THE BLACK SWAN

ALSO BY NASSIM NICHOLAS TALEB

Fooled by Randomness

The Black Swan

The Impact of the Highly Improbable

Nassim Nicholas Taleb

ALLEN LANE
an imprint of
PENGUIN BOOKS

ALLEN LANE

Published by the Penguin Group
Penguin Books Ltd, 80 Strand, London WC2R 0RL, England
Penguin Group (USA) Inc., 375 Hudson Street, New York, New York 10014, USA
Penguin Group (Canada), 90 Eglinton Avenue East, Suite 700, Toronto, Ontario, Canada M4P 2Y3
(a division of Pearson Penguin Canada Inc.)
Penguin Ireland, 25 St Stephen's Green, Dublin 2, Ireland (a division of Penguin Books Ltd)
Penguin Group (Australia), 250 Camberwell Road,
Camberwell, Victoria 3124, Australia (a division of Pearson Australia Group Pty Ltd)
Penguin Books India Pvt Ltd, 11 Community Centre,
Panchsheel Park, New Delhi – 110 017, India
Penguin Group (NZ), 67 Apollo Drive, Rosedale, North Shore 0632, New Zealand
(a division of Pearson New Zealand Ltd)
Penguin Books (South Africa) (Pty) Ltd, 24 Sturdee Avenue,
Rosebank, Johannesburg 2196, South Africa

Penguin Books Ltd, Registered Offices: 80 Strand, London WC2R 0RL, England

www.penguin.com

First published in the United States of America by The Random House Publishing Group 2007
First published in Great Britain by Allen Lane 2007
1

Printed in Great Britain by Clays Ltd, St Ives plc

A CIP catalogue record for this book is available from the British Library

978-1-846-14085-3

www.greenpenguin.co.uk

Mixed Sources
Product group from well-managed forests and other controlled sources
www.fsc.org Cert no. SA-COC-1592
© 1996 Forest Stewardship Council
FSC

Penguin Books is committed to a sustainable future
for our business, our readers and our planet.
The book in your hands is made from paper
certified by the Forest Stewardship Council.

To Benoît Mandelbrot,
a Greek among Romans

CONTENTS

PROLOGUE

ON THE PLUMAGE OF BIRDS

Before the discovery of Australia, people in the Old World were convinced that *all* swans were white, an unassailable belief as it seemed completely confirmed by empirical evidence. The sighting of the first black swan might have been an interesting surprise for a few ornithologists (and others extremely concerned with the coloring of birds), but that is not where the significance of the story lies. It illustrates a severe limitation to our learning from observations or experience and the fragility of our knowledge. One single observation can invalidate a general statement derived from millennia of confirmatory sightings of millions of white swans. All you need is one single (and, I am told, quite ugly) black bird.*

I push one step beyond this philosophical-logical question into an empirical reality, and one that has obsessed me since childhood. What we call here a Black Swan (and capitalize it) is an event with the following three attributes.

First, it is an *outlier,* as it lies outside the realm of regular expectations, because nothing in the past can convincingly point to its possibility. Second, it carries an extreme impact. Third, in spite of its outlier status,

* The spread of camera cell phones has afforded me a large collection of pictures of black swans sent by traveling readers. Last Christmas I also got a case of Black Swan Wine (not my favorite), a videotape (I don't watch videos), and two books. I prefer the pictures.

human nature makes us concoct explanations for its occurrence *after* the fact, making it explainable and predictable.

I stop and summarize the triplet: rarity, extreme impact, and retrospective (though not prospective) predictability.* A small number of Black Swans explain almost everything in our world, from the success of ideas and religions, to the dynamics of historical events, to elements of our own personal lives. Ever since we left the Pleistocene, some ten millennia ago, the effect of these Black Swans has been increasing. It started accelerating during the industrial revolution, as the world started getting more complicated, while ordinary events, the ones we study and discuss and try to predict from reading the newspapers, have become increasingly inconsequential.

Just imagine how little your understanding of the world on the eve of the events of 1914 would have helped you guess what was to happen next. (Don't cheat by using the explanations drilled into your cranium by your dull high school teacher.) How about the rise of Hitler and the subsequent war? How about the precipitous demise of the Soviet bloc? How about the rise of Islamic fundamentalism? How about the spread of the Internet? How about the market crash of 1987 (and the more unexpected recovery)? Fads, epidemics, fashion, ideas, the emergence of art genres and schools. All follow these Black Swan dynamics. Literally, just about everything of significance around you might qualify.

This combination of low predictability and large impact makes the Black Swan a great puzzle; but that is not yet the core concern of this book. Add to this phenomenon the fact that we tend to act as if it does not exist! I don't mean just you, your cousin Joey, and me, but almost all "social scientists" who, for over a century, have operated under the false belief that their tools could measure uncertainty. For the applications of the sciences of uncertainty to real-world problems has had ridiculous effects; I have been privileged to see it in finance and economics. Go ask your portfolio manager for his definition of "risk," and odds are that he will supply you with a *measure* that *excludes* the possibility of the Black Swan—hence one that has no better predictive value for assessing the total risks than astrology (we will see how they dress up the intellectual fraud with mathematics). This problem is endemic in social matters.

* The highly expected *not happening* is also a Black Swan. Note that, by symmetry, the occurrence of a highly improbable event is the equivalent of the nonoccurrence of a highly probable one.

The central idea of this book concerns our blindness with respect to randomness, particularly the large deviations: Why do we, scientists or nonscientists, hotshots or regular Joes, tend to see the pennies instead of the dollars? Why do we keep focusing on the minutiae, not the possible significant large events, in spite of the obvious evidence of their huge influence? And, if you follow my argument, why does reading the newspaper actually *decrease* your knowledge of the world?

It is easy to see that life is the cumulative effect of a handful of significant shocks. It is not so hard to identify the role of Black Swans, from your armchair (or bar stool). Go through the following exercise. Look into your own existence. Count the significant events, the technological changes, and the inventions that have taken place in our environment since you were born and compare them to what was expected before their advent. How many of them came on a schedule? Look into your own personal life, to your choice of profession, say, or meeting your mate, your exile from your country of origin, the betrayals you faced, your sudden enrichment or impoverishment. How often did these things occur according to plan?

What You Do Not Know

Black Swan logic makes *what you don't know* far more relevant than what you do know. Consider that many Black Swans can be caused and exacerbated *by their being unexpected.*

Think of the terrorist attack of September 11, 2001: had the risk been reasonably *conceivable* on September 10, it would not have happened. If such a possibility were deemed worthy of attention, fighter planes would have circled the sky above the twin towers, airplanes would have had locked bulletproof doors, and the attack would not have taken place, period. Something else might have taken place. What? I don't know.

Isn't it strange to see an event happening precisely because it was not supposed to happen? What kind of defense do we have against that? Whatever you come to know (that New York is an easy terrorist target, for instance) may become inconsequential if your enemy knows that you know it. It may be odd that, in such a strategic game, what you know can be truly inconsequential.

This extends to all businesses. Think about the "secret recipe" to making a killing in the restaurant business. If it were known and obvious, then someone next door would have already come up with the idea and it would have become generic. The next killing in the restaurant industry needs to be an idea that is not easily conceived of by the current population of restaurateurs. It has to be at some distance from expectations. The more unexpected the success of such a venture, the smaller the number of competitors, and the more successful the entrepreneur who implements the idea. The same applies to the shoe and the book businesses—or any kind of entrepreneurship. The same applies to scientific theories—nobody has interest in listening to trivialities. The payoff of a human venture is, in general, inversely proportional to what it is expected to be.

Consider the Pacific tsunami of December 2004. Had it been expected, it would not have caused the damage it did—the areas affected would have been less populated, an early warning system would have been put in place. What you know cannot really hurt you.

Experts and "Empty Suits"

The inability to predict outliers implies the inability to predict the course of history, given the share of these events in the dynamics of events.

But we act as though we are able to predict historical events, or, even worse, as if we are able to change the course of history. We produce thirty-year projections of social security deficits and oil prices without realizing that we cannot even predict these for next summer—our cumulative prediction errors for political and economic events are so monstrous that every time I look at the empirical record I have to pinch myself to verify that I am not dreaming. What is surprising is not the magnitude of our forecast errors, but our absence of awareness of it. This is all the more worrisome when we engage in deadly conflicts: wars are fundamentally unpredictable (and we do not know it). Owing to this misunderstanding of the causal chains between policy and actions, we can easily trigger Black Swans thanks to aggressive ignorance—like a child playing with a chemistry kit.

Our inability to predict in environments subjected to the Black Swan, coupled with a general lack of the awareness of this state of affairs, means that certain professionals, while believing they are experts, are in fact not. Based on their empirical record, they do not know more about their subject matter than the general population, but they are much better at narrating—or, worse, at smoking you with complicated mathematical models. They are also more likely to wear a tie.

Black Swans being unpredictable, we need to adjust to their existence (rather than naïvely try to predict them). There are so many things we can do if we focus on antiknowledge, or what we do not know. Among many other benefits, you can set yourself up to collect serendipitous Black Swans (of the positive kind) by maximizing your exposure to them. Indeed, in some domains—such as scientific discovery and venture capital investments—there is a disproportionate payoff from the unknown, since you typically have little to lose and plenty to gain from a rare event. We will see that, contrary to social-science wisdom, almost no discovery, no technologies of note, came from design and planning—they were just Black Swans. The strategy for the discoverers and entrepreneurs is to rely less on top-down planning and focus on maximum tinkering and recognizing opportunities when they present themselves. So I disagree with the followers of Marx and those of Adam Smith: the reason free markets work is because they allow people to be lucky, thanks to aggressive trial and error, not by giving rewards or "incentives" for skill. The strategy is, then, to tinker as much as possible and try to collect as many Black Swan opportunities as you can.

Learning to Learn

Another related human impediment comes from excessive focus on what we do know: we tend to learn the precise, not the general.

What did people learn from the 9/11 episode? Did they learn that some events, owing to their dynamics, stand largely outside the realm of the predictable? No. Did they learn the built-in defect of conventional wisdom? No. What did they figure out? They learned precise rules for avoiding Islamic prototerrorists and tall buildings. Many keep reminding me that it is important for us to be practical and take tangible steps rather than to "theorize" about knowledge. The story of the Maginot Line shows how we are conditioned to be specific. The French, after the Great War, built a wall along the previous German invasion route to prevent reinvasion—Hitler just (almost) effortlessly went around it. The French had been excellent students of history; they just learned with too much precision. They were too practical and exceedingly focused for their own safety.

We do not spontaneously learn that *we don't learn that we don't learn*. The problem lies in the structure of our minds: we don't learn rules, just facts, and only facts. Metarules (such as the rule that we have a tendency to not learn rules) we don't seem to be good at getting. We scorn the abstract; we scorn it with passion.

Why? It is necessary here, as it is my agenda in the rest of this book, both to stand conventional wisdom on its head and to show how inapplicable it is to our modern, complex, and increasingly *recursive* environment.*

But there is a deeper question: What are our minds made for? It looks as if we have the wrong user's manual. Our minds do not seem made to think and introspect; if they were, things would be easier for us today, but then we would not be here today and I would not have been here to talk about it—my counterfactual, introspective, and hard-thinking ancestor would have been eaten by a lion while his nonthinking but faster-reacting cousin would have run for cover. Consider that thinking is time-consuming and generally a great waste of energy, that our predecessors spent more than a hundred million years as nonthinking mammals and that in the blip in our history during which we have used our brain we have used it on subjects too peripheral to matter. Evidence shows that we do much less thinking than we believe we do—except, of course, when we think about it.

A NEW KIND OF INGRATITUDE

It is quite saddening to think of those people who have been mistreated by history. There were the *poètes maudits,* like Edgar Allan Poe or Arthur Rimbaud, scorned by society and later worshipped and force-fed to schoolchildren. (There are even schools named after high school dropouts.) Alas, this recognition came a little too late for the poet to get a serotonin kick out of it, or to prop up his romantic life on earth. But there are even more mistreated heroes—the very sad category of those who we do not know were heroes, who saved our lives, who helped us avoid disasters. They left no traces and did not even know that they were making a contribution. We remember the martyrs who died for a cause that we knew about, never those no less effective in their contribution but whose cause we were never

* *Recursive* here means that the world in which we live has an increasing number of feedback loops, causing events to be the cause of more events (say, people buy a book *because* other people bought it), thus generating snowballs and arbitrary and unpredictable planet-wide winner-take-all effects. We live in an environment where information flows too rapidly, accelerating such epidemics. Likewise, events can happen *because* they are not supposed to happen. (Our intuitions are made for an environment with simpler causes and effects and slowly moving information.) This type of randomness did not prevail during the Pleistocene, as socioeconomic life was far simpler then.

aware of—precisely because they were successful. Our ingratitude toward the *poètes maudits* fades completely in front of this other type of thanklessness. This is a far more vicious kind of ingratitude: the feeling of uselessness on the part of the silent hero. I will illustrate with the following thought experiment.

Assume that a legislator with courage, influence, intellect, vision, and perseverance manages to enact a law that goes into universal effect and employment on September 10, 2001; it imposes the continuously locked bulletproof doors in every cockpit (at high costs to the struggling airlines)—just in case terrorists decide to use planes to attack the World Trade Center in New York City. I know this is lunacy, but it is just a thought experiment (I am aware that there may be no such thing as a legislator with intellect, courage, vision, and perseverance; this is the point of the thought experiment). The legislation is not a popular measure among the airline personnel, as it complicates their lives. But it would certainly have prevented 9/11.

The person who imposed locks on cockpit doors gets no statues in public squares, not so much as a quick mention of his contribution in his obituary. "Joe Smith, who helped avoid the disaster of 9/11, died of complications of liver disease." Seeing how superfluous his measure was, and how it squandered resources, the public, with great help from airline pilots, might well boot him out of office. *Vox clamantis in deserto*. He will retire depressed, with a great sense of failure. He will die with the impression of having done nothing useful. I wish I could go attend his funeral, but, reader, I can't find him. And yet, recognition can be quite a pump. Believe me, even those who genuinely claim that they do not believe in recognition, and that they separate labor from the fruits of labor, actually get a serotonin kick from it. See how the silent hero is rewarded: even his own hormonal system will conspire to offer no reward.

Now consider again the events of 9/11. In their aftermath, who got the recognition? Those you saw in the media, on television performing heroic acts, and those whom you saw trying to give you the impression that they were performing heroic acts. The latter category includes someone like the New York Stock Exchange chairman Richard Grasso, who "saved the stock exchange" and received a huge bonus for his contribution (the equivalent of several *thousand* average salaries). All he had to do was be there to ring the opening bell on television—the television that, we will see, is the carrier of unfairness and a major cause of Black Swan blindness.

Who gets rewarded, the central banker who avoids a recession or the one who comes to "correct" his predecessors' faults and happens to be there during some economic recovery? Who is more valuable, the politician who avoids a war or the one who starts a new one (and is lucky enough to win)?

It is the same logic reversal we saw earlier with the value of what we don't know; everybody knows that you need more prevention than treatment, but few reward acts of prevention. We glorify those who left their names in history books at the expense of those contributors about whom our books are silent. We humans are not just a superficial race (this may be curable to some extent); we are a very unfair one.

LIFE IS VERY UNUSUAL

This is a book about uncertainty; to this author, the rare event *equals* uncertainty. This may seem like a strong statement—that we need to principally study the rare and extreme events in order to figure out common ones—but I will make myself clear as follows. There are two possible ways to approach phenomena. The first is to rule out the extraordinary and focus on the "normal." The examiner leaves aside "outliers" and studies ordinary cases. The second approach is to consider that in order to understand a phenomenon, one needs first to consider the extremes—particularly if, like the Black Swan, they carry an extraordinary cumulative effect.

I don't particularly care about the usual. If you want to get an idea of a friend's temperament, ethics, and personal elegance, you need to look at him under the tests of severe circumstances, not under the regular rosy glow of daily life. Can you assess the danger a criminal poses by examining only what he does on an *ordinary* day? Can we understand health without considering wild diseases and epidemics? Indeed the normal is often irrelevant.

Almost everything in social life is produced by rare but consequential shocks and jumps; all the while almost everything studied about social life focuses on the "normal," particularly with "bell curve" methods of inference that tell you close to nothing. Why? Because the bell curve ignores large deviations, cannot handle them, yet makes us confident that we have tamed uncertainty. Its nickname in this book is GIF, Great Intellectual Fraud.

PLATO AND THE NERD

At the start of the Jewish revolt in the first century of our era, much of the Jews' anger was caused by the Romans' insistence on putting a statue of Caligula in their temple in Jerusalem in exchange for placing a statue of the Jewish god Yahweh in Roman temples. The Romans did not realize that what the Jews (and the subsequent Levantine monotheists) meant by *god* was abstract, all embracing, and had nothing to do with the anthropomorphic, too human representation that Romans had in mind when they said *deus*. Critically, the Jewish god did not lend himself to symbolic representation. Likewise, what many people commoditize and label as "unknown," "improbable,"or "uncertain" is not the same thing to me; it is not a concrete and precise category of knowledge, a *nerdified* field, but its opposite; it is the lack (and limitations) of knowledge. It is the exact contrary of knowledge; one should learn to avoid using terms made for knowledge to describe its opposite.

What I call *Platonicity,* after the ideas (and personality) of the philosopher Plato, is our tendency to mistake the map for the territory, to focus on pure and well-defined "forms," whether objects, like triangles, or social notions, like utopias (societies built according to some blueprint of what "makes sense"), even nationalities. When these ideas and crisp constructs inhabit our minds, we privilege them over other less elegant objects, those with messier and less tractable structures (an idea that I will elaborate progressively throughout this book).

Platonicity is what makes us think that we understand more than we actually do. But this does not happen everywhere. I am not saying that Platonic forms don't exist. Models and constructions, these intellectual maps of reality, are not always wrong; they are wrong only in some specific applications. The difficulty is that a) you do not know beforehand (only after the fact) *where* the map will be wrong, and b) the mistakes can lead to severe consequences. These models are like potentially helpful medicines that carry random but very severe side effects.

The *Platonic fold* is the explosive boundary where the Platonic mindset enters in contact with messy reality, where the gap between what you know and what you think you know becomes dangerously wide. It is here that the Black Swan is produced.

TOO DULL TO WRITE ABOUT

It was said that the artistic filmmaker Luchino Visconti made sure that when actors pointed at a closed box meant to contain jewels, there were real jewels inside. It could be an effective way to make actors live their part. I think that Visconti's gesture may also come out of a plain sense of aesthetics and a desire for authenticity—somehow it may not feel right to fool the viewer.

This is an essay expressing a primary idea; it is neither the recycling nor repackaging of other people's thoughts. An essay is an impulsive meditation, not science reporting. I apologize if I skip a few obvious topics in this book out of the conviction that what is too dull for me to write about might be too dull for the reader to read. (Also, to avoid dullness may help to filter out the nonessential.)

Talk is cheap. Someone who took too many philosophy classes in college (or perhaps not enough) might object that the sighting of a Black Swan does not invalidate the theory that *all swans are white* since such a black bird is not technically a swan since whiteness to him may be the essential property of a swan. Indeed those who read too much Wittgenstein (and writings about comments about Wittgenstein) may be under the impression that language problems are important. They may certainly be important to attain prominence in philosophy departments, but they are something we, practitioners and decision makers in the real world, *leave for the weekend.* As I explain in the chapter called "The Uncertainty of the Phony," for all of their intellectual appeal, these niceties have no serious implications Monday to Friday as opposed to more substantial (but neglected) matters. People in the classroom, not having faced many true situations of decision making under uncertainty, do not realize what is important and what is not—even those who are scholars of uncertainty (or *particularly* those who are scholars of uncertainty). What I call the practice of uncertainty can be piracy, commodity speculation, professional gambling, working in some branches of the Mafia, or just plain serial entrepreneurship. Thus I rail against "sterile skepticism," the kind we can do nothing about, and against the exceedingly theoretical language problems that have made much of modern philosophy largely irrelevant to what is derisively called the "general public." (In the past, for better or worse, those rare philosophers and thinkers who were not self-standing depended on a patron's support. Today academics in abstract disciplines depend on one another's opinion, without external checks, with the severe occasional pathological result of turning their pursuits into insular prowess-showing contests. Whatever the shortcomings of the old system, at least it enforced *some* standard of relevance.)

The philosopher Edna Ullmann-Margalit detected an inconsistency in this book and asked me to justify the use of the precise metaphor of a Black Swan to describe the unknown, the abstract, and imprecise uncertain—white ravens, pink elephants, or evaporating denizens of a remote planet orbiting Tau Ceti. Indeed, she caught me red handed. There is a contradiction; this book is a story, and I prefer to use stories and vignettes to illustrate our gullibility about stories and our preference for the dangerous compression of narratives.

You need a story to displace a story. Metaphors and stories are far more potent (alas) than ideas; they are also easier to remember and more fun to read. If I have to go after what I call the narrative disciplines, my best tool is a narrative.

Ideas come and go, stories stay.

THE BOTTOM LINE

The beast in this book is not just the bell curve and the self-deceiving statistician, nor the Platonified scholar who needs theories to fool himself with. It is the drive to "focus" on what makes sense to us. Living on our planet, today, requires a lot more imagination than we are made to have. We lack imagination and repress it in others.

Note that I am not relying in this book on the beastly method of collecting selective "corroborating evidence." For reasons I explain in Chapter 5, I call this overload of examples naïve empiricism—successions of anecdotes selected to fit a story do not constitute evidence. Anyone looking for confirmation will find enough of it to deceive himself—and no doubt his peers.* The Black Swan idea is based on the structure of randomness in empirical reality.

To summarize: in this (personal) essay, I stick my neck out and make a claim, against many of our habits of thought, that our world is dominated by the extreme, the unknown, and the very improbable (improbable ac-

* It is also naïve empiricism to provide, in support of some argument, series of eloquent confirmatory quotes by dead authorities. By searching, you can always find someone who made a well-sounding statement that confirms your point of view—and, on every topic, it is possible to find another dead thinker who said the exact opposite. Almost all of my non–Yogi Berra quotes are from people I disagree with.

cording our current knowledge)—and all the while we spend our time engaged in small talk, focusing on the known, and the repeated. This implies the need to use the extreme event as a starting point and not treat it as an exception to be pushed under the rug. I also make the bolder (and more annoying) claim that in spite of our progress and the growth in knowledge, or perhaps *because* of such progress and growth, the future will be increasingly less predictable, while both human nature and social "science" seem to conspire to hide the idea from us.

Chapters Map

The sequence of this book follows a simple logic; it flows from what can be labeled purely literary (in subject and treatment) to what can be deemed entirely scientific (in subject, though not in treatment). Psychology will be mostly present in Part One and in the early part of Part Two; business and natural science will be dealt with mostly in the second half of Part Two and in Part Three. Part One, "Umberto Eco's Antilibrary," is mostly about how we perceive historical and current events and what distortions are present in such perception. Part Two, "We Just Can't Predict," is about our errors in dealing with the future and the unadvertised limitations of some "sciences"—and what to do about these limitations. Part Three, "Those Gray Swans of Extremistan," goes deeper into the topic of extreme events, explains how the bell curve (that great intellectual fraud) is generated, and reviews the ideas in the natural and social sciences loosely lumped under the label "complexity." Part Four, "The End," will be very short.

I derived an unexpected amount of enjoyment writing this book—in fact, it just wrote itself—and I hope that the reader will experience the same. I confess that I got hooked on this withdrawal into pure ideas after the constraints of an active and transactional life. After this book is published, my aim is to spend time away from the clutter of public activities in order to think about my philosophical-scientific idea in total tranquillity.

Part 1

UMBERTO ECO'S ANTILIBRARY, OR HOW WE SEEK VALIDATION

The writer Umberto Eco belongs to that small class of scholars who are encyclopedic, insightful, and nondull. He is the owner of a large personal library (containing thirty thousand books), and separates visitors into two categories: those who react with "Wow! Signore *professore dottore* Eco, what a library you have! How many of these books have you read?" and the others—a very small minority—who get the point that a private library is not an ego-boosting appendage but a research tool. Read books are far less valuable than unread ones. The library should contain as much of *what you do not know* as your financial means, mortgage rates, and the currently tight real-estate market allow you to put there. You will accumulate more knowledge and more books as you grow older, and the growing number of unread books on the shelves will look at you menacingly. Indeed, the more you know, the larger the rows of unread books. Let us call this collection of unread books an *antilibrary.*

We tend to treat our knowledge as personal property to be protected and defended. It is an ornament that allows us to rise in the pecking order. So this tendency to offend Eco's library sensibility by focusing on the known is a human bias that extends to our mental operations. People don't walk around with anti-résumés telling you what they have not studied or experienced (it's the job of their competitors to do that), but it would be nice if they did. Just as we need to stand library logic on its head, we will work on standing knowledge itself on its head. Note that the Black

Swan comes from our misunderstanding of the likelihood of surprises, those unread books, because we take what we know a little too seriously.

Let us call an antischolar—someone who focuses on the unread books, and makes an attempt not to treat his knowledge as a treasure, or even a possession, or even a self-esteem enhancement device—a skeptical empiricist.

The chapters in this section address the question of how we humans deal with knowledge—and our preference for the anecdotal over the empirical. Chapter 1 presents the Black Swan as grounded in the story of my own obsession. I will make a central distinction between the two varieties of randomness in Chapter 3. After that, Chapter 4 briefly returns to the Black Swan problem in its original form: how we tend to generalize from what we see. Then I present the three facets of the same Black Swan problem: a) *The error of confirmation,* or how we are likely to undeservedly scorn the virgin part of the library (the tendency to look at what confirms our knowledge, not our ignorance), in Chapter 5; b) *the narrative fallacy,* or how we fool ourselves with stories and anecdotes (Chapter 6); c) how emotions get in the way of our inference (Chapter 7); and d) *the problem of silent evidence,* or the tricks history uses to hide Black Swans from us (Chapter 8). Chapter 9 discusses the lethal fallacy of building knowledge from the world of games.

Chapter One

THE APPRENTICESHIP OF AN EMPIRICAL SKEPTIC

Anatomy of a Black Swan—The triplet of opacity—Reading books backward—The rearview mirror—Everything becomes explainable—Always talk to the driver (with caution)—History doesn't crawl; it jumps—"It was so unexpected"—Sleeping for twelve hours

This is not an autobiography, so I will skip the scenes of war. Actually, even if it were an autobiography, I would still skip the scenes of war. I cannot compete with action movies or memoirs of adventurers more accomplished than myself, so I will stick to my specialties of chance and uncertainty.

ANATOMY OF A BLACK SWAN

For more than a millennium the eastern Mediterranean seaboard called Syria Libanensis, or Mount Lebanon, had been able to accommodate at least a dozen different sects, ethnicities, and beliefs—it worked like magic. The place resembled major cities of the eastern Mediterranean (called the Levant) more than it did the other parts in the interior of the Near East (it was easier to move by ship than by land through the mountainous terrain). The Levantine cities were mercantile in nature; people dealt with one another according to a clear protocol, preserving a peace conducive to commerce, and they socialized quite a bit across communities. This millennium of peace was interrupted only by small occasional friction *within* Moslem and Christian communities, rarely between Christians and Moslems. While the cities were mercantile and mostly Hellenistic, the mountains had been settled by all manner of religious minorities who claimed to have fled both the Byzantine and Moslem orthodoxies. A mountainous terrain is an ideal refuge from the mainstream, except that your enemy is the other refugee competing for the same type of rugged real estate. The mosaic of cultures and religions there was deemed an example of coexistence: Christians of all varieties (Maronites, Armenians, Greco-Syrian Byzantine Orthodox, even Byzantine Catholic, in addition to the few Roman Catholics left over from the Crusades); Moslems (Shiite and Sunni); Druzes; and a few Jews. It was taken for granted that people learned to be tolerant there; I recall how we were taught in school how far more civilized and wiser we were than those in the Balkan communities, where not only did the locals refrain from bathing but also fell prey to fractious fighting. Things appeared to be in a state of stable equilibrium, evolving out of a historical tendency for betterment and tolerance. The terms *balance* and *equilibrium* were often used.

Both sides of my family came from the Greco-Syrian community, the last Byzantine outpost in northern Syria, which included what is now called Lebanon. Note that the Byzantines called themselves "Romans"—*Roumi* (plural *Roum*) in the local languages. We originate from the olive-growing area at the base of Mount Lebanon—we chased the Maronite Christians into the mountains in the famous battle of Amioun, my ancestral village. Since the Arab invasion in the seventh century, we had been living in mercantile peace with the Moslems, with only some occasional harassment by the Lebanese Maronite Christians from the mountains. By some (literally) Byzantine arrangement between the Arab rulers and the Byzantine emperors, we managed to pay taxes to both sides and get protection from both. We thus managed to live in peace for more than a millennium almost devoid of bloodshed: our last true problem was the later troublemaking crusaders, not the Moslem Arabs. The Arabs, who seemed interested only in warfare (and poetry) and, later, the Ottoman Turks, who seemed only concerned with warfare (and pleasure), left to us the uninteresting pursuit of commerce and the less dangerous one of scholarship (like the translation of Aramaic and Greek texts).

By any standard the country called Lebanon, to which we found ourselves suddenly incorporated after the fall of the Ottoman Empire, in the early twentieth century, appeared to be a stable paradise; it was also cut in a way to be predominantly Christian. People were suddenly brainwashed to believe in the nation-state as an entity.* The Christians convinced themselves that they were at the origin and center of what is loosely called Western culture yet with a window on the East. In a classical case of static thinking, nobody took into account the differentials in birthrate between communities and it was assumed that a slight Christian majority would remain permanent. Levantines had been granted Roman citizenship, which allowed Saint Paul, a Syrian, to travel freely through the ancient world. People felt connected to everything they felt was worth connecting to; the place was exceedingly open to the world, with a vastly sophisticated lifestyle, a prosperous economy, and temperate weather just like California, with snow-covered mountains jutting above the Mediterranean. It attracted a collection of spies (both Soviet and Western), prostitutes (blondes), writers, poets, drug dealers, adventurers, compulsive gamblers, tennis players, après-skiers, and merchants—all professions that complement one another. Many people acted as if they were in an old James Bond movie, or the days when playboys smoked, drank, and, instead of going to the gym, cultivated relationships with good tailors.

The main attribute of paradise was there: cabdrivers were said to be polite (though, from what I remember, they were not polite to me). True, with hindsight, the place may appear more Elysian in the memory of people than it actually was.

I was too young to taste the pleasures of the place, as I became a rebellious idealist and, very early on, developed an ascetic taste, averse to the ostentatious signaling of wealth, allergic to Levantine culture's overt pursuit of luxury and its obsession with things monetary.

As a teenager, I could not wait to go settle in a metropolis with fewer James Bond types around. Yet I recall something that felt special in the intellectual air. I attended the French lycée that had one of the highest success rates for the French *baccalauréat* (the high school degree), even in the subject of the French language. French was spoken there with some purity: as in prerevolutionary Russia, the Levantine Christian and Jewish patrician class (from Istanbul to Alexandria) spoke and wrote formal French as a language of distinction. The most privileged were sent to school in

* It is remarkable how fast and how effectively you can construct a nationality with a flag, a few speeches, and a national anthem; to this day I avoid the label "Lebanese," preferring the less restrictive "Levantine" designation.

France, as both my grandfathers were—my paternal namesake in 1912 and my mother's father in 1929. Two thousand years earlier, by the same instinct of linguistic distinction, the snobbish Levantine patricians wrote in Greek, not the vernacular Aramaic. (The New Testament was written in the bad local patrician Greek of our capital, Antioch, prompting Nietzsche to shout that "God spoke bad Greek.") And, after Hellenism declined, they took up Arabic. So in addition to being called a "paradise," the place was also said to be a miraculous crossroads of what are superficially tagged "Eastern" and "Western" cultures.

On Walking Walks

My ethos was shaped when, at fifteen, I was put in jail for (allegedly) attacking a policeman with a slab of concrete during a student riot—an incident with strange ramifications since my grandfather was then the minister of the interior, and the person who signed the order to crush our revolt. One of the rioters was shot dead when a policeman who had been hit on the head with a stone panicked and randomly opened fire on us. I recall being at the center of the riot, and feeling a huge satisfaction upon my capture while my friends were scared of both prison and their parents. We frightened the government so much that we were granted amnesty.

There were some obvious benefits in showing one's ability to act on one's opinions, and not compromising an inch to avoid "offending" or bothering others. I was in a state of rage and didn't care what my parents (and grandfather) thought of me. This made them quite scared of *me,* so I could not afford to back down, or even blink. Had I concealed my participation in the riot (as many friends did) and been discovered, instead of being openly defiant, I am certain that I would have been treated as a black sheep. It is one thing to be cosmetically defiant of authority by wearing unconventional clothes—what social scientists and economists call "cheap signaling"—and another to prove willingness to translate belief into action.

My paternal uncle was not too bothered by my political ideas (these come and go); he was outraged that I used them as an excuse to dress sloppily. To him, inelegance on the part of a close family member was the mortal offense.

Public knowledge of my capture had another major benefit: it allowed me to avoid the usual outward signs of teenage rebellion. I discovered that it is much more effective to act like a nice guy and be "reasonable" if you prove willing to go beyond just verbiage. You can afford to be compassionate, lax, and courteous if, once in a while, when it is least expected of you, but completely justified, you sue someone, or savage an enemy, just to show that you can walk the walk.

"Paradise" Evaporated

The Lebanese "paradise" suddenly evaporated, after a few bullets and mortar shells. A few months after my jail episode, after close to thirteen centuries of remarkable ethnic coexistence, a Black Swan, coming out of nowhere, transformed the place from heaven to hell. A fierce civil war began between Christians and Moslems, including the Palestinian refugees who took the Moslem side. It was brutal, since the combat zones were in the center of the town and most of the fighting took place in residential areas (my high school was only a few hundred feet from the war zone). The conflict lasted more than a decade and a half. I will not get too descriptive. It may be that the invention of gunfire and powerful weapons turned what, in the age of the sword, would have been just tense conditions into a spiral of uncontrollable tit-for-tat warfare.

Aside from the physical destruction (which turned out to be easy to reverse with a few motivated contractors, bribed politicians, and naïve bondholders), the war removed much of the crust of sophistication that had made the Levantine cities a continuous center of great intellectual refinement for three thousand years. Christians had been leaving the area since Ottoman times—those who moved to the West took Western first names and melded in. Their exodus accelerated. The number of cultured people dropped below some critical level. Suddenly the place became a vacuum. Brain drain is hard to reverse, and some of the old refinement may be lost forever.

The Starred Night

The next time you experience a blackout, take some solace by looking at the sky. You will not recognize it. Beirut had frequent power shutdowns during the war. Before people bought their own generators, one side of the sky was clear at night, owing to the absence of light pollution. That was the side of town farthest from the combat zone. People deprived of television drove to watch the erupting lights of nighttime battles. They appeared to prefer the risk of being blown up by mortar shells to the boredom of a dull evening.

So you could see the stars with great clarity. I had been told in high school that the planets are in something called *equilibrium,* so we did not have to worry about the stars hitting us unexpectedly. To me, that eerily resembled the stories we were also told about the "unique historical stability" of Lebanon. The very idea of assumed equilibrium bothered me. I looked at the constellations in the sky and did not know what to believe.

HISTORY AND THE TRIPLET OF OPACITY

History is opaque. You see what comes out, not the script that produces events, the generator of history. There is a fundamental incompleteness in your grasp of such events, since you do not see what's inside the box, how the mechanisms work. What I call the generator of historical events is different from the events themselves, much as the minds of the gods cannot be read just by witnessing their deeds. You are very likely to be fooled about their intentions.

This disconnect is similar to the difference between the food you see on the table at the restaurant and the process you can observe in the kitchen. (The last time I brunched at a certain Chinese restaurant on Canal Street in downtown Manhattan, I saw a rat coming out of the kitchen.)

The human mind suffers from three ailments as it comes into contact with history, what I call the *triplet of opacity.* They are:

a. the illusion of understanding, or how everyone thinks he knows what is going on in a world that is more complicated (or random) than they realize;
b. the retrospective distortion, or how we can assess matters only after the fact, as if they were in a rearview mirror (history seems clearer and more organized in history books than in empirical reality); and
c. the overvaluation of factual information and the handicap of authoritative and learned people, particularly when they create categories—when they "Platonify."

Nobody Knows What's Going On

The first leg of the triplet is the pathology of thinking that the world in which we live is more understandable, more explainable, and therefore more predictable than it actually is.

I was constantly told by adults that the war, which ended up lasting close to seventeen years, was going to end in "only a matter of days." They seemed quite confident in their forecasts of duration, as can be evidenced by the number of people who sat waiting in hotel rooms and other temporary quarters in Cyprus, Greece, France, and elsewhere for the war to finish. One uncle kept telling me how, some thirty years earlier, when the rich Palestinians fled to Lebanon, they considered it a *very temporary* solution (most of those still alive are still there, six decades later). Yet when I asked him if it was going to be the same with our conflict, he replied, "No, of course not. This place is different; it has always been different." Somehow what he detected in others did not seem to apply to him.

This duration blindness in the middle-aged exile is quite a widespread disease. Later, when I decided to avoid the exile's obsession with his roots (exiles' roots penetrate their personalities a bit too deeply), I studied exile literature precisely to avoid the traps of a consuming and obsessive nostalgia. These exiles seemed to have become prisoners of their memory of idyllic origin—they sat together with other prisoners of the past and spoke about the old country, and ate their traditional food while some of their folk music played in the background. They continuously ran counterfactuals in their minds, generating alternative scenarios that could have happened and prevented these historical ruptures, such as "if the Shah had not named this incompetent man as prime minister, we would still be there." It was as if the historical rupture had a specific cause, and that the catastrophe could have been averted by removing *that* specific cause. So I pumped every displaced person I could find for information on their behavior during exile. Almost all act in the same way.

One hears endless stories of Cuban refugees with suitcases still half packed who came to Miami in the 1960s for "a matter of a few days" after the installation of the Castro regime. And of Iranian refugees in Paris and London who fled the Islamic Republic in 1978 thinking that their absence would be a brief vacation. A few are still waiting, more than a quarter century later, for the return. Many Russians who left in 1917, such as the writer Vladimir Nabokov, settled in Berlin, perhaps to be close enough for a quick return. Nabokov himself lived all his life in temporary housing, in both indigence and luxury, ending his days at the Montreux Palace hotel on Lake Geneva.

There was, of course, some wishful thinking in all of these forecasting errors, the blindness of hope, but there was a knowledge problem as well. The dynamics of the Lebanese conflict had been patently unpredictable, yet people's reasoning as they examined the events showed a constant: almost all those who cared seemed convinced that they understood what was going on. Every single day brought occurrences that lay completely outside their forecast, but they could not figure out that they had not forecast them. Much of what took place would have been deemed completely crazy with respect to the past. Yet it did not seem that crazy *after* the events. This retrospective plausibility causes a discounting of the rarity and conceivability of the event. I later saw the exact same illusion of understanding in business success and the financial markets.

History Does Not Crawl, It Jumps

Later, upon replaying the wartime events in my memory as I formulated my ideas on the perception of random events, I developed the governing impression that our minds are wonderful explanation machines, capable of making sense out of almost anything, capable of mounting explanations for all manner of phenomena, and generally incapable of accepting the idea of unpredictability. These events were unexplainable, but intelligent people thought they were capable of providing convincing explanations for them—after the fact. Furthermore, the more intelligent the person, the better sounding the explanation. What's more worrisome is that all these beliefs and accounts appeared to be logically coherent and devoid of inconsistencies.

So I left the place called Lebanon as a teenager, but, since a large number of my relatives and friends remained there, I kept coming back to visit, especially during the hostilities. The war was not continuous: there were periods of fighting interrupted by "permanent" solutions. I felt closer to my roots during times of trouble and experienced the urge to come back and show support to those left behind who were often demoralized by the departures—and envious of the fair-weather friends who could seek economic and personal safety only to return for vacations during these occasional lulls in the conflict. I was unable to work or read when I was outside Lebanon while people were dying, but, paradoxically, I was less concerned by the events and able to pursue my intellectual interests guilt-free when I was *inside* Lebanon. Interestingly, people partied quite heavily during the war and developed an even bigger taste for luxuries, making the visits quite attractive in spite of the fighting.

There were a few difficult questions. How could one have predicted that people who seemed a model of tolerance could become the purest of barbarians overnight? Why was the change so abrupt? I initially thought that perhaps the Lebanese war was truly not possible to predict, unlike other conflicts, and that the Levantines were too complicated a race to figure out. Later I slowly realized, as I started to consider all the big events in history, that their irregularity was not a local property.

The Levant has been something of a mass producer of consequential events nobody saw coming. Who predicted the rise of Christianity as a dominant religion in the Mediterranean basin, and later in the Western world? The Roman chroniclers of that period did not even take note of the new religion—historians of Christianity are baffled by the absence of contemporary mentions. Apparently, few of the big guns took the ideas of a seemingly heretical Jew seriously enough to think that he would leave traces for posterity. We only have a single contemporary reference to Jesus of Nazareth—in *The Jewish Wars* of Josephus—which itself may have been added later by a devout copyist. How about the competing religion that emerged seven centuries later; who forecast that a collection of horsemen would spread their empire and Islamic law from the Indian subcontinent to Spain in just a few years? Even more than the rise of Christianity, it was the spread of Islam (the third edition, so to speak) that carried full unpredictability; many historians looking at the record have been taken aback by the swiftness of the change. Georges Duby, for one, expressed his amazement about how quickly close to ten centuries of Levantine Hellenism were blotted out "with a strike of a sword." A later holder of the same history chair at the Collège de France, Paul Veyne, aptly talked about religions spreading "like bestsellers"—a comparison that indicates unpredictability. These kinds of discontinuities in the chronology of events did not make the historian's profession too easy: the studious examination of the past in the greatest of detail does not teach you much about the mind of History; it only gives you the illusion of understanding it.

History and societies do not crawl. They make jumps. They go from fracture to fracture, with a few vibrations in between. Yet we (and historians) like to believe in the predictable, small incremental progression.

It struck me, a belief that has never left me since, that we are just a great machine for looking backward, and that humans are great at self-delusion. Every year that goes by increases my belief in this distortion.

Dear Diary: On History Running Backward

Events present themselves to us in a distorted way. Consider the nature of information: of the millions, maybe even trillions, of small facts that prevail before an event occurs, only a few will turn out to be relevant later to your understanding of what happened. Because your memory is limited and filtered, you will be inclined to remember those data that subsequently match the facts, unless you are like the eponymous Funes in the short story by Jorge Luis Borges, "Funes, the Memorious," who forgets nothing and seems condemned to live with the burden of the accumulation of unprocessed information. (He does not manage to live too long.)

I had my first exposure to the retrospective distortion as follows. During my childhood I had been a voracious, if unsteady, reader, but I spent the first phase of the war in a basement, diving body and soul into all manner of books. School was closed and it was raining mortar shells. It is dreadfully boring to be in basements. My initial worries were mostly about how to fight boredom and what to read next*—though being forced to read for lack of other activities is not as enjoyable as reading out of one's own volition. I wanted to be a philosopher (I still do), so I felt that I needed to make an investment by forcibly studying others' ideas. Circumstances motivated me to study theoretical and general accounts of wars and conflicts, trying to get into the guts of History, to get into the workings of that big machine that generates events.

Surprisingly, the book that influenced me was not written by someone in the thinking business but by a journalist: William Shirer's *Berlin Diary: The Journal of a Foreign Correspondent, 1934–1941*. Shirer was a radio correspondent, famous for his book *The Rise and Fall of the Third Reich*. It occurred to me that the *Journal* offered an unusual perspective. I had already read (or read about) the works of Hegel, Marx, Toynbee, Aron, and Fichte on the philosophy of history and its properties and thought that I had a vague idea of the notions of dialectics, to the extent that there was

* Benoît Mandelbrot, who had a similar experience at about the same age, though close to four decades earlier, remembers his own war episode as long stretches of painful boredom punctuated by brief moments of extreme fear.

something to understand in these theories. I did not grasp much, except that history had some logic and that things developed through contradiction (or opposites) in a way that elevated mankind into higher forms of society—that kind of thing. This sounded awfully similar to the theorizing around me about the war in Lebanon. To this day I surprise people who put the ludicrous question to me about what books "shaped my thinking" by telling them that this book taught me (albeit inadvertently) the most about philosophy and theoretical history—and, we will see, about science as well, since I learned the difference between forward and backward processes.

How? Simply, the diary purported to describe the events *as they were taking place,* not after. I was in a basement with history audibly unfolding above me (the sound of mortar shells kept me up all night). I was a teenager attending the funerals of classmates. I was experiencing a nontheoretical unfolding of History and I was reading about someone apparently experiencing history as it went along. I made efforts to mentally produce a movielike representation of the future and realized it was not so obvious. I realized that if I were to start writing about the events later they would seem more . . . *historical.* There was a difference between the *before* and the *after.*

The journal was purportedly written without Shirer knowing what was going to happen next, when the information available to him was not corrupted by the subsequent outcomes. Some comments here and there were quite illuminating, particularly those concerning the French belief that Hitler was a transitory phenomenon, which explained their lack of preparation and subsequent rapid capitulation. At no time was the extent of the ultimate devastation deemed possible.

While we have a highly unstable memory, a diary provides indelible facts recorded more or less immediately; it thus allows the fixation of an unrevised perception and enables us to later study events in their own context. Again, it is the purported method of description of the event, not its execution, that was important. In fact, it is likely that Shirer and his editors did some cheating, since the book was published in 1941 and publishers, I am told, are in the business of delivering texts to the general public instead of providing faithful depictions of the authors' mind-sets stripped of retrospective distortions. (By "cheating," I mean removing at the time of publication elements that did not turn out to be relevant to what happened, thus enhancing those that may interest the public. Indeed the editing process can be severely distorting, particularly when the author is assigned what is called a "good editor.") Still, encountering Shirer's book provided me with an intuition about the workings of history. One would suppose that people living through the beginning of WWII had an inkling that something momentous was taking place. Not at all.*

Shirer's diary turned out to be a training program in the dynamics of uncertainty. I wanted to be a philosopher, not knowing at the time what most professional philosophers did for a living. The idea led me to adventure (rather to the adventurous practice of uncertainty) and also to mathematical and scientific pursuits instead.

Education in a Taxicab

I will introduce the third element of the triplet, the curse of learning, as follows. I closely watched my grandfather, who was minister of defense, and later minister of the interior and deputy prime minister in the early days of the war, before the fading of his political role. In spite of his position he did not seem to know what was going to happen any more than did his driver, Mikhail. But unlike my grandfather, Mikhail used to repeat "God knows" as his main commentary on events, transferring the task of understanding higher up.

I noticed that very intelligent and informed persons were at no advantage over cabdrivers in their predictions, but there was a crucial difference. Cabdrivers did not believe that they understood as much as learned people—really, they were not the experts and they knew it. Nobody knew anything, but elite thinkers thought that they knew more than the rest because they were elite thinkers, and if you're a member of the elite, you automatically know more than the nonelite.

It is not just knowledge but information that can be of dubious value. It came to my notice that almost everybody was acquainted with current events in their smallest details. The overlap between newspapers was so

* The historian Niall Ferguson showed that, despite all the standard accounts of the buildup to the Great War, which describe "mounting tensions" and "escalating crises," the conflict came as a surprise. Only retrospectively was it seen as unavoidable by backward-looking historians. Ferguson used a clever empirical argument to make his point: he looked at the prices of imperial bonds, which normally include investors' anticipation of government's financing needs and decline in expectation of conflicts since wars cause severe deficits. But bond prices did not reflect the anticipation of war. Note that this study illustrates, in addition, how working with prices can provide a good understanding of history.

large that you would get less and less information the more you read. Yet everyone was so eager to become familiar with every fact that they read every freshly printed document and listened to every radio station as if the great answer was going to be revealed to them in the next bulletin. People became encyclopedias of who had met with whom and which politician said what to which other politician (and with what tone of voice: "Was he more friendly than usual?"). Yet to no avail.

CLUSTERS

I also noticed during the Lebanese war that journalists tended to cluster not necessarily around the same opinions but frequently around the same framework of analyses. They assign the same importance to the same sets of circumstances and cut reality into the same categories—once again the manifestation of Platonicity, the desire to cut reality into crisp shapes. What Robert Fisk calls "hotel journalism" further increased the mental contagion. While Lebanon in earlier journalism was part of the Levant, i.e., the eastern Mediterranean, it now suddenly became part of the Middle East, as if someone had managed to transport it closer to the sands of Saudi Arabia. The island of Cyprus, around sixty miles from my village in northern Lebanon, and with almost identical food, churches, and habits, suddenly became part of Europe (of course the natives on both sides became subsequently conditioned). While in the past a distinction had been drawn between Mediterranean and non-Mediterranean (i.e., between the olive oil and the butter), in the 1970s the distinction suddenly became that between Europe and non-Europe. Islam being the wedge between the two, one does not know where to place the indigenous Arabic-speaking Christians (or Jews) in that story. Categorizing is necessary for humans, but it becomes pathological when the category is seen as definitive, preventing people from considering the fuzziness of boundaries, let alone revising their categories. Contagion was the culprit. If you selected one hundred independent-minded journalists capable of seeing factors in isolation from one another, you would get one hundred different opinions. But the process of having these people report in lockstep caused the dimensionality of the opinion set to shrink considerably—they converged on opinions and used the same items as causes. For instance, to depart from Lebanon for a moment, all reporters now refer to the "roaring eighties," assuming that there was something particularly distinct about that exact decade. And during the Internet bubble of the late 1990s, journalists agreed on crazy indicators as explanatory of the quality of worthless companies that everyone wanted very badly.*

If you want to see what I mean by the arbitrariness of categories, check the situation of polarized politics. The next time a Martian visits earth, try to explain to him why those who favor allowing the elimination of a fetus in the mother's womb also oppose capital punishment. Or try to explain to him why those who accept abortion are supposed to be favorable to high taxation but against a strong military. Why do those who prefer sexual freedom need to be against individual economic liberty?

I noticed the absurdity of clustering when I was quite young. By some farcical turn of events, in that civil war of Lebanon, Christians became pro–free market and the capitalistic system—i.e., what a journalist would call "the Right"—and the Islamists became socialists, getting support from Communist regimes (*Pravda,* the organ of the Communist regime, called them "oppression fighters," though subsequently when the Russians invaded Afghanistan, it was the Americans who sought association with bin Laden and his Moslem peers).

The best way to prove the arbitrary character of these categories, and the contagion effect they produce, is to remember how frequently these clusters reverse in history. Today's alliance between Christian fundamentalists and the Israeli lobby would certainly seem puzzling to a nineteenth-century intellectual—Christians used to be anti-Semites and Moslems were the protectors of the Jews, whom they preferred to Christians. Libertarians used to be left-wing. What is interesting to me as a probabilist is that some random event makes one group that initially supports an issue ally itself with another group that supports another issue, thus causing the two items to fuse and unify . . . until the surprise of the separation.

Categorizing always produces reduction in true complexity. It is a manifestation of the Black Swan generator, that unshakable Platonicity that I defined in the Prologue. Any reduction of the world around us can have explosive consequences since it rules out some sources of uncertainty; it drives us to a misunderstanding of the fabric of the world. For instance, you may think that radical Islam (and its values) are your allies against the threat of Communism, and so you may help them develop, until they send two planes into downtown Manhattan.

* We will see in Chapter 10 some clever quantitative tests done to prove such herding; they show that, in many subject matters, the distance between opinions is remarkably narrower than the distance between the average of opinions and truth.

It was a few years after the beginning of the Lebanese war, as I was attending the Wharton School, at the age of twenty-two, that I was hit with the idea of efficient markets—an idea that holds that there is no way to derive profits from traded securities since these instruments have automatically incorporated all the available information. Public information can therefore be useless, particularly to a businessman, since prices can already "include" all such information, and news shared with millions gives you no real advantage. Odds are that one or more of the hundreds of millions of other readers of such information will already have bought the security, thus pushing up the price. I then completely gave up reading newspapers and watching television, which freed up a considerable amount of time (say one hour or more a day, enough time to read more than a hundred additional books per year, which, after a couple of decades, starts mounting). But this argument was not quite the entire reason for my dictum in this book to avoid the newspapers, as we will see further benefits in avoiding the toxicity of information. It was initially a great excuse to avoid keeping up with the minutiae of business, a perfect alibi since I found nothing interesting about the details of the business world—inelegant, dull, pompous, greedy, unintellectual, selfish, and boring.

Where Is the Show?

Why someone with plans to become a "philosopher" or a "scientific philosopher of history" would wind up in business school, and the Wharton School no less, still escapes me. There I saw that it was not merely some inconsequential politician in a small and antique country (and his philosophical driver Mikhail) who did not know what was going on. After all, people in small countries are supposed to *not know* what is going on. What I saw was that in one of the most prestigious business schools in the world, in the most potent country in the history of the world, the executives of the most powerful corporations were coming to describe what they did for a living, and it was possible that they too did not know what was going on. As a matter of fact, in my mind it was far more than a possibility. I felt in my spine the weight of the epistemic arrogance of the human race.*

I became obsessive. At the time, I started becoming conscious of my

* I then realized that the great strength of the free-market system is the fact that company executives don't need to know what's going on.

subject—the *highly improbable consequential event.* And it was not only well-dressed, testosterone-charged corporate executives who were usually fooled by this concentrated luck, but persons of great learning. This awareness turned my Black Swan from a problem of lucky or unlucky people in business into a problem of knowledge and science. My idea is that not only are some scientific results useless in real life, because they underestimate the impact of the highly improbable (or lead us to ignore it), but that many of them may be actually creating Black Swans. These are not just taxonomic errors that can make you flunk a class in ornithology. I started to see the consequences of the idea.

8¾ LBS LATER

Four and a half years after my graduation from Wharton (and 8¾ pounds heavier), on October 19, 1987, I walked home from the offices of the investment bank Credit Suisse First Boston in Midtown Manhattan to the Upper East Side. I walked slowly, as I was in a bewildered state.

That day saw a traumatic financial event: the largest market drop in (modern) history. It was all the more traumatic in that it took place at a time when we thought we had become sufficiently sophisticated with all these intelligent-talking Platonified economists (with their phony bell curve–based equations) to prevent, or at least forecast and control, big shocks. The drop was not even the response to any discernible news. The occurrence of the event lay outside anything one could have imagined on the previous day—had I pointed out its possibility, I would have been called a lunatic. It qualified as a Black Swan, but I did not know the expression then.

I ran into a colleague of mine, Demetrius, on Park Avenue, and, as I started talking to him, an anxiety-ridden woman, losing all inhibitions, jumped into the conversation: "Hey, do the two of you know what's going on?" People on the sidewalk looked dazed. Earlier I had seen a few adults silently sobbing in the trading room of First Boston. I had spent the day at the epicenter of the events, with shell-shocked people running around like rabbits in front of headlights. When I got home, my cousin Alexis called to tell me that his neighbor committed suicide, jumping from his upper-floor apartment. It did not even feel eerie. It felt like Lebanon, with a twist: having seen both, I was struck that financial distress could be more demoralizing than war (just consider that financial problems and the accompanying humiliations can lead to suicide, but war doesn't appear to do so directly).

I feared a Pyrrhic victory: I had been vindicated intellectually, but I was afraid of being too right and seeing the system crumble under my feet. I did not really want to be *that* right. I will always remember the late Jimmy P. who, seeing his net worth in the process of melting down, kept half-jokingly begging the price on the screen to stop moving.

But I realized then and there that I did not give a hoot about the money. I experienced the strangest feeling I have ever had in my life, this deafening trumpet signaling to me that *I was right,* so loudly that it made my bones vibrate. I have never had it since and will never be able to explain it to those who have never experienced it. It was a physical sensation, perhaps a mixture of joy, pride, and terror.

And I felt vindicated? How?

During the one or two years after my arrival at Wharton, I had developed a precise but strange specialty: betting on rare and unexpected events, those that were on the *Platonic fold,* and considered "inconceivable" by the Platonic "experts." Recall that the Platonic fold is where our representation of reality ceases to apply—but we do not know it.

For I was early to embrace, as a day job, the profession of "quantitative finance." I became a "quant" and trader at the same time—a quant is a brand of industrial scientist who applies mathematical models of uncertainty to financial (or socioeconomic) data and complex financial instruments. Except that I was a quant exactly in reverse: I studied the flaws and the limits of these models, looking for the *Platonic fold* where they break down. Also I engaged in speculative trading, not "just tawk," which was rare for quants since they were prevented from "taking risks," their role being confined to analysis, not decision making. I was convinced that I was totally incompetent in predicting market prices—but that others were generally incompetent also but did not know it, or did not know that they were taking massive risks. Most traders were just "picking pennies in front of a streamroller," exposing themselves to the high-impact rare event yet sleeping like babies, unaware of it. Mine was the only job you could do if you thought of yourself as risk-hating, risk-aware, and highly ignorant.

Also, the technical baggage that comes with being a quant (a mixture of applied mathematics, engineering, and statistics), in addition to the immersion in practice, turned out to be very useful for someone wanting to be a philosopher.* First, when you spend a couple of decades doing mass-scale empirical work with data and taking risks based on such studies, you can easily spot elements in the texture of the world that the Platonified "thinker" is too brainwashed, or threatened, to see. Second, it allowed me to become formal and systematic in my thinking instead of wallowing in the anecdotal. Finally, both the philosophy of history and epistemology (the philosophy of knowledge) seemed inseparable from the empirical study of times series data, which is a succession of numbers in time, a sort of historical document containing numbers instead of words. And numbers are easy to process on computers. Studying historical data makes you conscious that history runs forward, not backward, and that it is messier than narrated accounts. Epistemology, the philosophy of history, and statistics aim at understanding truths, investigating the mechanisms that generate them, and separating regularity from the coincidental in historical matters. They all address the question of what one knows, except that they are all to be found in different buildings, so to speak.

The Four-Letter Word of Independence

That night, on October 19, 1987, I slept for twelve hours straight.

It was hard to tell my friends, all hurt in some manner by the crash, about this feeling of vindication. Bonuses at the time were a fraction of what they are today, but if my employer, First Boston, and the financial system survived until year-end, I would get the equivalent of a fellowship. This is sometimes called "f*** you money," which, in spite of its coarseness, means that it allows you to act like a Victorian gentleman, free from slavery. It is a psychological buffer: the capital is not so large as to make you spoiled-rich, but large enough to give you the freedom to choose a

* I specialized in complicated financial instruments called "derivatives," those that required advanced mathematics—but for which the errors for using the wrong mathematics were the greatest. The subject was new and attractive enough for me to get a doctorate in it.

Note that I was not able to build a career just by betting on Black Swans—there were not enough tradable opportunities. I could, on the other hand, avoid being exposed to them by protecting my portfolio against large losses. So, in order to eliminate the dependence on randomness, I focused on technical inefficiencies between complicated instruments, and on exploiting these opportunities without exposure to the rare event, before they disappeared as my competitors became technologically advanced. Later on in my career I discovered the easier (and less randomness laden) business of protecting, insurance-style, large portfolios against the Black Swan.

new occupation without excessive consideration of the financial rewards. It shields you from prostituting your mind and frees you from outside authority—any outside authority. (Independence is person-specific: I have always been taken aback at the high number of people in whom an astonishingly high income led to additional sycophancy as they became more dependent on their clients and employers and more addicted to making even more money.) While not substantial by some standards, it literally cured me of all financial ambition—it made me feel ashamed whenever I diverted time away from study for the pursuit of material wealth. Note that the designation *f*** you* corresponds to the exhilarating ability to pronounce that compact phrase *before* hanging up the phone.

These were the days when it was extremely common for traders to break phones when they lost money. Some resorted to destroying chairs, tables, or whatever would make noise. Once, in the Chicago pits, another trader tried to strangle me and it took four security guards to drag him away. He was irate because I was standing in what he deemed his "territory." Who would want to leave such an environment? Compare it to lunches in a drab university cafeteria with gentle-mannered professors discussing the latest departmental intrigue. So I stayed in the quant and trading businesses (I'm still there), but organized myself to do minimal but intense (and entertaining) work, focus only on the most technical aspects, never attend business "meetings," avoid the company of "achievers" and people in suits who don't read books, and take a sabbatical year for every three on average to fill up gaps in my scientific and philosophical culture. To slowly distill my single idea, I wanted to become a flâneur, a professional meditator, sit in cafés, lounge, unglued to desks and organization structures, sleep as long as I needed, read voraciously, and not owe any explanation to anybody. I wanted to be left alone in order to build, small steps at a time, an entire system of thought based on my Black Swan idea.

Limousine Philosopher

The war in Lebanon and the crash of 1987 seemed identical phenomena. It became obvious to me that nearly everyone had a mental blindspot in acknowledging the role of such events: it was as if they were not able to see these mammoths, or that they rapidly forgot about them. The answer was looking straight at me: it was a psychological, perhaps even biological, *blindness;* the problem lay not in the nature of events, but in the way we perceived them.

I end this autobiographical prelude with the following story. I had no defined specialty (outside of my day job), and wanted none. When people at cocktail parties asked me what I did for a living, I was tempted to answer, "I am a *skeptical empiricist* and a flâneur-reader, someone committed to getting very deep into an idea," but I made things simple by saying that I was a limousine driver.

Once, on a transatlantic flight, I found myself upgraded to first class next to an expensively dressed, high-powered lady dripping with gold and jewelry who continuously ate nuts (low-carb diet, perhaps), insisted on drinking only Evian, all the while reading the European edition of *The Wall Street Journal*. She kept trying to start a conversation in broken French, since she saw me reading a book (in French) by the sociologist-philosopher Pierre Bourdieu—which, ironically, dealt with the marks of social distinction. I informed her (in English) that I was a limousine driver, proudly insisting that I only drove "very upper-end" cars. An icy silence lasted the whole flight, and, although I could feel the tension, it allowed me to read in peace.

Chapter Two

YEVGENIA'S BLACK SWAN

Pink glasses and success—How Yevgenia stops marrying philosophers—I told you so

Five years ago, Yevgenia Nikolayevna Krasnova was an obscure and unpublished novelist, with an unusual background. She was a neuroscientist with an interest in philosophy (her first three husbands had been philosophers), and she got it into her stubborn Franco-Russian head to express her research and ideas in literary form. She dressed up her theories as stories, and mixed them with all manner of autobiographical commentary. She avoided the journalistic prevarications of contemporary narrative nonfiction ("On a clear April morning, John Smith left his house. . . ."). Foreign dialogue was always written in the original language, with translations appended like movie subtitles. She refused to dub into bad English conversations that took place in bad Italian.*

No publisher would have given her the time of day, except that there was, at the time, some interest in those rare scientists who could manage to express themselves in semi-understandable sentences. A few publishers agreed to speak with her; they hoped that she would grow up and write a "popular science book on consciousness." She received enough attention

* Her third husband was an Italian philosopher.

to get the courtesy of rejection letters and occasional insulting comments instead of the far more insulting and demeaning silence.

Publishers were confused by her manuscript. She could not even answer their first question: "Is this fiction or nonfiction?" Nor could she respond to the "Who is this book written for?" on the publishers' book proposal forms. She was told, "You need to understand who your audience is" and "amateurs write for themselves, professionals write for others." She was also told to conform to a precise genre because "bookstores do not like to be confused and need to know where to place a book on the shelves." One editor protectively added, "This, my dear friend, will only sell ten copies, including those bought by your ex-husbands and family members."

She had attended a famous writing workshop five years earlier and came out nauseated. "Writing well" seemed to mean obeying arbitrary rules that had grown into gospel, with the confirmatory reinforcement of what we call "experience." The writers she met were learning to retrofit what was deemed successful: they all tried to imitate stories that had appeared in past issues of *The New Yorker*—not realizing that most of what is new, by definition, cannot be modeled on past issues of *The New Yorker.* Even the idea of a "short story" was a me-too concept to Yevgenia. The workshop instructor, gentle but firm in his delivery, told her that her case was utterly hopeless.

Yegvenia ended up posting the entire manuscript of her main book, *A Story of Recursion,* on the Web. There it found a small audience, which included the shrewd owner of a small unknown publishing house, who wore pink-rimmed glasses and spoke primitive Russian (convinced that he was fluent). He offered to publish her, and agreed to her condition to keep her text completely unedited. He offered her a fraction of the standard royalty rate in return for her editorial stricture—he had so little to lose. She accepted since she had no choice.

It took five years for Yevgenia to graduate from the "egomaniac without anything to justify it, stubborn and difficult to deal with" category to "persevering, resolute, painstaking, and fiercely independent." For her book slowly caught fire, becoming one of the great and strange successes in literary history, selling millions of copies and drawing so-called critical acclaim. The start-up house has since become a big corporation, with a (polite) receptionist to greet visitors as they enter the main office. Her book has been translated into forty languages (even French). You see her picture everywhere. She is said to be a pioneer of something called the

Consilient School. Publishers now have a theory that "truck drivers who read books do not read books written for truck drivers" and hold that "readers despise writers who pander to them." A scientific paper, it is now understood, can hide trivialities or irrelevance with equations and jargon; consilient prose, by exposing an idea in raw form, allows it to be judged by the public.

Today, Yevgenia has stopped marrying philosophers (they argue too much), and she hides from the press. In classrooms, literary scholars discuss the many clues indicating the inevitability of the new style. The distinction between fiction and nonfiction is considered too archaic to withstand the challenges of modern society. It was so evident that we needed to remedy the fragmentation between art and science. After the fact, her talent was so obvious.

Many of the editors she later met blamed her for not coming to them, convinced that they would have immediately seen the merit in her work. In a few years, a literary scholar will write the essay "From Kundera to Krasnova," showing how the seeds of her work can be found in Kundera—a precursor who mixed essay and metacommentary (Yevgenia never read Kundera, but did see the movie version of one of his books—there was no commentary in the movie). A prominent scholar will show how the influence of Gregory Bateson, who injected autobiographical scenes into his scholarly research papers, is visible on every page (Yevgenia has never heard of Bateson).

Yevgenia's book is a Black Swan.

Chapter Three

THE SPECULATOR AND THE PROSTITUTE

On the critical difference between speculators and prostitutes—Fairness, unfairness, and Black Swans—Theory of knowledge and professional incomes—How Extremistan is not the best place to visit, except, perhaps, if you are a winner

Yevgenia's rise from the second basement to superstar is possible in only one environment, which I call Extremistan.* I will soon introduce the central distinction between the Black Swan–generating province of Extremistan and the tame, quiet, and uneventful province of Mediocristan.

THE BEST (WORST) ADVICE

When I play back in my mind all the "advice" people have given me, I see that only a couple of ideas have stuck with me for life. The rest has been mere words, and I am glad that I did not heed most of it. Most consisted of recommendations such as "be measured and reasonable in your statements," contradicting the Black Swan idea, since empirical reality is not "measured," and its own version of "reasonableness" does not corre-

* To those readers who Googled Yevgenia Krasnova, I am sorry to say that she is (officially) a fictional character.

spond to the conventional middlebrow definition. To be genuinely empirical is to reflect reality as faithfully as possible; to be honorable implies not fearing the appearance and consequences of being outlandish. The next time someone pesters you with unneeded advice, gently remind him of the fate of the monk whom Ivan the Terrible put to death for delivering uninvited (and moralizing) advice. It works as a short-term cure.

The most important piece of advice was, in retrospect, bad, but it was also, paradoxically, the most consequential, as it pushed me deeper into the dynamics of the Black Swan. It came when I was twenty-two, one February afternoon, in the corridor of a building at 3400 Walnut Street in Philadelphia, where I lived. A second-year Wharton student told me to get a profession that is "scalable," that is, one in which you are not paid by the hour and thus subject to the limitations of the amount of your labor. It was a very simple way to discriminate among professions and, from that, to generalize a separation between types of uncertainty—and it led me to the major philosophical problem, the problem of induction, which is the technical name for the Black Swan. It allowed me to turn the Black Swan from a logical impasse into an easy-to-implement solution, and, as we will see in the next chapters, to ground it in the texture of empirical reality.

How did career advice lead to such ideas about the nature of uncertainty? Some professions, such as dentists, consultants, or massage professionals, cannot be scaled: there is a cap on the number of patients or clients you can see in a given period of time. If you are a prostitute, you work by the hour and are (generally) paid by the hour. Furthermore, your presence is (I assume) necessary for the service you provide. If you open a fancy restaurant, you will at best steadily fill up the room (unless you franchise it). In these professions, no matter how highly paid, your income is subject to gravity. Your revenue depends on your continuous efforts more than on the quality of your decisions. Moreover, this kind of work is largely predictable: it will vary, but not to the point of making the income of a single day more significant than that of the rest of your life. In other words, it will not be Black Swan driven. Yevgenia Nikolayevna would not have been able to cross the chasm between underdog and supreme hero overnight had she been a tax accountant or a hernia specialist (but she would not have been an underdog either).

Other professions allow you to add zeroes to your output (and your income), if you do well, at little or no extra effort. Now being lazy, considering laziness as an asset, and eager to free up the maximum amount of time in my day to meditate and read, I immediately (but mistakenly) drew a conclusion. I separated the "idea" person, who sells an intellectual product in the form of a transaction or a piece of work, from the "labor" person, who sells you his work.

If you are an idea person, you do not have to work hard, only think intensely. You do the same work whether you produce a hundred units or a thousand. In quant trading, the same amount of work is involved in buying a hundred shares as in buying a hundred thousand, or even a million. It is the same phone call, the same computation, the same legal document, the same expenditure of brain cells, the same effort in verifying that the transaction is right. Furthermore, you can work from your bathtub or from a bar in Rome. You can use leverage as a replacement for work! Well, okay, I was a little wrong about trading: one cannot work from a bathtub, but, when done right, the job allows considerable free time.

The same property applies to recording artists or movie actors: you let the sound engineers and projectionists do the work; there is no need to show up at every performance in order to perform. Similarly, a writer expends the same effort to attract one single reader as she would to capture several hundred million. J. K. Rowling, the author of the Harry Potter books, does not have to write each book again every time someone wants to read it. But this is not so for a baker: he needs to bake every single piece of bread in order to satisfy each additional customer.

So the distinction between writer and baker, speculator and doctor, fraudster and prostitute, is a helpful way to look at the world of activities. It separates those professions in which one can add zeroes of income with no greater labor from those in which one needs to add labor and time (both of which are in limited supply)—in other words, those subjected to gravity.

BEWARE THE SCALABLE

But why was the advice from my fellow student bad?

If the advice was helpful, and it was, in creating a classification for ranking uncertainty and knowledge, it was a mistake as far as choices of profession went. It might have paid off for me, but only because I was lucky and happened to be "in the right place at the right time," as the saying goes. If I myself had to give advice, I would recommend someone pick a profession that is *not* scalable! A scalable profession is good only if you are successful; they are more competitive, produce monstrous inequalities, and are far more random, with huge disparities between efforts and rewards—a few can take a large share of the pie, leaving others out entirely at no fault of their own.

One category of profession is driven by the mediocre, the average, and the middle-of-the-road. In it, the mediocre is collectively consequential. The other has either giants or dwarves—more precisely, a very small number of giants and a huge number of dwarves.

Let us see what is behind the formation of unexpected giants—the Black Swan formation.

The Advent of Scalability

Consider the fate of Giaccomo, an opera singer at the end of the nineteenth century, before sound recording was invented. Say he performs in a small and remote town in central Italy. He is shielded from those big egos at La Scala in Milan and other major opera houses. He feels safe as his vocal cords will always be in demand somewhere in the district. There is no way for him to export his singing, and there is no way for the big guns to export theirs and threaten his local franchise. It is not yet possible for him to store his work, so his presence is needed at every performance, just as a barber is (still) needed today for every haircut. So the total pie is unevenly split, but only mildly so, much like your calorie consumption. It is cut in a few pieces and everyone has a share; the big guns have larger audiences and get more invitations than the small guy, but this is not too worrisome. Inequalities exist, but let us call them *mild*. There is no scalability yet, no way to double the largest in-person audience without having to sing twice.

Now consider the effect of the first music recording, an invention that introduced a great deal of injustice. Our ability to reproduce and repeat performances allows me to listen on my laptop to hours of background music of the pianist Vladimir Horowitz (now extremely dead) performing Rachmaninoff's *Preludes,* instead of to the local Russian émigré musician (still living), who is now reduced to giving piano lessons to generally untalented children for close to minimum wage. Horowitz, though dead, is putting the poor man out of business. I would rather listen to Vladimir Horowitz or Arthur Rubinstein for $10.99 a CD than pay $9.99 for one by some unknown (but very talented) graduate of the Juilliard School or the Prague Conservatory. If you ask me why I select Horowitz, I will answer that it is because of the order, rhythm, or passion, when in fact there are probably a legion of people I have never heard about, and will never hear about—those who did not make it to the stage, but who might play just as well.

Some people naïvely believe that the process of unfairness started with the gramophone, according to the logic that I just presented. I disagree. I am convinced that the process started much, much earlier, with our DNA, which stores information about our selves and allows us to repeat our performance without our being there by spreading our genes down the generations. Evolution is *scalable:* the DNA that wins (whether by luck or survival advantage) will reproduce itself, like a bestselling book or a successful record, and become pervasive. Other DNA will vanish. Just consider the difference between us humans (excluding financial economists and businessmen) and other living beings on our planet.

Furthermore, I believe that the big transition in social life came not with the gramophone, but when someone had the great but unjust idea to invent the alphabet, thus allowing us to store information and reproduce it. It accelerated further when another inventor had the even more dangerous and iniquitous notion of starting a printing press, thus promoting texts across boundaries and triggering what ultimately grew into a winner-take-all ecology. Now, what was so unjust about the spread of books? The alphabet allowed stories and ideas to be replicated with high fidelity and without limit, without any additional expenditure of energy on the author's part for the subsequent performances. He didn't even have to be alive for them—death is often a good career move for an author. This implies that those who, for some reason, start getting some attention can quickly reach more minds than others and displace the competitors from the bookshelves. In the days of bards and troubadours, everyone had an audience. A storyteller, like a baker or a coppersmith, had a market, and the assurance that none from far away could dislodge him from his territory. Today, a few take almost everything; the rest, next to nothing.

By the same mechanism, the advent of the cinema displaced neighborhood actors, putting the small guys out of business. But there is a difference. In pursuits that have a technical component, like being a pianist or a brain surgeon, talent is easy to ascertain, with subjective opinion playing a relatively small part. The inequity comes when someone perceived as being marginally better gets the whole pie.

In the arts—say the cinema—things are far more vicious. What we call "talent" generally comes from success, rather than its opposite. A great deal of empiricism has been done on the subject, most notably by Art De

Vany, an insightful and original thinker who singlemindedly studied wild uncertainty in the movies. He showed that, sadly, much of what we ascribe to skills is an after-the-fact attribution. The movie makes the actor, he claims—and a large dose of nonlinear luck makes the movie.

The success of movies depends severely on contagions. Such contagions do not just apply to the movies: they seem to affect a wide range of cultural products. It is hard for us to accept that people do not fall in love with works of art only for their own sake, but also in order to feel that they belong to a community. By imitating, we get closer to others—that is, other imitators. It fights solitude.

This discussion shows the difficulty in predicting outcomes in an environment of concentrated success. So for now let us note that the division between professions can be used to understand the division between types of random variables. Let us go further into the issue of knowledge, of inference about the unknown and the properties of the known.

SCALABILITY AND GLOBALIZATION

Whenever you hear a snotty (and frustrated) European middlebrow presenting his stereotypes about Americans, he will often describe them as "uncultured," "unintellectual," and "poor in math" because, unlike his peers, Americans are not into equation drills and the constructions middlebrows call "high culture"—like knowledge of Goethe's inspirational (and central) trip to Italy, or familiarity with the Delft school of painting. Yet the person making these statements is likely to be addicted to his iPod, wear blue jeans, and use Microsoft Word to jot down his "cultural" statements on his PC, with some Google searches here and there interrupting his composition. Well, it so happens that America is currently far, far more creative than these nations of museumgoers and equation solvers. It is also far more tolerant of bottom-up tinkering and undirected trial and error. And globalization has allowed the United States to specialize in the creative aspect of things, the production of concepts and ideas, that is, the scalable part of the products, and, increasingly, by exporting jobs, separate the less scalable components and assign them to those happy to be paid by the hour. There is more money in designing a shoe than in actually making it: Nike, Dell, and Boeing can get paid for just thinking, organizing, and leveraging their know-how and ideas while subcontracted factories in developing countries do the grunt work and engineers in cultured and mathematical states do the noncreative technical grind. The American economy has leveraged itself heavily on the idea generation, which explains why losing manufacturing jobs can be coupled with a rising standard of living. Clearly the drawback of a world economy where the payoff goes to ideas is higher inequality among the idea generators together with a greater role for both opportunity and luck—but I will leave the socioeconomic discussion for Part Three and focus here on knowledge.

TRAVELS INSIDE MEDIOCRISTAN

This scalable/nonscalable distinction allows us to make a clear-cut differentiation between two varieties of uncertainties, two types of randomness.

Let's play the following thought experiment. Assume that you round up a thousand people randomly selected from the general population and have them stand next to one another in a stadium. You can even include Frenchmen (but please, not too many out of consideration for the others in the group), Mafia members, non-Mafia members, and vegetarians.

Imagine the heaviest person you can think of and add him to that sample. Assuming he weighs three times the average, between four hundred and five hundred pounds, he will rarely represent more than a very small fraction of the weight of the entire population (in this case, about a half of a percent).

You can get even more aggressive. If you picked the heaviest biologically possible human on the planet (who yet can still be called a human), he would not represent more than, say, 0.6 percent of the total, a very negligible increase. And if you had ten thousand persons, his contribution would be vanishingly small.

In the utopian province of Mediocristan, particular events don't contribute much individually—only collectively. I can state the supreme law of Mediocristan as follows: *When your sample is large, no single instance will significantly change the aggregate or the total.* The largest observation will remain impressive, but eventually insignificant, to the sum.

I'll borrow another example from my friend Bruce Goldberg: your caloric consumption. Look at how much you consume per year—if you are classified as human, close to eight hundred thousand calories. No single day, not even Thanksgiving at your great-aunt's, will represent a large share of that. Even if you tried to kill yourself by eating, that day's calories would not seriously affect your yearly consumption.

Now, if I told you that it is possible to run into someone who weighs several thousand tons, or stands several hundred miles tall, you would be perfectly justified in having my frontal lobe examined, or in suggesting that I switch to science-fiction writing. But you cannot so easily rule out extreme variations with a different brand of quantities, to which we turn next.

The Strange Country of Extremistan

Consider by comparison the net worth of the thousand people you lined up in the stadium. Add to them the wealthiest person to be found on the planet—say, Bill Gates, the founder of Microsoft. Assume his net worth to be close to $80 billion—with the total capital of the others around a few million. How much of the total wealth would he represent? 99.9 percent? Indeed, all the others would represent no more than a rounding error for his net worth, the variation of his personal portfolio over the past second. For someone's weight to represent such a share, he would need to weigh fifty million pounds!

Try it again with, say, book sales. Line up a thousand authors (or people begging to get published, but calling themselves authors instead of waiters), and check their book sales. Then add the living writer who (currently) has the most readers. J. K. Rowling, the author of the Harry Potter series, with several hundred million books sold, will dwarf the remaining thousand authors with, say, collectively, a few hundred thousand readers at most.

Try it also with academic citations (the mention of one academic by another academic in a formal publication), media references, income, company size, and so on. Let us call these *social* matters, as they are man-made, as opposed to physical ones, like the size of waistlines.

In Extremistan, inequalities are such that one single observation can disproportionately impact the aggregate, or the total.

So while weight, height, and calorie consumption are from Mediocristan, wealth is not. Almost all social matters are from Extremistan. Another way to say it is that social quantities are informational, not physical: you cannot touch them. Money in a bank account is something important, but certainly *not physical.* As such it can take any value without necessitating the expenditure of energy. It is just a number!

Note that before the advent of modern technology, wars used to belong to Mediocristan. It is hard to kill many people if you need to slaughter them one at the time. Today, with tools of mass destruction, all it takes is a button, a nutcase, or a small error to wipe out the planet.

Look at the implication for the Black Swan. Extremistan can produce Black Swans, and does, since a few occurrences have had huge influences on history. This is the main idea of this book.

Extremistan and Knowledge

While this distinction (between Mediocristan and Extremistan) has severe ramifications for both social fairness and the dynamics of events, let us see its application to knowledge, which is where most of its value lies. If a Martian came to earth and engaged in the business of measuring the heights of the denizens of this happy planet, he could safely stop at a hundred humans to get a good picture of the average height. If you live in Mediocristan, you can be comfortable with what you have measured—provided that you know for sure that it comes from Mediocristan. You can also be comfortable with *what you have learned* from the data. The epistemological consequence is that with Mediocristan-style randomness it is not *possible** to have a Black Swan surprise such that a single event can dominate a phenomenon. *Primo,* the first hundred days should reveal all you need to know about the data. *Secondo,* even if you do have a surprise, as we saw in the case of the heaviest human, it would not be consequential.

If you are dealing with quantities from Extremistan, you will have trouble figuring out the average from any sample since it can depend so much on one single observation. The idea is not more difficult than that. In Extremistan, one unit can easily affect the total in a disproportionate way. In this world, you should always be suspicious of the knowledge you derive from data. This is a very simple test of uncertainty that allows you to distinguish between the two kinds of randomness. Capish?

What you can know from data in Mediocristan augments very rapidly with the supply of information. But knowledge in Extremistan grows slowly and erratically with the addition of data, some of it extreme, possibly at an unknown rate.

* I emphasize *possible* because the chance of these occurrences is typically in the order of one in several trillion trillion, as close to impossible as it gets.

Wild and Mild

If we follow my distinction of scalable versus nonscalable, we can see clear differences shaping up between Mediocristan and Extremistan. Here are a few examples.

Matters that seem to belong to Mediocristan (subjected to what we call type 1 randomness): height, weight, calorie consumption, income for a baker, a small restaurant owner, a prostitute, or an orthodontist; gambling profits (in the very special case, assuming the person goes to a casino and maintains a constant betting size), car accidents, mortality rates, "IQ" (as measured).

Matters that seem to belong to Extremistan (subjected to what we call type 2 randomness): wealth, income, book sales per author, book citations per author, name recognition as a "celebrity," number of references on Google, populations of cities, uses of words in a vocabulary, numbers of speakers per language, damage caused by earthquakes, deaths in war, deaths from terrorist incidents, sizes of planets, sizes of companies, stock ownership, height between species (consider elephants and mice), financial markets (but your investment manager does not know it), commodity prices, inflation rates, economic data. The Extremistan list is much longer than the prior one.

The Tyranny of the Accident

Another way to rephrase the general distinction is as follows: Mediocristan is where we must endure the tyranny of the collective, the routine, the obvious, and the predicted; Extremistan is where we are subjected to the tyranny of the singular, the accidental, the unseen, and the unpredicted. As hard as you try, you will never lose a lot of weight in a single day; you need the collective effect of many days, weeks, even months. Likewise, if you work as a dentist, you will never get rich in a single day—but you can do very well over thirty years of motivated, diligent, disciplined, and regular attendance to teeth-drilling sessions. If you are subject to Extremistan-based speculation, however, you can gain or lose your fortune in a single minute.

Table 1 summarizes the differences between the two dynamics, to which I will refer in the rest of the book; confusing the left column with the right one can lead to dire (or extremely lucky) consequences.

TABLE 1

Mediocristan	*Extremistan*
Nonscalable	Scalable
Mild or type 1 randomness	Wild (even superwild) or type 2 randomness
The most typical member is mediocre	The most "typical" is either giant or dwarf, i.e., there is no typical member
Winners get a small segment of the total pie	Winner-take-almost-all effects
Example: audience of an opera singer before the gramophone	Today's audience for an artist
More likely to be found in our ancestral environment	More likely to be found in our modern environment
Impervious to the Black Swan	Vulnerable to the Black Swan
Subject to gravity	There are no physical constraints on what a number can be
Corresponds (generally) to physical quantities, i.e., height	Corresponds to numbers, say, wealth
As close to utopian equality as reality can spontaneously deliver	Dominated by extreme winner-take-all inequality
Total is not determined by a single instance or observation	Total will be determined by a small number of extreme events
When you observe for a while you can get to know what's going on	It takes a long time to know what's going on
Tyranny of the collective	Tyranny of the accidental
Easy to predict from what you see and extend to what you do not see	Hard to predict from past information
History crawls	History makes jumps
Events are distributed* according to the "bell curve" (the GIF) or its variations	The distribution is either Mandelbrotian "gray" Swans (tractable scientifically) or totally intractable Black Swans

* What I call "probability distribution" here is the model used to calculate the odds of different events, how they are distributed. When I say that an event is distributed according to the "bell curve," I mean that the Gaussian bell curve (after C. F. Gauss; more on him later) can help provide probabilities of various occurrences.

This framework, showing that Extremistan is where most of the Black Swan action is, is only a rough approximation—please do not Platonify it; don't simplify it beyond what's necessary.

Extremistan does not always imply Black Swans. Some events can be rare and consequential, but somewhat predictable, particularly to those who are prepared for them and have the tools to understand them (instead of listening to statisticians, economists, and charlatans of the bell-curve variety). They are near–Black Swans. They are somewhat tractable scientifically—knowing about their incidence should lower your surprise; these events are rare but expected. I call this special case of "gray" swans Mandelbrotian randomness. This category encompasses the randomness that produces phenomena commonly known by terms such as *scalable, scale-invariant, power laws, Pareto-Zipf laws, Yule's law, Paretian-stable processes, Levy-stable,* and *fractal laws,* and we will leave them aside for now since they will be covered in some depth in Part Three. They are scalable, according to the logic of this chapter, but you can know a little more about *how* they scale since they share much with the laws of nature.

You can still experience severe Black Swans in Mediocristan, though not easily. How? You may forget that something is random, think that it is deterministic, then have a surprise. Or you can tunnel and miss on a source of uncertainty, whether mild or wild, owing to lack of imagination—most Black Swans result from this "tunneling" disease, which I will discuss in Chapter 9.

This has been a "literary" overview of the central distinction of this book, offering a trick to distinguish between what can belong in Mediocristan and what belongs in Extremistan. I said that I will get into a more thorough examination in Part Three, so let us focus on epistemology for now and see how the distinction affects our knowledge.

Chapter Four

ONE THOUSAND AND ONE DAYS, OR HOW NOT TO BE A SUCKER

Surprise, surprise—Sophisticated methods for learning from the future—Sextus was always ahead—The main idea is not to be a sucker—Let us move to Mediocristan, if we can find it

Which brings us to the Black Swan Problem in its original form.

Imagine someone of authority and rank, operating in a place where rank matters—say, a government agency or a large corporation. He could be a verbose political commentator on Fox News stuck in front of you at the health club (impossible to avoid looking at the screen), the chairman of a company discussing the "bright future ahead," a Platonic medical doctor who has categorically ruled out the utility of mother's milk (because he did not see anything special in it), or a Harvard Business School professor who does not laugh at your jokes. He takes what he knows a little too seriously.

Say that a prankster surprises him one day by surreptitiously sliding a thin feather up his nose during a moment of relaxation. How would his dignified pompousness fare after the surprise? Contrast his authoritative demeanor with the shock of being hit by something totally unexpected that he does not understand. For a brief moment, before he regains his bearings, you will see disarray in his face.

I confess having developed an incorrigible taste for this kind of prank during my first sleepaway summer camp. Introduced into the nostril of a sleeping camper, a feather would induce sudden panic. I spent part of my childhood practicing variations on the prank: in place of a thin feather you can roll the corner of a tissue to make it long and narrow. I got some practice on my younger brother. An equally effective prank would be to drop an ice cube down someone's collar when he expects it least, say during an official dinner. I had to stop these pranks as I got deeper into adulthood, of course, but I am often involuntarily hit with such an image when bored out of my wits in meetings with serious-looking businesspersons (dark suits and standardized minds) theorizing, explaining things, or talking about random events with plenty of "because" in their conversation. I zoom in on one of them and imagine the ice cube sliding down his back—it would be less fashionable, though certainly more spectacular, if you put a living mouse there, particularly if the person is ticklish and is wearing a tie, which would block the rodent's normal route of exit.*

Pranks can be compassionate. I remember in my early trading days, at age twenty-five or so, when money was starting to become easy. I would take taxis, and if the driver spoke skeletal English and looked particularly depressed, I'd give him a $100 bill as a tip, just to give him a little jolt and get a kick out of his surprise. I'd watch him unfold the bill and look at it with some degree of consternation ($1 million certainly would have been better but it was not within my means). It was also a simple hedonic experiment: it felt elevating to make someone's day with the trifle of $100. I eventually stopped; we all become stingy and calculating when our wealth grows and we start taking money seriously.

I don't need much help from fate to get larger-scale entertainment: reality provides such forced revisions of beliefs at quite a high frequency. Many are quite spectacular. In fact, the entire knowledge-seeking enterprise is based on taking conventional wisdom and accepted scientific beliefs and shattering them into pieces with new counterintuitive evidence, whether at a micro scale (every scientific discovery is an attempt to produce a micro–Black Swan) or at a larger one (as with Poincaré's and Einstein's relativity). Scientists may be in the business of laughing at their predecessors, but owing to an array of human mental dispositions, few realize that someone will laugh at their beliefs in the (disappointingly near) future. In this case, my readers and I are laughing at the *present* state of social knowledge. These big guns do not see the inevitable overhaul of

* I am safe since I never wear ties (except at funerals).

their work coming, which means that you can usually count on them to be in for a surprise.

HOW TO LEARN FROM THE TURKEY

The überphilosopher Bertrand Russell presents a particularly toxic variant of my surprise jolt in his illustration of what people in his line of business call the Problem of Induction or Problem of Inductive Knowledge (capitalized for its seriousness)—certainly the mother of all problems in life. How can we *logically* go from specific instances to reach general conclusions? How do we know what we know? How do we know that what we have observed from given objects and events suffices to enable us to figure out their other properties? There are traps built into any kind of knowledge gained from observation.

Consider a turkey that is fed every day. Every single feeding will firm up the bird's belief that it is the general rule of life to be fed every day by friendly members of the human race "looking out for its best interests," as a politician would say. On the afternoon of the Wednesday before Thanksgiving, something *unexpected* will happen to the turkey. It will incur a revision of belief.*

The rest of this chapter will outline the Black Swan problem in its original form: How can we know the future, given knowledge of the past; or, more generally, how can we figure out properties of the (infinite) unknown based on the (finite) known? Think of the feeding again: What can a turkey learn about what is in store for it tomorrow from the events of yesterday? A lot, perhaps, but certainly a little less than it thinks, and it is just that "little less" that may make all the difference.

The turkey problem can be generalized to any situation where *the same hand that feeds you can be the one that wrings your neck*. Consider the case of the increasingly integrated German Jews in the 1930s—or my description in Chapter 1 of how the population of Lebanon got lulled into a false sense of security by the appearance of mutual friendliness and tolerance.

Let us go one step further and consider induction's most *worrisome* aspect: learning backward. Consider that the turkey's experience may have, rather than no value, a *negative* value. It learned from observation, as we

* Since Russell's original example used a chicken, this is the enhanced North American adaptation.

FIGURE 1: ONE THOUSAND AND ONE DAYS OF HISTORY

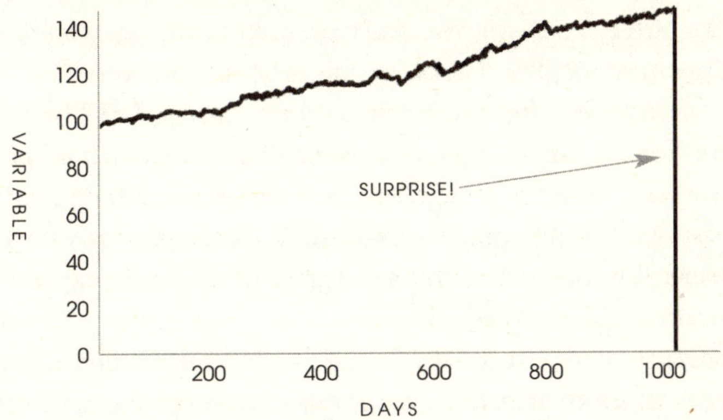

A turkey before and after Thanksgiving. The history of a process over a thousand days tells you nothing about what is to happen next. This naïve projection of the future from the past can be applied to anything.

are all advised to do (hey, after all, this is what is believed to be the scientific method). Its confidence increased as the number of friendly feedings grew, and it felt increasingly safe even though the slaughter was more and more imminent. Consider that the feeling of safety reached its maximum when the risk was at the highest! But the problem is even more general than that; it strikes at the nature of empirical knowledge itself. Something has worked in the past, until—well, it unexpectedly no longer does, and what we have learned from the past turns out to be at best irrelevant or false, at worst viciously misleading.

Figure 1 provides the prototypical case of the problem of induction as encountered in real life. You observe a hypothetical variable for one thousand days. It could be anything (with a few mild transformations): book sales, blood pressure, crimes, your personal income, a given stock, the interest on a loan, or Sunday attendance at a specific Greek Orthodox church. You subsequently derive *solely from past data* a few conclusions concerning the properties of the pattern with projections for the next thousand, even five thousand, days. On the one thousand and first day—boom! A big change takes place that is completely unprepared for by the past.

Consider the surprise of the Great War. After the Napoleonic conflicts, the world had experienced a period of peace that would lead any observer to believe in the disappearance of severely destructive conflicts. Yet, surprise! It turned out to be the deadliest conflict, up until then, in the history of mankind.

Note that after the event you start predicting the possibility of other outliers happening locally, that is, in the process you were just surprised by, *but not elsewhere*. After the stock market crash of 1987 half of America's traders braced for another one every October—not taking into account that there was no antecedent for the first one. We worry too late—ex post. Mistaking a naïve observation of the past as something definitive or representative of the future is the one and only cause of our inability to understand the Black Swan.

It would appear to a quoting dilettante—i.e., one of those writers and scholars who fill up their texts with phrases from some dead authority—that, as phrased by Hobbes, "from like antecedents flow like consequents." Those who believe in the unconditional benefits of past experience should consider this pearl of wisdom allegedly voiced by a famous ship's captain:

> *But in all my experience, I have never been in any accident . . . of any sort worth speaking about. I have seen but one vessel in distress in all my years at sea. I never saw a wreck and never have been wrecked nor was I ever in any predicament that threatened to end in disaster of any sort.*
>
> E. J. Smith, 1907, Captain, RMS *Titanic*

Captain Smith's ship sank in 1912 in what became the most talked-about shipwreck in history.*

* Statements like those of Captain Smith are so common that it is not even funny. In September 2006, a fund called Amaranth, ironically named after a flower that "never dies," had to shut down after it lost close to $7 billion in a few days, the most impressive loss in trading history (another irony: I shared office space with the traders). A few days prior to the event, the company made a statement to the effect that investors should not worry because they had twelve risk managers—people who use models of the past to produce risk measures on the odds of such an event. Even if they had one hundred and twelve risk managers, there would be no meaningful difference; they still would have blown up. Clearly you cannot manufacture more information than the past can deliver; if you buy one hundred copies of *The New York Times*, I am not too certain that it would help you gain incremental knowledge of the future. We just don't know how much information there is in the past.

Trained to Be Dull

Similarly, think of a bank chairman whose institution makes steady profits over a long time, only to lose everything in a single reversal of fortune. Traditionally, bankers of the lending variety have been pear-shaped, clean-shaven, and dress in possibly the most comforting and boring manner, in dark suits, white shirts, and red ties. Indeed, for their lending business, banks hire dull people and train them to be even more dull. But this is for show. If they look conservative, it is because their loans only go bust on rare, very rare, occasions. There is no way to gauge the effectiveness of their lending activity by observing it over a day, a week, a month, or . . . even a century! In the summer of 1982, large American banks lost close to all their past earnings (cumulatively), about everything they ever made in the history of American banking—everything. They had been lending to South and Central American countries that all defaulted at the same time—"an event of an exceptional nature." So it took just one summer to figure out that this was a sucker's business and that all their earnings came from a very risky game. All that while the bankers led everyone, especially themselves, into believing that they were "conservative." They are not conservative; just phenomenally skilled at self-deception by burying the possibility of a large, devastating loss under the rug. In fact, the travesty repeated itself a decade later, with the "risk-conscious" large banks once again under financial strain, many of them near-bankrupt, after the real-estate collapse of the early 1990s in which the now defunct savings and loan industry required a taxpayer-funded bailout of more than half a trillion dollars. The Federal Reserve bank protected them at our expense: when "conservative" bankers make profits, they get the benefits; when they are hurt, we pay the costs.

After graduating from Wharton, I initially went to work for Bankers Trust (now defunct). There, the chairman's office, rapidly forgetting about the story of 1982, broadcast the results of every quarter with an announcement explaining how smart, profitable, conservative (and good looking) they were. It was obvious that their profits were simply cash borrowed from destiny with some random payback time. I have no problem with risk taking, just please, please, do not call yourself conservative and act superior to other businesses who are not as vulnerable to Black Swans.

Another recent event is the almost-instant bankruptcy, in 1998, of a financial investment company (hedge fund) called Long-Term Capital Management (LTCM), which used the methods and risk expertise of two "Nobel economists," who were called "geniuses" but were in fact using phony, bell curve–style mathematics while managing to convince themselves that it was great science and thus turning the entire financial establishment into suckers. One of the largest trading losses ever in history took place in almost the blink of an eye, with no warning signal (more, much more on that in Chapter 17).*

A Black Swan Is Relative to Knowledge

From the standpoint of the turkey, the nonfeeding of the one thousand and first day is a Black Swan. For the butcher, it is not, since its occurrence is not unexpected. So you can see here that the Black Swan is a sucker's problem. In other words, it occurs relative to your expectation. You realize that you can eliminate a Black Swan by science (if you're able), or by keeping an open mind. Of course, like the LTCM people, you can create Black Swans with science, by giving people confidence that the Black Swan cannot happen—this is when science turns normal citizens into suckers.

Note that these events do not have to be *instantaneous* surprises. Some of the historical fractures I mention in Chapter 1 have lasted a few decades, like, say, the computer that brought consequential effects on society without its invasion of our lives being noticeable from day to day. Some Black Swans can come from the slow building up of incremental changes in the same direction, as with books that sell large amounts over years, never showing up on the bestseller lists, or from technologies that creep up on us slowly, but surely. Likewise, the growth of Nasdaq stocks in the late 1990s took a few years—but the growth would seem sharper if you were to plot it on a long historical line. Matters should be seen on some relative, not absolute, timescale: earthquakes last minutes, 9/11 lasted hours, but historical changes and technological implementations

* The main tragedy of the high impact–low probability event comes from the mismatch between the time taken to compensate someone and the time one needs to be comfortable that he is not making a bet against the rare event. People have an incentive to bet against it, or to game the system since they can be paid a bonus reflecting their yearly performance when in fact all they are doing is producing illusory profits that they will lose back one day. Indeed, the tragedy of capitalism is that since the quality of the returns is not observable from past data, owners of companies, namely shareholders, can be taken for a ride by the managers who show returns and cosmetic profitability but in fact might be taking hidden risks.

are Black Swans that can take decades. In general, positive Black Swans take time to show their effect while negative ones happen very quickly—it is much easier and much faster to destroy than to build. (During the Lebanese war, my parents' house in Amioun and my grandfather's house in a nearby village were destroyed in just a few hours, dynamited by my grandfather's enemies who controlled the area. It took seven thousand times longer—two years—to rebuild them. This asymmetry in timescales explains the difficulty in reversing time.)

A BRIEF HISTORY OF THE BLACK SWAN PROBLEM

This turkey problem (a.k.a. the problem of induction) is a very old one, but for some reason it is likely to be called "Hume's problem" by your local philosophy professor.

People imagine us skeptics and empiricists to be morose, paranoid, and tortured in our private lives, which may be the exact opposite of what history (and my private experience) reports. Like many of the skeptics I hang around with, Hume was jovial and a bon vivant, eager for literary fame, salon company, and pleasant conversation. His life was not devoid of anecdotes. He once fell into a swamp near the house he was building in Edinburgh. Owing to his reputation among the locals as an atheist, a woman refused to pull him out of it until he recited the Lord's Prayer and the Belief, which, being practical-minded, he did. But not before he argued with her about whether Christians were obligated to help their enemies. Hume looked unprepossessing. "He exhibited that preoccupied stare of the thoughtful scholar that so commonly impresses the undiscerning as imbecile," writes a biographer.

Strangely, Hume during his day was not mainly known for the works that generated his current reputation—he became rich and famous through writing a bestselling history of England. Ironically, when Hume was alive, his philosophical works, to which we now attach his fame, "fell deadborn off the presses," while the works for which he was famous at the time are now harder to find. Hume wrote with such clarity that he puts to shame almost all current thinkers, and certainly the entire German graduate curriculum. Unlike Kant, Fichte, Schopenhauer, and Hegel, Hume is the kind of thinker who is *sometimes* read by the person mentioning his work.

I often hear "Hume's problem" mentioned in connection with the problem of induction, but the problem is old, older than the interesting

Scotsman, perhaps as old as philosophy itself, maybe as old as olive-grove conversations. Let us go back into the past, as it was formulated with no less precision by the ancients.

Sextus the (Alas) Empirical

The violently antiacademic writer, and antidogma activist, Sextus Empiricus operated close to a millennium and a half before Hume, and formulated the turkey problem with great precision. We know very little about him; we do not know whether he was a philosopher or more of a copyist of philosophical texts by authors obscure to us today. We surmise that he lived in Alexandria in the second century of our era. He belonged to a school of medicine called "empirical," since its practitioners doubted theories and causality and relied on past experience as guidance in their treatment, though not putting much trust in it. Furthermore, they did not trust that anatomy revealed function too obviously. The most famous proponent of the empirical school, Menodotus of Nicomedia, who merged empiricism and philosophical skepticism, was said to keep medicine an art, not a "science," and insulate its practice from the problems of dogmatic science. The practice of medicine explains the addition of *empiricus* ("the empirical") to Sextus's name.

Sextus represented and jotted down the ideas of the school of the Pyrrhonian skeptics who were after some form of intellectual therapy resulting from the suspension of belief. Do you face the possibility of an adverse event? Don't worry. Who knows, it may turn out to be good for you. Doubting the consequences of an outcome will allow you to remain imperturbable. The Pyrrhonian skeptics were docile citizens who followed customs and traditions whenever possible, but taught themselves to systematically doubt everything, and thus attain a level of serenity. But while conservative in their habits, they were rabid in their fight against dogma.

Among the surviving works of Sextus's is a diatribe with the beautiful title *Adversos Mathematicos,* sometimes translated as *Against the Professors.* Much of it could have been written last Wednesday night!

Where Sextus is mostly interesting for my ideas is in his rare mixing of philosophy and decision making in his practice. He was a doer, hence classical scholars don't say nice things about him. The methods of empirical medicine, relying on seemingly purposeless trial and error, will be central to my ideas on planning and prediction, on how to benefit from the Black Swan.

In 1998, when I went out on my own, I called my research laboratory and trading firm Empirica, not for the same antidogmatist reasons, but on account of the far more depressing reminder that it took at least another fourteen centuries after the works of the school of empirical medicine before medicine changed and finally became adogmatic, suspicious of theorizing, profoundly skeptical, and evidence-based! Lesson? That awareness of a problem does not mean much—particularly when you have special interests and self-serving institutions in play.

Algazel

The third major thinker who dealt with the problem was the eleventh-century Arabic-language skeptic Al-Ghazali, known in Latin as Algazel. His name for a class of dogmatic scholars was *ghabi,* literally "the imbeciles," an Arabic form that is funnier than "moron" and more expressive than "obscurantist." Algazel wrote his own *Against the Professors,* a diatribe called *Tahafut al falasifa,* which I translate as "The Incompetence of Philosophy." It was directed at the school called *falasifah*—the Arabic intellectual establishment was the direct heir of the classical philosophy of the academy, and they managed to reconcile it with Islam through rational argument.

Algazel's attack on "scientific" knowledge started a debate with Averroës, the medieval philosopher who ended up having the most profound influence of any medieval thinker (on Jews and Christians, though not on Moslems). The debate between Algazel and Averroës was finally, but sadly, won by both. In its aftermath, many Arab religious thinkers integrated and exaggerated Algazel's skepticism of the scientific method, preferring to leave causal considerations to God (in fact it was a stretch of his idea). The West embraced Averroës's rationalism, built upon Aristotle's, which survived through Aquinas and the Jewish philosophers who called themselves Averroan for a long time. Many thinkers blame the Arabs' later abandonment of scientific method on Algazel's huge influence. He ended up fueling Sufi mysticism, in which the worshipper attempts to enter into communion with God, severing all connections with earthly matters. All of this came from the Black Swan problem.

The Skeptic, Friend of Religion

While the ancient skeptics advocated learned ignorance as the first step in honest inquiries toward truth, later medieval skeptics, both Moslems and Christians, used skepticism as a tool to avoid accepting what today we call science. Belief in the importance of the Black Swan problem, worries about induction, and skepticism can make some religious arguments more appealing, though in stripped-down, anticlerical, theistic form. This idea of relying on faith, not reason, was known as fideism. So there is a tradition of Black Swan skeptics who found solace in religion, best represented by Pierre Bayle, a French-speaking Protestant erudite, philosopher, and theologian, who, exiled in Holland, built an extensive philosophical architecture related to the Pyrrhonian skeptics. Bayle's writings exerted some considerable influence on Hume, introducing him to ancient skepticism—to the point where Hume took ideas wholesale from Bayle. Bayle's *Dictionnaire historique et critique* was the most read piece of scholarship of the eighteenth century, but like many of my French heroes (such as Frédéric Bastiat), Bayle does not seem to be part of the French curriculum and is nearly impossible to find in the original French language. Nor is the fourteenth-century Algazelist Nicolas of Autrecourt.

Indeed, it is not a well-known fact that the most complete exposition of the ideas of skepticism, until recently, remains the work of a powerful Catholic bishop who was an august member of the French Academy. Pierre-Daniel Huet wrote his *Philosophical Treatise on the Weaknesses of the Human Mind* in 1690, a remarkable book that tears through dogmas and questions human perception. Huet presents arguments against causality that are quite potent—he states, for instance, that any event can have an infinity of possible causes.

Both Huet and Bayle were erudites and spent their lives reading. Huet, who lived into his nineties, had a servant follow him with a book to read aloud to him during meals and breaks and thus avoid lost time. He was deemed the most read person in his day. Let me insist that erudition is important to me. It signals genuine intellectual curiosity. It accompanies an open mind and the desire to probe the ideas of others. Above all, an erudite can be dissatisfied with his own knowledge, and such dissatisfaction is a wonderful shield against Platonicity, the simplifications of the five-minute manager, or the philistinism of the overspecialized scholar. Indeed, scholarship without erudition can lead to disasters.

I Don't Want to Be a Turkey

But promoting philosophical skepticism is not quite the mission of this book. If awareness of the Black Swan problem can lead us into withdrawal and extreme skepticism, I take here the exact opposite direction. I am interested in deeds and true empiricism. So, this book was not written by a Sufi mystic, or even by a skeptic in the ancient or medieval sense, or even (we will see) in a philosophical sense, but by a practitioner whose principal aim is to not be a sucker in things that matter, period.

Hume was radically skeptical in the philosophical cabinet, but abandoned such ideas when it came to daily life, since he could not handle them. I am doing here the exact opposite: I am skeptical in matters that have implications for daily life. In a way, all I care about is making a decision without being the turkey.

Many middlebrows have asked me over the past twenty years, "How do you, Taleb, cross the street given your extreme risk consciousness?" or have stated the more foolish "You are asking us to take *no* risks." Of course I am not advocating total risk phobia (we will see that I favor an aggressive type of risk taking): all I will be showing you in this book is how to avoid crossing the street *blindfolded.*

They Want to Live in Mediocristan

I have just presented the Black Swan problem in its historical form: the central difficulty of generalizing from available information, or of learning from the past, the known, and the seen. I have also presented the list of those who, I believe, are the most relevant historical figures.

You can see that it is extremely convenient for us to assume that we live in Mediocristan. Why? Because it allows you to rule out these Black Swan surprises! The Black Swan problem either does not exist or is of small consequence if you live in Mediocristan!

Such an assumption magically drives away the problem of induction, which since Sextus Empiricus has been plaguing the history of thinking. The statistician can do away with epistemology.

Wishful thinking! We do not live in Mediocristan, so the Black Swan needs a different mentality. As we cannot push the problem under the rug, we will have to dig deeper into it. This is not a terminal difficulty—and we can even benefit from it.

. . .

Now, there are other themes arising from our blindness to the Black Swan:

a. We focus on preselected segments of the seen and generalize from it to the unseen: the error of confirmation.
b. We fool ourselves with stories that cater to our Platonic thirst for distinct patterns: the narrative fallacy.
c. We behave as if the Black Swan does not exist: human nature is not programmed for Black Swans.
d. What we see is not necessarily all that is there. History hides Black Swans from us and gives us a mistaken idea about the odds of these events: this is the distortion of silent evidence.
e. We "tunnel": that is, we focus on a few well-defined sources of uncertainty, on too specific a list of Black Swans (at the expense of the others that do not easily come to mind).

I will discuss each of the points in the next five chapters. Then, in the conclusion of Part One, I will show how, in effect, they are the *same* topic.

Chapter Five

CONFIRMATION SHMONFIRMATION!

I have so much evidence—Can Zoogles be (sometimes) Boogles?—Corroboration shmorroboration—Popper's idea

As much as it is ingrained in our habits and conventional wisdom, confirmation can be a dangerous error.

Assume I told you that I had evidence that the football player O. J. Simpson (who was accused of killing his wife in the 1990s) was not a criminal. Look, the other day I had breakfast with him and *he didn't kill anybody*. I am serious, I did not see him kill a single person. Wouldn't that *confirm* his innocence? If I said such a thing you would certainly call a shrink, an ambulance, or perhaps even the police, since you might think that I spent too much time in trading rooms or in cafés thinking about this Black Swan topic, and that my logic may represent such an immediate danger to society that I myself need to be locked up immediately.

You would have the same reaction if I told you that I took a nap the other day on the railroad track in New Rochelle, New York, and was not killed. Hey, look at me, I am alive, I would say, and that is evidence that lying on train tracks is risk-free. Yet consider the following. Look again at Figure 1 in Chapter 4; someone who observed the turkey's first thousand days (but not the shock of the thousand and first) would tell you, and rightly so, that there is *no evidence* of the possibility of large events, i.e.,

Black Swans. You are likely to confuse that statement, however, particularly if you do not pay close attention, with the statement that there is *evidence of no possible* Black Swans. Even though it is in fact vast, the logical distance between the two assertions will seem very narrow in your mind, so that one can be easily substituted for the other. Ten days from now, if you manage to remember the first statement at all, you will be likely to retain the second, inaccurate version—that there is *proof of no Black Swans*. I call this confusion the round-trip fallacy, since these statements are not *interchangeable*.

Such confusion of the two statements partakes of a trivial, very trivial (but crucial), logical error—but we are not immune to trivial, logical errors, nor are professors and thinkers particularly immune to them (complicated equations do not tend to cohabit happily with clarity of mind). Unless we concentrate very hard, we are likely to unwittingly simplify the problem because our minds routinely do so without our knowing it.

It is worth a deeper examination here.

Many people confuse the statement "almost all terrorists are Moslems" with "almost all Moslems are terrorists." Assume that the first statement is true, that 99 percent of terrorists are Moslems. This would mean that only about .001 percent of Moslems are terrorists, since there are more than one billion Moslems and only, say, ten thousand terrorists, one in a hundred thousand. So the logical mistake makes you (unconsciously) overestimate the odds of a randomly drawn individual Moslem person (between the age of, say, fifteen and fifty) being a terrorist by close to fifty thousand times!

The reader might see in this round-trip fallacy the unfairness of stereotypes—minorities in urban areas in the United States have suffered from the same confusion: even if most criminals come from their ethnic subgroup, most of their ethnic subgroup are not criminals, but they still suffer from discrimination by people who should know better.

"I never meant to say that the Conservatives are generally stupid. I meant to say that stupid people are generally Conservative," John Stuart Mill once complained. This problem is chronic: if you tell people that the key to success is not always skills, they think that you are telling them that it is never skills, always luck.

Our inferential machinery, that which we use in daily life, is not made for a complicated environment in which a statement changes markedly when its wording is slightly modified. Consider that in a primitive environment there is no consequential difference between the statements *most killers are wild animals* and *most wild animals are killers.* There is an error here, but it is almost inconsequential. Our statistical intuitions have not evolved for a habitat in which these subtleties can make a big difference.

Zoogles Are Not All Boogles

All zoogles are boogles. You saw a boogle. Is it a zoogle? Not necessarily, *since not all boogles are zoogles;* adolescents who make a mistake in answering this kind of question on their SAT test might not make it to college. Yet another person can get very high scores on the SATs and still feel a chill of fear when someone from the wrong side of town steps into the elevator. This inability to automatically transfer knowledge and sophistication from one situation to another, or from theory to practice, is a quite disturbing attribute of human nature.

Let us call it the *domain specificity* of our reactions. By domain-specific I mean that our reactions, our mode of thinking, our intuitions, depend on the context in which the matter is presented, what evolutionary psychologists call the "domain" of the object or the event. The classroom is a domain; real life is another. We react to a piece of information not on its logical merit, but on the basis of which framework surrounds it, and how it registers with our social-emotional system. Logical problems approached one way in the classroom might be treated differently in daily life. Indeed they *are* treated differently in daily life.

Knowledge, even when it is exact, does not often lead to appropriate actions because we tend to forget what we know, or forget how to process it properly if we do not pay attention, even when we are experts. Statisticians, it has been shown, tend to leave their brains in the classroom and engage in the most trivial inferential errors once they are let out on the streets. In 1971, the psychologists Danny Kahneman and Amos Tversky plied professors of statistics with statistical questions not phrased as statistical questions. One was similar to the following (changing the example for clarity): Assume that you live in a town with two hospitals—one large, the other small. On a given day 60 percent of those born in one of the two hospitals are boys. Which hospital is it likely to be? Many statisticians made the equivalent of the mistake (during a casual conversation) of choosing the larger hospital, when in fact the very basis of statistics is that large samples are more stable and should fluctuate less from the long-term average—here, 50 percent for each of the sexes—than smaller samples.

These statisticians would have flunked their own exams. During my days as a quant I counted hundreds of such severe inferential mistakes made by statisticians who forgot that they were statisticians.

For another illustration of the way we can be ludicrously domain-specific in daily life, go to the luxury Reebok Sports Club in New York City, and look at the number of people who, after riding the escalator for a couple of floors, head directly to the StairMasters.

This domain specificity of our inferences and reactions works both ways: some problems we can understand in their applications but not in textbooks; others we are better at capturing in the textbook than in the practical application. People can manage to effortlessly solve a problem in a social situation but struggle when it is presented as an abstract logical problem. We tend to use different mental machinery—so-called modules—in different situations: our brain lacks a central all-purpose computer that starts with logical rules and applies them equally to all possible situations.

And as I've said, we can commit *a logical mistake in reality but not in the classroom*. This asymmetry is best visible in cancer detection. Take doctors examining a patient for signs of cancer; tests are typically done on patients who want to know if they are cured or if there is "recurrence." (In fact, recurrence is a misnomer; it simply means that the treatment did not kill all the cancerous cells and that these undetected malignant cells have started to multiply out of control.) It is not feasible, in the present state of technology, to examine every single one of the patient's cells to see if all of them are nonmalignant, so the doctor takes a sample by scanning the body with as much precision as possible. Then she makes an assumption about what she did not see. I was once taken aback when a doctor told me after a routine cancer checkup, "Stop worrying, we have evidence of cure." "Why?" I asked. "There is evidence of *no* cancer" was the reply. "How do you know?" I asked. He replied, "The scan is negative." Yet he went around calling himself doctor!

An acronym used in the medical literature is NED, which stands for No Evidence of Disease. There is no such thing as END, Evidence of No Disease. Yet my experience discussing this matter with plenty of doctors, even those who publish papers on their results, is that many slip into the round-trip fallacy during conversation.

Doctors in the midst of the scientific arrogance of the 1960s looked down at mothers' milk as something primitive, as if it could be replicated by their laboratories—not realizing that mothers' milk might include useful components that could have eluded their scientific understanding—a simple confusion of *absence of evidence* of the benefits of mothers' milk with *evidence of absence* of the benefits (another case of Platonicity as "it did not make sense" to breast-feed when we could simply use bottles). Many people paid the price for this naïve inference: those who were not breast-fed as infants turned out to be at an increased risk of a collection of health problems, including a higher likelihood of developing certain types of cancer—there had to be in mothers' milk some necessary nutrients that still elude us. Furthermore, benefits to mothers who breast-feed were also neglected, such as a reduction in the risk of breast cancer.

Likewise with tonsils: the removal of tonsils may lead to a higher incidence of throat cancer, but for decades doctors never suspected that this "useless" tissue might actually have a use that escaped their detection. The same with the dietary fiber found in fruits and vegetables: doctors in the 1960s found it useless because they saw no immediate evidence of its necessity, and so they created a malnourished generation. Fiber, it turns out, acts to slow down the absorption of sugars in the blood and scrapes the intestinal tract of precancerous cells. Indeed medicine has caused plenty of damage throughout history, owing to this simple kind of inferential confusion.

I am not saying here that doctors should not have beliefs, only that some kinds of definitive, closed beliefs need to be avoided—this is what Menodotus and his school seemed to be advocating with their brand of skeptical-empirical medicine that avoided theorizing. Medicine has gotten better—but many kinds of knowledge have not.

Evidence

By a mental mechanism I call naïve empiricism, we have a natural tendency to look for instances that confirm our story and our vision of the world—these instances are always easy to find. Alas, with tools, and fools, anything can be easy to find. You take past instances that corroborate your theories and you treat them as *evidence*. For instance, a diplomat will show you his "accomplishments," not what he failed to do. Mathematicians will try to convince you that their science is useful to society by pointing out instances where it proved helpful, not those where it was a waste of time, or, worse, those numerous mathematical applications that inflicted a severe cost on society owing to the highly unempirical nature of elegant mathematical theories.

Even in testing a hypothesis, we tend to look for instances where the hypothesis proved true. Of course we can easily find confirmation; all we have to do is look, or have a researcher do it for us. I can *find confirmation* for just about anything, the way a skilled London cabbie can find traffic to increase the fare, even on a holiday.

Some people go further and give me examples of events that we have been able to foresee with some success—indeed there are a few, like landing a man on the moon and the economic growth of the twenty-first century. One can find plenty of "counterevidence" to the points in this book, the best being that newspapers are excellent at predicting movie and theater schedules. Look, I predicted yesterday that the sun would rise today, and it did!

NEGATIVE EMPIRICISM

The good news is that there is a way around this naïve empiricism. I am saying that a series of corroborative facts is not *necessarily* evidence. Seeing white swans does not confirm the nonexistence of black swans. There is an exception, however: I know what statement is wrong, but not necessarily what statement is correct. If I see a black swan I can certify that *all swans are not white*! If I see someone kill, then I can be practically certain that he is a criminal. If I don't see him kill, I cannot be certain that he is innocent. The same applies to cancer detection: the finding of a single malignant tumor proves that you have cancer, but the absence of such a finding cannot allow you to say with certainty that you are cancer-free.

We can get closer to the truth by negative instances, not by verification! It is misleading to build a general rule from observed facts. Contrary to conventional wisdom, our body of knowledge does not increase from a series of confirmatory observations, like the turkey's. But there are some things I can remain skeptical about, and others I can safely consider certain. This makes the consequences of observations one-sided. It is not much more difficult than that.

This asymmetry is immensely practical. It tells us that we do not have to be complete skeptics, just semiskeptics. The subtlety of real life over the books is that, in your decision making, you need be interested only in one side of the story: if you seek *certainty* about whether the patient has cancer, not *certainty* about whether he is healthy, then you might be satisfied with negative inference, since it will supply you the certainty you seek. So we can learn a lot from data—but not as much as we expect. Sometimes a lot of data can be meaningless; at other times one single piece of information can be very meaningful. It is true that a thousand days cannot prove you right, but one day can prove you to be wrong.

The person who promoted this idea of one-sided semiskepticism is Sir Doktor Professor Karl Raimund Popper, who may be the only philosopher of science who is actually read and discussed by actors in the real world (though not as enthusiastically by professional philosophers). As I am writing these lines, a black-and-white picture of him is hanging on the wall of my study. It was a gift I got in Munich from the essayist Jochen Wegner, who, like me, considers Popper to be about all "we've got" among modern thinkers—well, almost. He writes to us, not to other philosophers. "We" are the empirical decision makers who hold that uncertainty is our discipline, and that understanding how to act under conditions of incomplete information is the highest and most urgent human pursuit.

Popper generated a large-scale theory around this asymmetry, based on a technique called "falsification" (to falsify is to prove wrong) meant to distinguish between science and nonscience, and people immediately started splitting hairs about its technicalities, even though it is not the most interesting, or the most original, of Popper's ideas. This idea about the asymmetry of knowledge is so liked by practitioners, because it is obvious to them; it is the way they run their business. The philosopher *maudit* Charles Sanders Peirce, who, like an artist, got only posthumous respect, also came up with a version of this Black Swan solution when Popper was wearing diapers—some people even called it the Peirce-Popper approach. Popper's far more powerful and original idea is the "open" society, one that relies on skepticism as a modus operandi, refusing and resisting definitive truths. He accused Plato of closing our minds, according to the arguments I described in the Prologue. But Popper's biggest idea was his insight concerning the fundamental, severe, and incurable unpredictability of the world, and that I will leave for the chapter on prediction.*

Of course, it is not so easy to "falsify," i.e., to state that something is wrong with full certainty. Imperfections in your testing method may yield a mistaken "no." The doctor discovering cancer cells might have faulty

* Neither Peirce nor Popper was the first to come up with this asymmetry. The philosopher Victor Brochard mentioned the importance of negative empiricism in 1878, as if it were a matter held by the empiricists to be the sound way to do business—ancients understood it implicitly. Out-of-print books deliver many surprises.

equipment causing optical illusions; or he could be a bell-curve-using economist disguised as a doctor. An eyewitness to a crime might be drunk. *But it remains the case that you know what is wrong with a lot more confidence than you know what is right.* All pieces of information are not equal in importance.

Popper introduced the mechanism of conjectures and refutations, which works as follows: you formulate a (bold) conjecture and you start looking for the observation that would prove you wrong. This is the alternative to our search for confirmatory instances. If you think the task is easy, you will be disappointed—few humans have a natural ability to do this. I confess that I am not one of them; it does not come naturally to me.

Counting to Three

Cognitive scientists have studied our natural tendency to look only for corroboration; they call this vulnerability to the corroboration error the *confirmation bias*. There are some experiments showing that people focus only on the books read in Umberto Eco's library. You can test a given rule either directly, by looking at instances where it works, or indirectly, by focusing on where it does not work. As we saw earlier, disconfirming instances are far more powerful in establishing truth. Yet we tend to not be aware of this property.

The first experiment I know of concerning this phenomenon was done by the psychologist P. C. Wason. He presented subjects with the three-number sequence 2, 4, 6, and asked them to try to guess the rule generating it. Their method of guessing was to produce other three-number sequences, to which the experimenter would respond "yes" or "no" depending on whether the new sequences were consistent with the rule. Once confident with their answers, the subjects would formulate the rule. (Note the similarity of this experiment to the discussion in Chapter 1 of the way history presents itself to us: assuming history is generated according to some logic, we see only the events, never the rules, but need to guess how it works.) The correct rule was "numbers in ascending order," nothing more. Very few subjects discovered it because in order to do so they had to offer a series in descending order (that the experimenter would say "no" to). Wason noticed that the subjects had a rule in mind, but gave him examples aimed at confirming it instead of trying to supply series that were inconsistent with their hypothesis. Subjects tenaciously kept trying to confirm the rules that *they* had made up.

This experiment inspired a collection of similar tests, of which another example: Subjects were asked which questions to ask to find out whether a person was extroverted or not, purportedly for another type of experiment. It was established that subjects supplied mostly questions for which a "yes" answer would *support* the hypothesis.

But there are exceptions. Among them figure chess grand masters, who, it has been shown, actually do focus on where a speculative move might be weak; rookies, by comparison, look for confirmatory instances instead of falsifying ones. But don't play chess to practice skepticism. Scientists believe that it is the search for their own weaknesses that makes them good chess players, not the practice of chess that turns them into skeptics. Similarly, the speculator George Soros, when making a financial bet, keeps looking for instances that would prove his initial theory wrong. This, perhaps, is true self-confidence: the ability to look at the world without the need to find signs that stroke one's ego.*

Sadly, the notion of corroboration is rooted in our intellectual habits and discourse. Consider this comment by the writer and critic John Updike: "When Julian Jaynes . . . speculates that until late in the second millennium B.C. men had no consciousness but were automatically obeying the voices of gods, we are astounded but compelled to follow this remarkable thesis through all the corroborative evidence." Jaynes's thesis may be right, but, Mr. Updike, the central problem of knowledge (and the point of this chapter) is that there is no such animal as *corroborative* evidence.

Saw Another Red Mini!

The following point further illustrates the absurdity of confirmation. If you believe that witnessing an additional white swan will bring confirmation that there are no black swans, then you should also accept the statement, on purely logical grounds, that the sighting of a red Mini Cooper should confirm that there are *no black swans*.

Why? Just consider that the statement "all swans are white" implies

* This confirmation problem pervades our modern life, since most conflicts have at their root the following mental bias: when Arabs and Israelis watch news reports they see different stories in the same succession of events. Likewise, Democrats and Republicans look at different parts of the same data and never converge to the same opinions. Once your mind is inhabited with a certain view of the world, you will tend to only consider instances proving you to be right. Paradoxically, the more information you have, the more justified you will feel in your views.

that *all nonwhite objects are not swans.* What confirms the latter statement should confirm the former. Therefore, the sighting of a nonwhite object that is not a swan should bring such confirmation. This argument, known as Hempel's raven paradox, was rediscovered by my friend the (thinking) mathematician Bruno Dupire during one of our intense meditating walks in London—one of those intense walk-discussions, intense to the point of our not noticing the rain. He pointed to a red Mini and shouted, "Look, Nassim, look! No Black Swan!"

Not Everything

We are not naïve enough to believe that someone will be immortal because we have never seen him die, or that someone is innocent of murder because we have never seen him kill. The problem of naïve generalization does not plague us everywhere. But such smart pockets of inductive skepticism tend to involve events that we have encountered in our natural environment, matters from which we have learned to avoid foolish generalization.

For instance, when children are presented with the picture of a single member of a group and are asked to guess the properties of other unseen members, they are capable of selecting *which* attributes to generalize. Show a child a photograph of someone overweight, tell her that he is a member of a tribe, and ask her to describe the rest of the population: she will (most likely) not jump to the conclusion that all the members of the tribe are weight-challenged. But she would respond differently to generalizations involving skin color. If you show her people of dark complexion and ask her to describe their co-tribesmen, she will assume that they too have dark skin.

So it seems that we are endowed with specific and elaborate inductive instincts showing us the way. Contrary to the opinion held by the great David Hume, and that of the British empiricist tradition, that *belief arises from custom,* as they assumed that we learn generalizations solely from experience and empirical observations, it was shown from studies of infant behavior that we come equipped with mental machinery that causes us to *selectively* generalize from experiences (i.e., to selectively acquire inductive learning in some domains but remain skeptical in others). By doing so, we are not learning from a mere thousand days, but benefiting, thanks to evolution, from the learning of our ancestors—which found its way into our biology.

Back to Mediocristan

And we may have learned things wrong from our ancestors. I speculate here that we probably inherited the instincts adequate for survival in the East African Great Lakes region where we presumably hail from, but these instincts are certainly not well adapted to the present, post-alphabet, intensely informational, and statistically complex environment.

Indeed our environment is a bit more complex than we (and our institutions) seem to realize. How? The modern world, being Extremistan, is dominated by rare—very rare—events. It can deliver a Black Swan after thousands and thousands of white ones, so we need to withhold judgment for longer than we are inclined to. As I said in Chapter 3, it is impossible—biologically impossible—to run into a human several hundred miles tall, so our intuitions rule these events out. But the sales of a book or the magnitude of social events do not follow such strictures. It takes a lot more than a thousand days to accept that a writer is ungifted, a market will not crash, a war will not happen, a project is hopeless, a country is "our ally," a company will not go bust, a brokerage-house security analyst is not a charlatan, or a neighbor will not attack us. In the distant past, humans could make inferences far more accurately and quickly.

Furthermore, the sources of Black Swans today have multiplied beyond measurability.* In the primitive environment they were limited to newly encountered wild animals, new enemies, and abrupt weather changes. These events were repeatable enough for us to have built an innate fear of them. This instinct to make inferences rather quickly, and to "tunnel" (i.e., focus on a small number of sources of uncertainty, or causes of known Black Swans) remains rather ingrained in us. This instinct, in a word, is our predicament.

* Clearly, weather-related and geodesic events (such as tornadoes and earthquakes) have not changed much over the past millennium, but what have changed are the socioeconomic consequences of such occurrences. Today, an earthquake or hurricane commands more and more severe economic consequences than it did in the past because of the interlocking relationships between economic entities and the intensification of the "network effects" that we will discuss in Part Three. Matters that used to have mild effects now command a high impact. Tokyo's 1923 earthquake caused a drop of about a third in Japan's GNP. Extrapolating from the tragedy of Kobe in 1994, we can easily infer that the consequences of another such earthquake in Tokyo would be far costlier than that of its predecessor.

Chapter Six

THE NARRATIVE FALLACY

The cause of the because—How to split a brain—Effective methods of pointing at the ceiling—Dopamine will help you win—I will stop riding motorcycles (but not today)—Both empirical and psychologist? Since when?

ON THE CAUSES OF MY REJECTION OF CAUSES

During the fall of 2004, I attended a conference on aesthetics and science in Rome, perhaps the best possible location for such a meeting since aesthetics permeates everything there, down to one's personal behavior and tone of voice. At lunch, a prominent professor from a university in southern Italy greeted me with extreme enthusiasm. I had listened earlier that morning to his impassioned presentation; he was so charismatic, so convinced, and so convincing that, although I could not understand much of what he said, I found myself fully agreeing with everything. I could only make out a sentence here and there, since my knowledge of Italian worked better in cocktail parties than in intellectual and scholarly venues. At some point during his speech, he turned all red with anger—thus convincing me (and the audience) that he was definitely right.

He assailed me during lunch to congratulate me for showing the effects of those causal links that are more prevalent in the human mind than in reality. The conversation got so animated that we stood together near the buffet table, blocking the other delegates from getting close to the food. He was speaking accented French (with his hands), I was answering in primitive Italian (with my hands), and we were so vivacious that the other guests were afraid to interrupt a conversation of such importance and animation. He was emphatic about my previous book on randomness, a sort of angry trader's reaction against blindness to luck in life and in the markets, which had been published there under the musical title *Giocati dal caso*. I had been lucky to have a translator who knew almost more about the topic than I did, and the book found a small following among Italian intellectuals. "I am a huge fan of your ideas, but I feel slighted. These are truly mine too, and you wrote the book that I (almost) planned to write," he said. "You are a lucky man; you presented in such a comprehensive way the effect of chance on society and the overestimation of cause and effect. You show how stupid we are to systematically try to *explain* skills."

He stopped, then added, in a calmer tone: "But, *mon cher ami*, let me tell you *quelque chose* [uttered very slowly, with his thumb hitting his index and middle fingers]: had you grown up in a Protestant society where people are told that efforts are linked to rewards and individual responsibility is emphasized, you would never have seen the world in such a manner. You were able to see luck and separate cause and effect *because* of your Eastern Orthodox Mediterranean heritage." He was using the French *à cause*. And he was so convincing that, for a minute, I agreed with his interpretation.

We like stories, we like to summarize, and we like to simplify, i.e., to reduce the dimension of matters. The first of the problems of human nature that we examine in this section, the one just illustrated above, is what I call the *narrative fallacy*. (It is actually a fraud, but, to be more polite, I will call it a fallacy.) The fallacy is associated with our vulnerability to overinterpretation and our predilection for compact stories over raw truths. It severely distorts our mental representation of the world; it is particularly acute when it comes to the rare event.

Notice how my thoughtful Italian fellow traveler shared my militancy against overinterpretation and against the overestimation of cause, yet was unable to see me and my work without a reason, a cause, tagged to both, as anything other than part of a story. He had to *invent* a cause. Furthermore, he was not aware of his having fallen into the causation trap, nor was I immediately aware of it myself.

The narrative fallacy addresses our limited ability to look at sequences of facts without weaving an explanation into them, or, equivalently, forcing a logical link, an *arrow of relationship,* upon them. Explanations bind facts together. They make them all the more easily remembered; they help them *make more sense*. Where this propensity can go wrong is when it increases our *impression* of understanding.

This chapter will cover, just like the preceding one, a single problem, but seemingly in different disciplines. The problem of narrativity, although extensively studied in one of its versions by psychologists, is not so "psychological": something about the way disciplines are designed masks the point that it is more generally a problem of *information*. While narrativity comes from an ingrained biological need to reduce dimensionality, robots would be prone to the same process of reduction. Information *wants* to be reduced.

To help the reader locate himself: in studying the problem of induction in the previous chapter, we examined what could be inferred about the unseen, what lies *outside* our information set. Here, we look at the seen, what lies *within* the information set, and we examine the distortions in the act of processing it. There is plenty to say on this topic, but the angle I take concerns narrativity's simplification of the world around us and its effects on our perception of the Black Swan and wild uncertainty.

SPLITTING BRAINS

Ferreting out antilogics is an exhilarating activity. For a few months, you experience the titillating sensation that you've just entered a new world. After that, the novelty fades, and your thinking returns to business as usual. The world is dull again until you find another subject to be excited about (or manage to put another hotshot in a state of total rage).

For me, one such antilogic came with the discovery—thanks to the literature on cognition—that, counter to what everyone believes, *not theorizing* is an act—that theorizing can correspond to the absence of willed activity, the "default" option. It takes considerable effort to see facts (and remember them) while withholding judgment and resisting explanations. And this theorizing disease is rarely under our control: it is largely anatomical, part of our biology, so fighting it requires fighting one's own self. So the ancient skeptics' precepts to withhold judgment go against our nature. Talk is cheap, a problem with advice-giving philosophy we will see in Chapter 13.

Try to be a true skeptic with respect to your interpretations and you will be worn out in no time. You will also be humiliated for resisting to theorize. (There are tricks to achieving true skepticism; but you have to go through the back door rather than engage in a frontal attack on yourself.) Even from an anatomical perspective, it is impossible for our brain to see anything in raw form without some interpretation. We may not even always be conscious of it.

Post hoc rationalization. In an experiment, psychologists asked women to select from among twelve pairs of nylon stockings the ones they preferred. The researchers then asked the women their reasons for their choices. Texture, "feel," and color featured among the selected reasons. All the pairs of stockings were, in fact, identical. The women supplied backfit, *post hoc* explanations. Does this suggest that we are better at explaining than at understanding? Let us see.

A series of famous experiments on split-brain patients gives us convincing physical—that is, biological—evidence of the automatic aspect of the act of interpretation. There appears to be a sense-making organ in us—though it may not be easy to zoom in on it with any precision. Let us see how it is detected.

Split-brain patients have no connection between the left and the right sides of their brains, which prevents information from being shared between the two cerebral hemispheres. These patients are jewels, rare and invaluable for researchers. You literally have two different persons, and you can communicate with each one of them separately; the differences between the two individuals give you some indication about the specialization of each of the hemispheres. This splitting is usually the result of surgery to remedy more serious conditions like severe epilepsy; no, scientists in Western countries (and most Eastern ones) are no longer allowed to cut human brains in half, even if it is for the pursuit of knowledge and wisdom.

Now, say that you induced such a person to perform an act—raise his finger, laugh, or grab a shovel—in order to ascertain how he ascribes a reason to his act (when in fact you know that there is no reason for it other than your inducing it). If you ask the right hemisphere, here isolated from the left side, to perform the action, then ask the other hemisphere for an explanation, the patient will invariably offer some interpretation: "I was pointing at the ceiling in order to . . . ," "I saw something interesting on the wall," or, if you ask this author, I will offer my usual "because I am originally from the Greek Orthodox village of Amioun, northern Lebanon," et cetera.

Now, if you do the opposite, namely instruct the isolated left hemisphere of a right-handed person to perform an act and ask the right hemisphere for the reasons, you will be plainly told, "I don't know." Note that the left hemisphere is where language and deduction generally reside. I warn the reader hungry for "science" against attempts to build a neural map: all I'm trying to show is the biological basis of this tendency toward causality, not its precise location. There are reasons for us to be suspicious of these "right brain/left brain" distinctions and subsequent pop-science generalizations about personality. Indeed, the idea that the left brain controls language may not be so accurate: the left brain seems more precisely to be where pattern recognition resides, and it may control language only insofar as language has a pattern-recognition attribute. Another difference between the hemispheres is that the right brain deals with novelty. It tends to see series of facts (the particular, or the trees) while the left one perceives the patterns, the gestalt (the general, or the forest).

To see an illustration of our biological dependence on a story, consider the following experiment. First, read this:

A BIRD IN THE
THE HAND IS WORTH
TWO IN THE BUSH

Do you see anything unusual? Try again.*

The Sydney-based brain scientist Alan Snyder (who has a Philadelphia accent) made the following discovery. If you inhibit the left hemisphere of a right-handed person (more technically, by directing low-frequency magnetic pulses into the left frontotemporal lobes), you lower his rate of error in reading the above caption. Our propensity to impose meaning and concepts blocks our awareness of the details making up the concept. However, if you zap people's left hemispheres, they become more realistic—they can draw better and with more verisimilitude. Their minds become better at seeing the objects themselves, cleared of theories, narratives, and prejudice.

Why is it hard to avoid interpretation? It is key that, as we saw with the vignette of the Italian scholar, brain functions often operate outside our awareness. You interpret pretty much as you perform other activities deemed automatic and outside your control, like breathing.

* The word *the* is written twice.

What makes nontheorizing *cost* you so much more energy than theorizing? First, there is the impenetrability of the activity. I said that much of it takes place outside of our awareness: if you don't know that you are making the inference, how can you stop yourself unless you stay in a continuous state of alert? And if you need to be continuously on the watch, doesn't that cause fatigue? Try it for an afternoon and see.

A Little More Dopamine

In addition to the story of the left-brain interpreter, we have more physiological evidence of our ingrained pattern seeking, thanks to our growing knowledge of the role of neurotransmitters, the chemicals that are assumed to transport signals between different parts of the brain. It appears that pattern perception increases along with the concentration in the brain of the chemical dopamine. Dopamine also regulates moods and supplies an internal reward system in the brain (not surprisingly, it is found in slightly higher concentrations in the left side of the brains of right-handed persons than on the right side). A higher concentration of dopamine appears to lower skepticism and result in greater vulnerability to pattern detection; an injection of L-dopa, a substance used to treat patients with Parkinson's disease, seems to increase such activity and lowers one's suspension of belief. The person becomes vulnerable to all manner of fads, such as astrology, superstitions, economics, and tarot-card reading.

Actually, as I am writing this, there is news of a pending lawsuit by a patient going after his doctor for more than $200,000—an amount he allegedly lost while gambling. The patient claims that the treatment of his Parkinson's disease caused him to go on wild betting sprees in casinos. It turns out that one of the side effects of L-dopa is that a small but significant minority of patients become compulsive gamblers. Since such gambling is associated with their seeing what they believe to be clear patterns in random numbers, this illustrates the *relation between knowledge and randomness*. It also shows that some aspects of what we call "knowledge" (and what I call narrative) are an ailment.

Once again, I warn the reader that I am not focusing on dopamine as the *reason* for our overinterpreting; rather, my point is that there is a physical and neural correlate to such operation and that our minds are largely victims of our physical embodiment. Our minds are like inmates, captive to our biology, unless we manage a cunning escape. It is the lack of our control of such inferences that I am stressing. Tomorrow, someone may discover another chemical or organic basis for our perception of patterns, or counter what I said about the left-brain interpreter by showing the role of a more complex structure; but it would not negate the idea that perception of causation has a biological foundation.

Andrey Nikolayevich's Rule

There is another, even deeper reason for our inclination to narrate, and it is not psychological. It has to do with the effect of order on information storage and retrieval in any system, and it's worth explaining here because of what I consider the central problems of probability and information theory.

The first problem is that information is *costly to obtain.*

The second problem is that information is also *costly to store*—like real estate in New York. The more orderly, less random, patterned, and *narratized* a series of words or symbols, the easier it is to store that series in one's mind or jot it down in a book so your grandchildren can read it someday.

Finally, information is costly to manipulate and retrieve.

With so many brain cells—one hundred billion (and counting)—the attic is quite large, so the difficulties probably do not arise from storage-capacity limitations, but may be just indexing problems. Your conscious, or working, memory, the one you are using to read these lines and make sense of their meaning, is considerably smaller than the attic. Consider that your working memory has difficulty holding a mere phone number longer than seven digits. Change metaphors slightly and imagine that your consciousness is a desk in the Library of Congress: no matter how many books the library holds, and makes available for retrieval, the size of your desk sets some processing limitations. Compression is vital to the performance of conscious work.

Consider a collection of words glued together to constitute a 500-page book. If the words are purely random, picked up from the dictionary in a totally unpredictable way, you will not be able to summarize, transfer, or reduce the dimensions of that book without losing something significant from it. You need 100,000 words to carry the exact message of a random 100,000 words with you on your next trip to Siberia. Now consider the opposite: a book filled with the repetition of the following sentence: "The chairman of [*insert here your company name*] is a lucky fellow who happened to be in the right place at the right time and claims credit for the company's success, without making a single allowance for luck," running ten times per page for 500 pages. The entire book can be accurately compressed, as I have just done, into 34 words (out of 100,000); you could reproduce it with total fidelity out of such a kernel. By finding the pattern, the logic of the series, you no longer need to memorize it all. You just store the pattern. And, as we can see here, a pattern is obviously more compact than raw information. You looked into the book and found a *rule*. It is along these lines that the great probabilist Andrey Nikolayevich Kolmogorov defined the degree of randomness; it is called "Kolmogorov complexity."

We, members of the human variety of primates, have a hunger for rules because we need to reduce the dimension of matters so they can get into our heads. Or, rather, sadly, so we can *squeeze* them into our heads. The more random information is, the greater the dimensionality, and thus the more difficult to summarize. The more you summarize, the more order you put in, the less randomness. Hence *the same condition that makes us simplify pushes us to think that the world is less random than it actually is*.

And the Black Swan is what we leave out of simplification.

Both the artistic and scientific enterprises are the product of our need to reduce dimensions and inflict some order on things. Think of the world around you, laden with trillions of details. Try to describe it and you will find yourself tempted to weave a thread into what you are saying. A novel, a story, a myth, or a tale, all have the same function: they spare us from the complexity of the world and shield us from its randomness. Myths impart order to the disorder of human perception and the perceived "chaos of human experience."*

Indeed, many severe psychological disorders accompany the feeling of loss of control of—being able to "make sense" of—one's environment.

Platonicity affects us here once again. The very same desire for order, interestingly, applies to scientific pursuits—it is just that, unlike art, the (stated) purpose of science is to get to the truth, not to give you a feeling of organization or make you feel better. We tend to use knowledge as therapy.

* The Parisian novelist Georges Perec tried to break away from narrative and attempted to write a book as large as the world. He had to settle for an exhaustive account of what happened on the Place Saint-Sulpice between October 18 and October 20, 1974. Even so, his account was not so exhaustive, and he ended up with a narrative.

A Better Way to Die

To view the potency of narrative, consider the following statement: "The king died and the queen died." Compare it to "The king died, and then the queen died of grief." This exercise, presented by the novelist E. M. Forster, shows the distinction between mere succession of information and a plot. But notice the hitch here: although we added information to the second statement, we effectively reduced the dimension of the total. The second sentence is, in a way, much lighter to carry and easier to remember; we now have one single piece of information in place of two. As we can remember it with less effort, we can also sell it to others, that is, market it better as a packaged idea. This, in a nutshell, is the definition and function of a *narrative*.

To see how the narrative can lead to a mistake in the assessment of the odds, do the following experiment. Give someone a well-written detective story—say, an Agatha Christie novel with a handful of characters who can all be plausibly deemed guilty. Now question your subject about the probabilities of each character's being the murderer. Unless she writes down the percentages to keep an exact tally of them, they should add up to well over 100 percent (even well over 200 percent for a good novel). The better the detective writer, the higher that number.

REMEMBRANCE OF THINGS NOT QUITE PAST

Our tendency to perceive—to impose—*narrativity* and *causality* are symptoms of the same disease—dimension reduction. Moreover, like causality, narrativity has a chronological dimension and leads to the perception of the flow of time. Causality makes time flow in a single direction, and so does narrativity.

But memory and the arrow of time can get mixed up. Narrativity can viciously affect the remembrance of past events as follows: we will tend to more easily remember those facts from our past that fit a narrative, while we tend to neglect others that do not *appear* to play a causal role in that narrative. Consider that we recall events in our memory all the while knowing the answer of what happened subsequently. It is literally impossible to ignore posterior information when solving a problem. This simple inability to remember not the true sequence of events but a reconstructed one will make history appear in hindsight to be far more explainable than it actually was—or is.

Conventional wisdom holds that memory is like a serial recording device like a computer diskette. In reality, memory is dynamic—not static—like a paper on which new texts (or new versions of the same text) will be continuously recorded, thanks to the power of posterior information. (In a remarkable insight, the nineteenth-century Parisian poet Charles Baudelaire compared our memory to a palimpsest, a type of parchment on which old texts can be erased and new ones written over them.) Memory is more of a self-serving dynamic revision machine: you remember the last time you remembered the event and, without realizing it, *change the story at every subsequent remembrance*.

So we pull memories along causative lines, revising them involuntarily and unconsciously. We continuously renarrate past events in the light of what appears to make what we think of as logical sense after these events occur.

By a process called reverberation, a memory corresponds to the strengthening of connections from an increase of brain activity in a given sector of the brain—the more activity, the stronger the memory. While we believe that the memory is fixed, constant, and connected, all this is very far from truth. What makes sense according to information obtained subsequently will be remembered more vividly. We invent some of our memories—a sore point in courts of law since it has been shown that plenty of people have invented child-abuse stories by dint of listening to theories.

The Madman's Narrative

We have far too many possible ways to interpret past events for our own good.

Consider the behavior of paranoid people. I have had the privilege to work with colleagues who have hidden paranoid disorders that come to the surface on occasion. When the person is highly intelligent, he can astonish you with the most far-fetched, yet completely plausible interpretations of the most innocuous remark. If I say to them, "I am afraid that . . . ," in reference to an undesirable state of the world, they may interpret it literally, that I am experiencing actual fright, and it triggers an episode of fear on the part of the paranoid person. Someone hit with such a disorder can muster the most insignificant of details and construct an elaborate and coherent theory of why there is a conspiracy against him. And if you gather, say, ten paranoid people, all in the same state of episodic delusion, the ten of them will provide ten distinct, yet coherent, interpretations of events.

When I was about seven, my schoolteacher showed us a painting of an assembly of impecunious Frenchmen in the Middle Ages at a banquet held by one of their benefactors, some benevolent king, as I recall. They were holding the soup bowls to their lips. The schoolteacher asked me why they had their noses in the bowls and I answered, "Because they were not taught manners." She replied, "Wrong. The reason is that they are hungry." I felt stupid at not having thought of this, but I could not understand what made one explanation more likely than the other, or why we weren't both wrong (there was no, or little, silverware at the time, which seems the most likely explanation).

Beyond our perceptional distortions, there is a problem with logic itself. How can someone have no clue yet be able to hold a set of perfectly sound and coherent viewpoints that match the observations and abide by every single possible rule of logic? Consider that two people can hold incompatible beliefs based on the exact same data. Does this mean that there are possible families of explanations and that each of these can be equally perfect and sound? Certainly not. One may have a million ways to explain things, but the true explanation is unique, whether or not it is within our reach.

In a famous argument, the logician W. V. Quine showed that there exist families of logically consistent interpretations and theories that can match a given series of facts. Such insight should warn us that mere absence of nonsense may not be sufficient to make something true.

Quine's problem is related to his finding difficulty in translating statements between languages, simply because one could interpret any sentence in an infinity of ways. (Note here that someone splitting hairs could find a self-canceling aspect to Quine's own writing. I wonder how he expects us to understand this very point in a noninfinity of ways).

This does not mean that we cannot talk about causes; there are ways to escape the narrative fallacy. How? By making conjectures and running experiments, or as we will see in Part Two (alas), by making testable predictions.* The psychology experiments I am discussing here do so: they select a population and run a test. The results should hold in Tennessee, in China, even in France.

* Such tests avoid both the narrative fallacy and much of the confirmation bias, since testers are obliged to take into account the failures as well as the successes of their experiments.

Narrative and Therapy

If narrativity causes us to see past events as more predictable, more expected, and less random than they actually were, then we should be able to make it work for us as therapy against some of the stings of randomness.

Say some unpleasant event, such as a car accident for which you feel indirectly responsible, leaves you with a bad lingering aftertaste. You are tortured by the thought that you caused injuries to your passengers; you are continuously aware that you could have avoided the accident. Your mind keeps playing alternative scenarios branching out of a main tree: if you did not wake up three minutes later than usual, you would have avoided the car accident. It was not your intension to injure your passengers, yet your mind is inhabited with remorse and guilt. People in professions with high randomness (such as in the markets) can suffer more than their share of the toxic effect of look-back stings: I should have sold my portfolio at the top; I could have bought that stock years ago for pennies and I would now be driving a pink convertible; et cetera. If you are a professional, you can feel that you "made a mistake," or, worse, that "mistakes were made," when you failed to do the equivalent of buying the winning lottery ticket for your investors, and feel the need to apologize for your "reckless" investment strategy (that is, what seems reckless in retrospect).

How can you get rid of such a persistent throb? Don't try to willingly avoid thinking about it: this will almost surely backfire. A more appropriate solution is to make the event appear more unavoidable. Hey, it was bound to take place and it seems futile to agonize over it. How can you do so? Well, *with a narrative*. Patients who spend fifteen minutes every day writing an account of their daily troubles feel indeed better about what has befallen them. You feel less guilty for not having avoided certain events; you feel less responsible for it. Things appear as if they were bound to happen.

If you work in a randomness-laden profession, as we see, you are likely to suffer burnout effects from that constant second-guessing of your past actions in terms of what played out subsequently. Keeping a diary is the least you can do in these circumstances.

TO BE WRONG WITH INFINITE PRECISION

We harbor a crippling dislike for the abstract.

One day in December 2003, when Saddam Hussein was captured, Bloomberg News flashed the following headline at 13:01: U.S. TREASURIES RISE; HUSSEIN CAPTURE MAY NOT CURB TERRORISM.

Whenever there is a market move, the news media feel obligated to give the "reason." Half an hour later, they had to issue a new headline. As these U.S. Treasury bonds fell in price (they fluctuate all day long, so there was nothing special about that), Bloomberg News had a new reason for the fall: Saddam's capture (the same Saddam). At 13:31 they issued the next bulletin: U.S. TREASURIES FALL; HUSSEIN CAPTURE BOOSTS ALLURE OF RISKY ASSETS.

So it was the same capture (the cause) explaining one event and its exact opposite. Clearly, this can't be; these two facts cannot be linked.

Do media journalists repair to the nurse's office every morning to get their daily dopamine injection so that they can narrate better? (Note the irony that the word *dope,* used to designate the illegal drugs athletes take to improve performance, has the same root as *dopamine.*)

It happens all the time: a cause is proposed to make you swallow the news and make matters more concrete. After a candidate's defeat in an election, you will be supplied with the "cause" of the voters' disgruntlement. Any conceivable cause can do. The media, however, go to great lengths to make the process "thorough" with their armies of fact-checkers. It is as if they wanted to be wrong with infinite precision (instead of accepting being approximately right, like a fable writer).

Note that in the absence of any other information about a person you encounter, you tend to fall back on her nationality and background as a salient attribute (as the Italian scholar did with me). How do I know that this attribution to the background is bogus? I did my own empirical test by checking how many traders with my background who experienced the same war became skeptical empiricists, and found none out of twenty-six. This nationality business helps you make a great story and satisfies your hunger for ascription of causes. It seems to be the dump site where all explanations go until one can ferret out a more obvious one (such as, say, some evolutionary argument that "makes sense"). Indeed, people tend to fool themselves with their self-narrative of "national identity," which, in a breakthrough paper in *Science* by sixty-five authors, was shown to be a total fiction. ("National traits" might be great for movies, they might help a lot with war, but they are Platonic notions that carry no empirical validity—yet, for example, both the English and the non-English erroneously believe in an English "national temperament.") Empirically, sex, social class, and profession seem to be better predictors of someone's behavior than nationality (a male from Sweden resembles a male from Togo more than a female from Sweden; a philosopher from Peru resembles a philosopher from Scotland more than a janitor from Peru; and so on).

The problem of overcausation does not lie with the journalist, but with the public. Nobody would pay one dollar to buy a series of abstract statistics reminiscent of a boring college lecture. We want to be told stories, and there is nothing wrong with that—except that we should check more thoroughly whether the story provides consequential distortions of reality. Could it be that fiction reveals truth while nonfiction is a harbor for the liar? Could it be that fables and stories are closer to the truth than is the thoroughly fact-checked ABC News? Just consider that the newspapers try to get impeccable facts, but weave them into a narrative in such a way as to convey the impression of causality (and knowledge). There are fact-checkers, not intellect-checkers. Alas.

But there is no reason to single out journalists. Academics in narrative disciplines do the same thing, but dress it up in a formal language—we will catch up to them in Chapter 10, on prediction.

Besides narrative and causality, journalists and public intellectuals of the sound-bite variety do not make the world simpler. Instead, they almost invariably make it look far more complicated than it is. The next time you are asked to discuss world events, plead ignorance, and give the arguments I offered in this chapter casting doubt on the visibility of the immediate cause. You will be told that "you overanalyze," or that "you are too complicated." All you will be saying is that you don't know!

Dispassionate Science

Now, if you think that science is an abstract subject free of sensationalism and distortions, I have some sobering news. Empirical researchers have found evidence that scientists too are vulnerable to narratives, emphasizing titles and "sexy" attention-grabbing punch lines over more substantive matters. They too are human and get their attention from sensational matters. The way to remedy this is through meta-analyses of scientific studies, in which an überresearcher peruses the entire literature, which includes the less-advertised articles, and produces a synthesis.

THE SENSATIONAL AND THE BLACK SWAN

Let us see how narrativity affects our understanding of the Black Swan. Narrative, as well as its associated mechanism of salience of the sensational fact, can mess up our projection of the odds. Take the following experiment conducted by Kahneman and Tversky, the pair introduced in the previous chapter: the subjects were forecasting professionals who were asked to imagine the following scenarios and estimate their odds.

a. A massive flood somewhere in America in which more than a thousand people die.
b. *An earthquake in California,* causing massive flooding, in which more than a thousand people die.

Respondents estimated the first event to be *less* likely than the second. An earthquake in California, however, is a readily imaginable *cause,* which greatly increases the mental availability—hence the assessed probability—of the flood scenario.

Likewise, if I asked you how many cases of lung cancer are likely to take place in the country, you would supply some number, say half a million. Now, if instead I asked you many cases of lung cancer are likely to take place *because* of smoking, odds are that you would give me a much higher number (I would guess more than twice as high). Adding the *because* makes these matters far more plausible, and far more *likely.* Cancer from smoking seems more likely than cancer without a cause attached to it—an unspecified cause means no cause at all.

I return to the example of E. M. Forster's plot from earlier in this chapter, but seen from the standpoint of probability. Which of these two statements seems more likely?

Joey seemed happily married. He killed his wife.

Joey seemed happily married. He killed his wife to get her inheritance.

Clearly the second statement seems more likely at first blush, which is a pure mistake of logic, since the first, being broader, can accommodate more causes, such as he killed his wife because he went mad, because she cheated with both the postman and the ski instructor, because he entered a state of delusion and mistook her for a financial forecaster.

All this can lead to pathologies in our decision making. How?

Just imagine that, as shown by Paul Slovic and his collaborators, people are more likely to pay for terrorism insurance than for plain insurance (which covers, among other things, terrorism).

The Black Swans we imagine, discuss, and worry about do not resemble those likely to be Black Swans. We worry about the wrong "improbable" events, as we will see next.

Black Swan Blindness

The first question about the paradox of the perception of Black Swans is as follows: How is it that *some* Black Swans are overblown in our minds when the topic of this book is that we mainly neglect Black Swans?

The answer is that there are two varieties of rare events: a) the *narrated* Black Swans, those that are present in the current discourse and that you are likely to hear about on television, and b) those nobody talks about, since they escape models—those that you would feel ashamed discussing in public because they do not seem plausible. I can safely say that it is entirely compatible with human nature that the incidences of Black Swans would be overestimated in the first case, but severely underestimated in the second one.

Indeed, lottery buyers overestimate their chances of winning because they visualize such a potent payoff—in fact, they are so blind to the odds that they treat odds of one in a thousand and one in a million almost in the same way.

Much of the empirical research agrees with this pattern of overestimation and underestimation of Black Swans. Kahneman and Tversky initially showed that people overreact to low-probability outcomes *when you discuss the event with them,* when you make them aware of it. If you ask someone, "What is the probability of death from a plane crash?" for instance, they will raise it. However, Slovic and his colleagues found, in insurance patterns, neglect of these highly improbable events in people's insurance purchases. They call it the "preference for insuring against probable small losses"—at the expense of the less probable but larger impact ones.

Finally, after years of searching for empirical tests of our scorn of the abstract, I found researchers in Israel that ran the experiments I had been waiting for. Greg Barron and Ido Erev provide experimental evidence that agents underweigh small probabilities when they engage in sequential experiments in which *they derive the probabilities themselves,* when they are not supplied with the odds. If you draw from an urn with a very small number of red balls and a high number of black ones, and if you do not have a clue about the relative proportions, you are likely to underestimate the number of red balls. It is only when you are supplied with their frequency—say, by telling you that 3 percent of the balls are red—that you overestimate it in your betting decision.

I've spent a lot of time wondering how we can be so myopic and short-termist yet survive in an environment that is not entirely from Mediocristan. One day, looking at the gray beard that makes me look ten years older than I am and thinking about the pleasure I derive from exhibiting it, I realized the following. Respect for elders in many societies might be a kind of compensation for our short-term memory. The word *senate* comes from *senatus,* "aged" in Latin; *sheikh* in Arabic means both a member of the ruling elite and "elder." Elders are repositories of complicated inductive learning that includes information about rare events. Elders can scare us with stories—which is why we become overexcited when we think of a *specific* Black Swan. I was excited to find out that this also holds true in the animal kingdom: a paper in *Science* showed that elephant matriarchs play the role of superadvisers on rare events.

We learn from repetition—at the expense of events that have not happened before. Events that are nonrepeatable are ignored before their occurrence, and overestimated after (for a while). After a Black Swan, such as September 11, 2001, people expect it to recur when in fact the odds of that happening have arguably been lowered. We like to think about *specific* and known Black Swans when in fact the very nature of randomness lies in its abstraction. As I said in the Prologue, it is the wrong definition of a god.

The economist Hyman Minsky sees the cycles of risk taking in the economy as following a pattern: stability and absence of crises encourage risk taking, complacency, and lowered awareness of the possibility of problems. Then a crisis occurs, resulting in people being shell-shocked and scared of investing their resources. Strangely, both Minsky and his school, dubbed Post-Keynesian, and his opponents, the libertarian "Austrian" economists, have the same analysis, except that the first group recommends governmental intervention to smooth out the cycle, while the second believes that civil servants should not be trusted to deal with such matters. While both schools of thought seem to fight each other, they both emphasize fundamental uncertainty and stand outside the mainstream economic departments (though they have large followings among businessmen and nonacademics). No doubt this emphasis on fundamental uncertainty bothers the Platonifiers.

All the tests of probability I discussed in this section are important; they show how we are fooled by the rarity of Black Swans but not by the role they play in the aggregate, their *impact*. In a preliminary study, the psychologist Dan Goldstein and I subjected students at the London Business School to examples from two domains, Mediocristan and Extremistan. We selected height, weight, and Internet hits per website. The subjects were good at guessing the role of rare events in Mediocristan-style environments. But their intuitions failed when it came to variables outside Mediocristan, showing that we are effectively not skilled at intuitively gauging the impact of the improbable, such as the contribution of a blockbuster to total book sales. In one experiment they underestimated by thirty-three times the *effect* of a rare event.

Next, let us see how this lack of understanding of abstract matters affects us.

The Pull of the Sensational

Indeed, abstract statistical information does not sway us as much as the anecdote—no matter how sophisticated the person. I will give a few instances.

The Italian Toddler. In the late 1970s, a toddler fell into a well in Italy. The rescue team could not pull him out of the hole and the child stayed at the bottom of the well, helplessly crying. Understandably, the whole of Italy was concerned with his fate; the entire country hung on the frequent news updates. The child's cries produced acute pains of guilt in the powerless rescuers and reporters. His picture was prominently displayed on magazines and newspapers, and you could hardly walk in the center of Milan without being reminded of his plight.

Meanwhile, the civil war was raging in Lebanon, with an occasional hiatus in the conflict. While in the midst of their mess, the Lebanese were also absorbed in the fate of that child. The *Italian* child. Five miles away, people were dying from the war, citizens were threatened with car bombs, but the fate of the Italian child ranked high among the interests of the population in the Christian quarter of Beirut. "Look how cute that poor thing is," I was told. And the entire town expressed relief upon his eventual rescue.

As Stalin, who knew something about the business of mortality, supposedly said, "One death is a tragedy; a million is a statistic." Statistics stay silent in us.

Terrorism kills, but the biggest killer remains the environment, responsible for close to 13 million deaths annually. But terrorism causes outrage, which makes us overestimate the likelihood of a potential terrorist attack—and react more violently to one when it happens. We feel the sting of man-made damage far more than that caused by nature.

Central Park. You are on a plane on your way to spend a long (bibulous) weekend in New York City. You are sitting next to an insurance salesman who, being a salesman, cannot stop talking. For him, *not talking* is the effortful activity. He tells you that his cousin (with whom he will celebrate the holidays) worked in a law office with someone whose brother-in-law's business partner's twin brother was mugged and killed in Central Park. Indeed, Central Park in glorious New York City. That was in 1989, if he remembers it well (the year is now 2007). The poor victim was only thirty-eight and had a wife and three children, one of whom had a birth defect and needed special care at Cornell Medical Center. Three children, one of whom needed special care, lost their father because of his foolish visit to Central Park.

Well, you are likely to avoid Central Park during your stay. You know you can get crime statistics from the Web or from any brochure, rather than anecdotal information from a verbally incontinent salesman. But you can't help it. For a while, the name Central Park will conjure up the image of that that poor, undeserving man lying on the polluted grass. It will take a lot of statistical information to override your hesitation.

Motorcycle Riding. Likewise, the death of a relative in a motorcycle accident is far more likely to influence your attitude toward motorcycles than volumes of statistical analyses. You can effortlessly look up accident statistics on the Web, but they do not easily come to mind. Note that I ride my red Vespa around town, since no one in my immediate environment has recently suffered an accident—although I am aware of this problem in logic, I am incapable of acting on it.

Now, I do not disagree with those recommending the use of a narrative to get attention. Indeed, our consciousness may be linked to our ability to concoct some form of story about ourselves. It is just that narrative can be lethal when used in the wrong places.

THE SHORTCUTS

Next I will go beyond narrative to discuss the more general attributes of thinking and reasoning behind our crippling shallowness. These defects in reasoning have been cataloged and investigated by a powerful research tradition represented by a school called the Society of Judgment and Decision Making (the only academic and professional society of which I am a member, and proudly so; its gatherings are the only ones where I do not have tension in my shoulders or anger fits). It is associated with the school of research started by Daniel Kahneman, Amos Tversky, and their friends, such as Robyn Dawes and Paul Slovic. It is mostly composed of empirical psychologists and cognitive scientists whose methodology hews strictly to running very precise, controlled experiments (physics-style) on humans and making catalogs of how people react, with minimal theorizing. They look for regularities. Note that empirical psychologists use the bell curve to gauge errors in their testing methods, but as we will see more technically in Chapter 15, this is one of the rare adequate applications of the bell curve in social science, owing to the nature of the experiments. We have seen such types of experiments earlier in this chapter with the flood in California, and with the identification of the confirmation bias in Chapter 5. These researchers have mapped our activities into (roughly) a dual mode of thinking, which they separate as "System 1" and "System 2," or the *experiential* and the *cognitive*. The distinction is straightforward.

System 1, the experiential one, is effortless, automatic, fast, opaque (we do not know that we are using it), parallel-processed, and can lend itself to errors. It is what we call "intuition," and performs these quick acts of prowess that became popular under the name *blink,* after the title of Malcolm Gladwell's bestselling book. System 1 is highly emotional, precisely because it is quick. It produces shortcuts, called "heuristics," that allow us to function rapidly and effectively. Dan Goldstein calls these heuristics "fast and frugal." Others prefer to call them "quick and dirty." Now, these shortcuts are certainly virtuous, since they are rapid, but, at times, they can lead us into some severe mistakes. This main idea generated an entire school of research called the *heuristics and biases* approach (heuristics corresponds to the study of shortcuts, biases stand for mistakes).

System 2, the cogitative one, is what we normally call *thinking*. It is what you use in a classroom, as it is effortful (even for Frenchmen), reasoned, slow, logical, serial, progressive, and self-aware (you can follow the steps in your reasoning). It makes fewer mistakes than the experiential system, and, since you know how you derived your result, you can retrace your steps and correct them in an adaptive manner.

Most of our mistakes in reasoning come from using System 1 when we are in fact thinking that we are using System 2. How? Since we react without thinking and introspection, the main property of System 1 is our lack of awareness of using it!

Recall the round-trip error, our tendency to confuse "no evidence of Black Swans" with "evidence of no Black Swans"; it shows System 1 at work. You have to make an effort (System 2) to override your first reaction. Clearly Mother Nature makes you use the fast System 1 to get out of trouble, so that you do not sit down and cogitate whether there is truly a tiger attacking you or if it is an optical illusion. You run immediately, before you become "conscious" of the presence of the tiger.

Emotions are assumed to be the weapon System 1 uses to direct us and force us to act quickly. It mediates risk avoidance far more effectively than our cognitive system. Indeed, neurobiologists who have studied the emotional system show how it often reacts to the presence of danger long before we are consciously aware of it—we experience fear and start reacting a few milliseconds before we realize that we are facing a snake.

Much of the trouble with human nature resides in our inability to use much of System 2, or to use it in a prolonged way without having to take a long beach vacation. In addition, we often just forget to use it.

Beware the Brain

Note that neurobiologists make, roughly, a similar distinction to that between System 1 and System 2, except that they operate along anatomical lines. Their distinction differentiates between parts of the brain, the *cortical* part, which we are supposed to use for thinking, and which distinguishes us from other animals, and the fast-reacting *limbic* brain, which is the center of emotions, and which we share with other mammals.

As a skeptical empiricist, I do not want to be the turkey, so I do not want to focus solely on specific organs in the brain, since we do not observe brain functions very well. Some people try to identify what are called the neural correlates of, say, decision making, or more aggressively the neural "substrates" of, say, memory. The brain might be more complicated machinery than we think; its anatomy has fooled us repeatedly in the past. We can, however, assess regularities by running precise and thorough experiments on how people react under certain conditions, and keep a tally of what we see.

For an example that justifies skepticism about unconditional reliance on neurobiology, and vindicates the ideas of the empirical school of medicine to which Sextus belonged, let's consider the intelligence of birds. I kept reading in various texts that the cortex is where animals do their "thinking," and that the creatures with the largest cortex have the highest intelligence—we humans have the largest cortex, followed by bank executives, dolphins, and our cousins the apes. Well, it turns out that some birds, such as parrots, have a high level of intelligence, equivalent to that of dolphins, but that the intelligence of birds correlates with the size of another part of the brain, called the hyperstriatum. So neurobiology with its attribute of "hard science" can sometimes (though not always) fool you into a Platonified, reductive statement. I am amazed that the "empirics," skeptical about links between anatomy and function, had such insight—no wonder their school played a very small part in intellectual history. As a skeptical empiricist I prefer the experiments of empirical psychology to the theories-based MRI scans of neurobiologists, even if the former appear less "scientific" to the public.

How to Avert the Narrative Fallacy

I'll conclude by saying that our misunderstanding of the Black Swan can be largely attributed to our using System 1, i.e., narratives, and the sensational—as well as the emotional—which imposes on us a wrong map of the likelihood of events. On a day-to-day basis, we are not introspective enough to realize that we understand what is going on a little less than warranted from a dispassionate observation of our experiences. We also tend to forget about the notion of Black Swans immediately after one occurs—since they are too abstract for us—focusing, rather, on the precise and vivid events that easily come to our minds. We do worry about Black Swans, just the wrong ones.

Let me bring Mediocristan into this. In Mediocristan, narratives seem to work—the past is likely to yield to our inquisition. But not in Extremistan, where you do not have repetition, and where you need to remain suspicious of the sneaky past and avoid the easy and obvious narrative.

Given that I have lived largely deprived of information, I've often felt that I inhabit a different planet than my peers, which can sometimes be extremely painful. It's like they have a virus controlling their brains that prevents them from seeing things going forward—the Black Swan around the corner.

The way to avoid the ills of the narrative fallacy is to favor experimentation over storytelling, experience over history, and clinical knowledge over theories. Certainly the newspaper cannot perform an experiment, but it can choose one report over another—there is plenty of empirical research to present and interpret from—as I am doing in this book. Being empirical does not mean running a laboratory in one's basement: it is just a mind-set that favors a certain class of knowledge over others. I do not forbid myself from using the word *cause,* but the causes I discuss are either bold speculations (presented as such) or the result of experiments, not stories.

Another approach is to predict and keep a tally of the predictions.

Finally, there may be a way to use a narrative—but for a good purpose. Only a diamond can cut a diamond; we can use our ability to convince with a story that conveys the right message—what storytellers seem to do.

So far we have discussed two internal mechanisms behind our blindness to Black Swans, the confirmation bias and the narrative fallacy. The next chapters will look into an external mechanism: a defect in the way we receive and interpret recorded events, and a defect in the way we act on them.

Chapter Seven

LIVING IN THE ANTECHAMBER OF HOPE

How to avoid watercoolers—Select your brother-in-law—Yevgenia's favorite book—What deserts can and cannot deliver—On the avoidance of hope— El desierto de los tártaros*—The virtues of slow motion*

Assume that, like Yevgenia, your activities depend on a Black Swan surprise—i.e., you are a reverse turkey. Intellectual, scientific, and artistic activities belong to the province of Extremistan, where there is a severe concentration of success, with a very small number of winners claiming a large share of the pot. This seems to apply to all professional activities I find nondull and "interesting" (I am still looking for a single counter-example, a nondull activity that belongs to Mediocristan).

Acknowledging the role of this concentration of success, and acting accordingly, causes us to be punished twice: we live in a society where the reward mechanism is based on the illusion of the regular; our hormonal reward system also needs tangible and steady results. It too thinks that the world is steady and well behaved—it falls for the confirmation error. The world has changed too fast for our genetic makeup. We are alienated from our environment.

PEER CRUELTY

Every morning you leave your cramped apartment in Manhattan's East Village to go to your laboratory at the Rockefeller University in the East Sixties. You return in the late evening, and people in your social network ask you if you had a good day, just to be polite. At the laboratory, people are more tactful. Of course you did not have a good day; you found nothing. You are not a watch repairman. Your *finding nothing* is very valuable, since it is part of the process of discovery—hey, you know where *not* to look. Other researchers, knowing your results, would avoid trying your special experiment, provided a journal is thoughtful enough to consider your "found nothing" as information and publish it.

Meanwhile your brother-in-law is a salesman for a Wall Street firm, and keeps getting large commissions—large and steady commissions. "He is doing very well," you hear, particularly from your father-in-law, with a small pensive nanosecond of silence after the utterance—which makes you realize that he just made a comparison. It was involuntary, but he made one.

Holidays can be terrible. You run into your brother-in-law at family reunions and, invariably, detect unmistakable signs of frustration on the part of your wife, who, briefly, fears that she married a loser, before remembering the logic of your profession. But she has to fight her first impulse. Her sister will not stop talking about their renovations, their new wallpaper. Your wife will be a little more silent than usual on the drive home. This sulking will be made slightly worse because the car you are driving is rented, since you cannot afford to garage a car in Manhattan. What should you do? Move to Australia and thereby make family reunions less frequent, or switch brothers-in-laws by marrying someone with a less "successful" brother?

Or should you dress like a hippie and become defiant? That may work for an artist, but not so easily for a scientist or a businessman. You are trapped.

You work on a project that does not deliver immediate or steady results; all the while, people around you work on projects that do. You are in trouble. Such is the lot of scientists, artists, and researchers lost in society rather than living in an insulated community or an artist colony.

Positive lumpy outcomes, for which we either collect big or get nothing, prevail in numerous occupations, those invested with a sense of mission, such as doggedly pursuing (in a smelly laboratory) the elusive cure for cancer, writing a book that will change the way people view the world (while living hand to mouth), making music, or painting miniature icons on subway trains and considering it a higher form of art despite the diatribes of the antiquated "scholar" Harold Bloom.

If you are a researcher, you will have to publish inconsequential articles in "prestigious" publications so that others say hello to you once in a while when you run into them at conferences.

If you run a public corporation, things were great for you before you had shareholders, when you and your partners were the sole owners, along with savvy venture capitalists who understood uneven results and the lumpy nature of economic life. But now you have a slow-thinking thirty-year-old security analyst at a downtown Manhattan firm who "judges" your results and reads too much into them. He likes routine rewards, and the last thing you can deliver are routine rewards.

Many people labor in life under the impression that they are doing something right, yet they may not show solid results for a long time. They need a capacity for continuously adjourned gratification to survive a steady diet of peer cruelty without becoming demoralized. They look like idiots to their cousins, they look like idiots to their peers, they need courage to continue. No confirmation comes to them, no validation, no fawning students, no Nobel, no Shnobel. "How was your year?" brings them a small but containable spasm of pain deep inside, since almost all of their years will seem wasted to someone looking at their life from the outside. Then bang, the lumpy event comes that brings the grand vindication. Or it may never come.

Believe me, it is tough to deal with the social consequences of the appearance of continuous failure. We are social animals; hell is other people.

Where the Relevant Is the Sensational

Our intuitions are not cut out for nonlinearities. Consider our life in a primitive environment where process and result are closely connected. You are thirsty; drinking brings you adequate satisfaction. Or even in a not-so-primitive environment, when you engage in building, say, a bridge or a stone house, more work will lead to more apparent results, so your mood is propped up by visible continuous feedback.

In a primitive environment, the relevant *is* the sensational. This applies to our knowledge. When we try to collect information about the world around us, we tend to be guided by our biology, and our attention flows effortlessly toward the sensational—not the relevant so much as the sensational. Somehow the guidance system has gone wrong in the process of our coevolution with our habitat—it was transplanted into a world in which the relevant is often boring, nonsensational.

Furthermore, we think that if, say, two variables are causally linked, then a steady input in one variable should *always* yield a result in the other one. Our emotional apparatus is designed for linear causality. For instance, if you study every day, you expect to learn something in proportion to your studies. If you feel that you are not going anywhere, your emotions will cause you to become demoralized. But modern reality rarely gives us the privilege of a satisfying, linear, positive progression: you may think about a problem for a year and learn nothing; then, unless you are disheartened by the emptiness of the results and give up, something will come to you in a flash.

Researchers spent some time dealing with this notion of gratification; neurology has been enlightening us about the tension between the notions of immediate rewards and delayed ones. Would you like a massage today, or two next week? Well, the news is that the logical part of our mind, that "higher" one, which distinguishes us from animals, can override our animal instinct, which asks for immediate rewards. So we are a little better than animals, after all—but perhaps not by much. And not all of the time.

Nonlinearities

The situation can get a little more tragic—the world is more nonlinear than we think, and than scientists would like to think.

With linearities, relationships between variables are clear, crisp, and constant, therefore Platonically easy to grasp in a single sentence, such as "A 10 percent increase in money in the bank corresponds to a 10 percent increase in interest income and a 5 percent increase in obsequiousness on the part of the personal banker." If you have more money in the bank, you get more interest. Nonlinear relationships can vary; perhaps the best way to describe them is to say that they cannot be expressed verbally in a way that does justice to them. Take the relationship between pleasure and drinking water. If you are in a state of painful thirst, then a bottle of water increases your well-being significantly. More water means more pleasure. But what if I gave you a cistern of water? Clearly your well-being becomes rapidly insensitive to further quantities. As a matter of fact, if I gave you the choice between a bottle or a cistern you would prefer the bottle—so your enjoyment *declines* with additional quantities.

These nonlinear relationships are ubiquitous in life. Linear relationships are truly the exception; we only focus on them in classrooms and textbooks because they are easier to understand. Yesterday afternoon I tried to take a fresh look around me to catalog what I could see during my day that was linear. I could not find anything, no more than someone hunting for squares or triangles could find them in the rain forest—or, as we will see in Part Three, any more than someone looking for bell-shape randomness finding it in socioeconomic phenomena.

You play tennis every day with no improvement, then suddenly you start beating the pro.

Your child does not seem to have a learning impediment, but he does not seem to want to speak. The schoolmaster pressures you to start considering "other options," namely therapy. You argue with her to no avail (she is supposed to be the "expert"). Then, suddenly, the child starts composing elaborate sentences, perhaps a bit too elaborate for his age group. I will repeat that linear progression, a Platonic idea, is not the norm.

Process over Results

We favor the sensational and the extremely visible. This affects the way we judge heroes. There is little room in our consciousness for heroes who do not deliver visible results—or those heroes who focus on process rather than results.

However, those who claim that they value process over result are not telling the whole truth, assuming of course that they are members of the human species. We often hear the semi-lie that writers do not write for glory, that artists create for the sake of art, because the activity is "its own reward." True, these activities can generate a steady flow of autosatisfaction. But this does not mean that artists do not crave some form of attention, or that they would not be better off if they got some publicity; it does not mean that writers do not wake up early Saturday morning to check if *The New York Times Book Review* has featured their work, even if it is a very long shot, or that they do not keep checking their mailbox for that long-awaited reply from *The New Yorker*. Even a philosopher the caliber of Hume spent a few weeks sick in bed after the trashing of his masterpiece (what later became known as his version of the Black Swan problem) by some dim-thinking reviewer—whom he knew to be wrong and to have missed his whole point.

Where it gets painful is when you see one of your peers, whom you despise, heading to Stockholm for his Nobel reception.

Most people engaged in the pursuits that I call "concentrated" spend most of their time waiting for the big day that (usually) never comes.

True, this takes your mind away from the pettiness of life—the cappuccino that is too warm or too cold, the waiter too slow or too intrusive, the food too spicy or not enough, the overpriced hotel room that does not quite resemble the advertised picture—all these considerations disappear because you have your mind on much bigger and better things. But this does not mean that the person insulated from materialistic pursuits becomes impervious to other pains, those issuing from disrespect. Often these Black Swan hunters feel shame, or are made to feel shame, at not contributing. "You betrayed those who had high hopes for you," they are told, increasing their feeling of guilt. The problem of lumpy payoffs is not so much in the lack of income they entail, but the pecking order, the loss of dignity, the subtle humiliations near the watercooler.

It is my great hope someday to see science and decision makers rediscover what the ancients have always known, namely that our highest currency is respect.

Even economically, the individual Black Swan hunters are not the ones who make the bucks. The researcher Thomas Astebro has shown that returns on independent inventions (you take the cemetery into account) are far lower than those on venture capital. Some blindness to the odds or an obsession with their own positive Black Swan is necessary for entrepreneurs to function. The venture capitalist is the one who gets the shekels. The economist William Baumol calls this "a touch of madness." This may indeed apply to all concentrated businesses: when you look at the empirical record, you not only see that venture capitalists do better than entrepreneurs, but publishers do better than writers, dealers do better than artists, and science does better than scientists (about 50 percent of scientific and scholarly papers, costing months, sometimes years, of effort, are never truly read). The person involved in such gambles is paid in a currency other than material success: hope.

Human Nature, Happiness, and Lumpy Rewards

Let me distill the main idea behind what researchers call hedonic happiness.

Making $1 million in one year, but nothing in the preceding nine, does not bring the same pleasure as having the total evenly distributed over the same period, that is, $100,000 every year for ten years in a row. The same applies to the inverse order—making a bundle the first year, then nothing for the remaining period. Somehow, your pleasure system will be saturated rather quickly, and it will not carry forward the hedonic balance like a sum on a tax return. As a matter of fact, your happiness depends far more on the number of instances of positive feelings, what psychologists call "positive affect," than on their intensity when they hit. In other words, good news is good news first; *how* good matters rather little. So to have a pleasant life you should spread these small "affects" across time as evenly as possible. Plenty of mildly good news is preferable to one single lump of great news.

Sadly, it may be even worse for you to make $10 million, then lose back nine, than to making nothing at all! True, you may end up with a million (as compared to nothing), but it may be better had you got zilch. (This assumes, of course, that you care about financial rewards.)

So from a narrowly defined accounting point of view, which I may call here "hedonic calculus," it does not pay to shoot for one large win. Mother Nature destined us to derive enjoyment from a steady flow of pleasant small, but frequent, rewards. As I said, the rewards do not have to be large, just frequent—a little bit here, a little bit there. Consider that our major satisfaction for thousands of years came in the form of food and water (and something else more private), and that while we need these steadily, we quickly reach saturation.

The problem, of course, is that we do not live in an environment where results are delivered in a steady manner—Black Swans dominate much of human history. It is unfortunate that the right strategy for our current environment may not offer *internal* rewards and positive feedback.

The same property in reverse applies to our unhappiness. It is better to lump all your pain into a brief period rather than have it spread out over a longer one.

But some people find it possible to transcend the asymmetry of pains and joys, escape the hedonic deficit, set themselves outside that game—and live with hope. There is some good news, as we see next.

The Antechamber of Hope

For Yevgenia Krasnova, a person could love one book, at most a few—beyond this was a form of promiscuity. Those who talk about books as commodities are inauthentic, just as those who collect acquaintances can be superficial in their friendships. A novel you like resembles a friend. You read it and reread it, getting to know it better. Like a friend, you accept it the way it is; you do not judge it. Montaigne was asked "why" he and the writer Etienne de la Boétie were friends—the kind of question people ask you at a cocktail party as if you knew the answer, or as if there were an answer to know. It was typical of Montaigne to reply, "Parce que c'était lui, parce que c'était moi" (because it was him and because it was me). Likewise, Yevgenia claims that she likes that *one* book "because it is it and because I am me." Yevgenia once even walked out on a schoolteacher because he analyzed that book and thus violated her rule. One does not sit idle listening as people wax analytical about your friends. A very stubborn schoolchild she was.

This book she has as a friend is *Il deserto dei tartari,* by Dino Buzzati, a novel that was well known in Italy and France during her childhood, but that, strangely, nobody she knows in America had heard of. Its English title is mistranslated as *The Tartar Steppe* instead of *The Desert of the Tartars*.

Yevgenia encountered *Il deserto* when she was thirteen, in her parents' weekend country house in a small village two hundred kilometers outside Paris, where their Russian and French books multiplied without the constraints of the overfed Parisian apartment. She was so bored in the country that she could not even read. Then, one afternoon, she opened the book and was sucked into it.

Inebriated by Hope

Giovanni Drogo is a man of promise. He has just graduated from the military academy with the rank of junior officer, and active life is just starting. But things do not turn out as planned: his initial four-year assignment is a remote outpost, the Bastiani fortress, protecting the nation from the Tartars likely to invade from the border desert—not too desirable a position. The fortress is located a few days by horseback from the town; there is nothing but bareness around it—none of the social buzz that a man of his age could look forward to. Drogo thinks that his assignment in the outpost is temporary, a way for him to pay his dues before more appealing positions present themselves. Later, back in town, in his impeccably ironed uniform and with his athletic figure, few ladies will be able to resist him.

What is Drogo to do in this hole? He discovers a loophole, a way to be transferred after only four months. He decides to use the loophole.

At the very last minute, however, Drogo takes a glance at the desert from the window of the medical office and decides to extend his stay. Something in the walls of the fort and the silent landscape ensnares him. The appeal of the fort and waiting for the attackers, the big battle with the ferocious Tartars, gradually become his only reason to exist. The entire atmosphere of the fort is one of anticipation. The other men spend their time looking at the horizon and awaiting the big event of the enemy attack. They are so focused that, on rare occasions, they can detect the most insignificant stray animal that appears at the edge of the desert and mistake it for an enemy attack.

Sure enough, Drogo spends the rest of his life extending his stay, delaying the beginning of his life in the city—thirty-five years of pure hope, spent in the grip of the idea that one day, from the remote hills that no human has ever crossed, the attackers will eventually emerge and help him rise to the occasion.

At the end of the novel we see Drogo dying in a roadside inn as the event for which he has waited all his life takes place. He has missed it.

The Sweet Trap of Anticipation

Yevgenia read *Il deserto* numerous times; she even learned Italian (and perhaps married an Italian) so she could read it in the original. Yet she never had the heart to reread the painful ending.

I presented the Black Swan as the outlier, the important event that is not expected to happen. But consider the opposite: the unexpected event that *you very badly want to happen*. Drogo is obsessed and blinded by the possibility of an unlikely event; that rare occurrence is his raison d'être. At thirteen, when she encountered the book, little did Yevgenia know that she would spend an entire life playing Giovanni Drogo in the antechamber of hope, waiting for the big event, sacrificing for it, and refusing intermediate steps, the consolation prizes.

She did not mind the sweet trap of anticipation: to her it was a life worth living; it was worth living in the cathartic simplicity of a single purpose. Indeed, "be careful what you wish for": she may have been happier before the Black Swan of her success than after.

One of the attributes of a Black Swan is an asymmetry in consequences—either positive or negative. For Drogo the consequences were thirty-five years spent waiting in the antechamber of hope for just a few randomly distributed hours of glory—which he ended up missing.

When You Need the Bastiani Fortress

Note that there was no brother-in-law around in Drogo's social network. He was lucky to have companions in his mission. He was a member of a community at the gate of the desert intently looking together at the horizon. Drogo had the advantage of an association with peers and the avoidance of social contact with others outside the community. We are local animals, interested in our immediate neighborhood—even if people far away consider us total idiots. Those homo sapiens are abstract and remote and we do not care about them because we do not run into them in elevators or make eye contact with them. Our shallowness can sometimes work for us.

It may be a banality that we need others for many things, but we need them far more than we realize, particularly for dignity and respect. Indeed, we have very few historical records of people who have achieved anything extraordinary without such peer validation—but we have the freedom to choose our peers. If we look at the history of ideas, we see schools of thought occasionally forming, producing unusual work unpopular outside the school. You hear about the Stoics, the Academic Skeptics, the Cynics, the Pyrrhonian Skeptics, the Essenes, the Surrealists, the Dadaists, the anarchists, the hippies, the fundamentalists. A school allows someone with unusual ideas with the remote possibility of a payoff to find company and create a microcosm insulated from others. The members of the group can be ostracized together—which is better than being ostracized alone.

If you engage in a Black Swan–dependent activity, it is better to be part of a group.

EL DESIERTO DE LOS TÁRTAROS

Yevgenia met Nero Tulip in the lobby of the Hotel Danieli in Venice. He was a trader who lived between London and New York. At the time, traders from London went to Venice on Friday noon during the low season, just to talk to other traders (from London).

As Yevgenia and Nero stood engaged in an effortless conversation, she noticed that her husband was looking uncomfortably at them from the bar where he sat, trying to stay focused on the pontifications of one of his childhood friends. Yevgenia realized that she was going to see a bit more of Nero.

They met again in New York, first in a clandestine way. Her husband, being a philosophy professor, had too much time on his hands, so he started paying close attention to her schedule and became clingy. The clingier he got, the more stifled Yevgenia felt, which made him even clingier. She dumped him, called her lawyer who was by then expecting to hear from her, and saw more of Nero openly.

Nero had a stiff gait since he was recovering from a helicopter crash—he gets a little too arrogant after episodes of success and starts taking uncalculated physical risks, though he remains financially hyperconservative, even paranoid. He had spent months immobile in a London hospital, hardly able to read or write, trying to resist having to watch television, teasing the nurses, and waiting for his bones to heal. He can draw the ceiling with its fourteen cracks from memory, as well as the shabby white building across the street with its sixty-three windowpanes, all in need of professional cleaning.

Nero claimed that he was comfortable in Italian when he drank, so Yevgenia gave him a copy of *Il deserto*. Nero did not read novels—"Novels are fun to write, not read," he claimed. So he left the book by his bedside for a while.

Nero and Yevgenia were, in a sense, like night and day. Yevgenia went to bed at dawn, working on her manuscripts at night. Nero rose at dawn, like most traders, even on weekends. He then worked for an hour on his opus, *Treatise on Probability,* and never touched it again after that. He had been writing it for a decade and felt rushed to finish it only when his life was threatened. Yevgenia smoked; Nero was mindful of his health, spending at least an hour a day at the gym or in the pool. Yevgenia hung around intellectuals and bohemians; Nero often felt comfortable with street-smart traders and businessmen who had never been to college and spoke with cripplingly severe Brooklyn accents. Yevgenia never understood how a classicist and a polyglot like Nero could socialize with people like that. What was worse, she had this French Fifth Republic overt disdain for money, unless disguised by an intellectual or cultural façade, and she could hardly bear these Brooklyn fellows with thick hairy fingers and gigantic bank accounts. Nero's post-Brooklyn friends, in turn, found her snotty. (One of the effects of prosperity has been a steady migration of streetwise people from Brooklyn to Staten Island and New Jersey.)

Nero was also elitist, unbearably so, but in a different way. He separated those who could *connect the dots,* Brooklyn-born or not, from those who could not, regardless of their levels of sophistication and learning.

A few months later, after he was done with Yevgenia (with inordinate relief) he opened *Il deserto* and was sucked into it. Yevgenia had the prescience that, like her, Nero would identify with Giovanni Drogo, the main character of *Il deserto*. He did.

Nero, in turn, bought cases of the English (bad) translation of the book and handed copies to anyone who said a polite hello to him, including his New York doorman who could hardly speak English, let alone read it. Nero was so enthusiastic while explaining the story that the doorman got interested and Nero had to order the Spanish translation for him, *El desierto de los tártaros*.

Bleed or Blowup

Let us separate the world into two categories. Some people are like the turkey, exposed to a major blowup without being aware of it, while others play reverse turkey, prepared for big events that might surprise others. In some strategies and life situations, you gamble dollars to win a succession of pennies while appearing to be winning all the time. In others, you risk a succession of pennies to win dollars. In other words, you bet either that the Black Swan will happen or that it will never happen, two strategies that require completely different mind-sets.

We have seen that we (humans) have a marked preference for making a little bit of income at a time. Recall from Chapter 4 that in the summer of 1982, large American banks lost close to everything they had ever earned, and more.

So some matters that belong to Extremistan are extremely dangerous but do not appear to be so beforehand, since they hide and delay their risks—so suckers think they are "safe." It is indeed a property of Extremistan to look less risky, in the short run, than it really is.

Nero called the businesses exposed to such blowups dubious businesses, particularly since he distrusted whatever method was being used to compute the odds of a blowup. Recall from Chapter 4 that the accounting period upon which companies' performances are evaluated is too short to reveal whether or not they are doing a great job. And, owing to the shallowness of our intuitions, we formulate our risk assessments too quickly.

I will rapidly present Nero's idea. His premise was the following trivial point: some business bets in which one wins big but infrequently, yet loses small but frequently, are worth making if others are suckers for them and *if you have the personal and intellectual stamina*. But you need such stamina. You also need to deal with people in your entourage heaping all manner of insult on you, much of it blatant. People often accept that a financial strategy with a small chance of success is not necessarily a bad one as long as the success is large enough to justify it. For a spate of psychological reasons, however, people have difficulty carrying out such a strategy, simply because it requires a combination of belief, a capacity for delayed gratification, and the willingness to be spat upon by clients without blinking. And those who lose money for any reason start looking like guilty dogs, eliciting more scorn on the part of their entourage.

Against that background of potential blowup disguised as skills, Nero engaged in a strategy that he called "bleed." You lose steadily, daily, for a long time, except when some event takes place for which you get paid disproportionately well. No single event can make you blow up, on the other hand—some changes in the world can produce extraordinarily large profits that pay back such bleed for years, sometimes decades, sometimes even centuries.

Of all the people he knew, Nero was the least genetically designed for such a strategy. His brain disagreed so heavily with his body that he found himself in a state of continuous warfare. It was his body that was his problem, which accumulated physical fatigue from the neurobiological effect of exposure to the small continuous losses, Chinese-water-torture-style, throughout the day. Nero discovered that the losses went to his emotional brain, bypassing his higher cortical structures and slowly affecting his hippocampus and weakening his memory. The hippocampus is the structure where memory is supposedly controlled. It is the most plastic part of the brain; it is also the part that is assumed to absorb all the damage from repeated insults like the chronic stress we experience daily from small doses of negative feelings—as opposed to the invigorating "good stress" of the tiger popping up occasionally in your living room. You can rationalize all you want; the hippocampus takes the insult of chronic stress seriously, incurring irreversible atrophy. Contrary to popular belief, these small, seemingly harmless stressors do not strengthen you; they can amputate part of your self.

It was the exposure to a high level of information that poisoned Nero's life. He could sustain the pain if he saw only weekly performance numbers, instead of updates every minute. He did better emotionally with his own portfolio than with those of clients, since he was not obligated to monitor it continuously.

If his neurobiological system was a victim of the confirmation bias, reacting to the short term and the visible, he could trick his brain to escape its vicious effect by focusing only on the longer haul. He refused to look at any printout of his track record that was shorter than ten years. Nero came of age, intellectually speaking, with the stock market crash of 1987, in which he derived monstrous returns on what small equity he controlled. This episode would forever make his track record valuable, taken as a whole. In close to twenty years of trading, Nero had only four good years. For him, one was more than enough. All he needed was one good year per century.

Investors were no problem for him—they needed his trading as insurance and paid him well. He just had to exhibit a mild degree of contempt toward those he wanted to shed, which did not take much effort on his part. This effort was not contrived: Nero did not think much of them and let his body language express it freely, all the while maintaining an unfashionably high level of courtesy. He made sure, after a long string of losses, that they did not think he was apologetic—indeed, paradoxically, they became more supportive that way. Humans will believe anything you say provided you do not exhibit the smallest shadow of diffidence; like animals, they can detect the smallest crack in your confidence before you express it. The trick is to be as smooth as possible in personal manners. It is much easier to signal self-confidence if you are exceedingly polite and friendly; you can control people without having to offend their sensitivity. The problem with business people, Nero realized, is that if you act like a loser they will treat you as a loser—you set the yardstick yourself. There is no absolute measure of good or bad. It is not what you are telling people, it is how you are saying it.

But you need to remain understated and maintain an Olympian calm in front of others.

When he worked as a trader for an investment bank, Nero had to face the typical employee-evaluation form. The form was supposed to keep track of "performance," supposedly as a check against employees slacking off. Nero found the evaluation absurd because it did not so much judge the quality of a trader's performance as encourage him to game the system by working for short-term profits at the expense of possible blowups—like banks that give foolish loans that have a small probability of blowing up, because the loan officer is shooting for his next quarterly evaluation. So one day early in his career, Nero sat down and listened very calmly to the evaluation of his "supervisor." When Nero was handed the evaluation form he tore it into small pieces in front of him. He did this very slowly, accentuating the contrast between the nature of the act and the tranquillity with which he tore the paper. The boss watched him blank with fear, eyes popping out of his head. Nero focused on his undramatic, slow-motion act, elated by both the feeling of standing up for his beliefs and the aesthetics of its execution. The combination of elegance and dignity was exhilarating. He knew that he would either be fired or left alone. He was left alone.

Chapter Eight

GIACOMO CASANOVA'S UNFAILING LUCK: THE PROBLEM OF SILENT EVIDENCE

The Diagoras problem—How Black Swans make their way out of history books—Methods to help you avoid drowning—The drowned do not usually vote—We should all be stockbrokers—Do silent witnesses count?—Casanova's étoile—New York is "so invincible"

Another fallacy in the way we understand events is that of silent evidence. History hides both Black Swans and its Black Swan–generating ability from us.

THE STORY OF THE DROWNED WORSHIPPERS

More than two thousand years ago, the Roman orator, belletrist, thinker, Stoic, manipulator-politician, and (usually) virtuous gentleman, Marcus Tullius Cicero, presented the following story. One Diagoras, a nonbeliever in the gods, was shown painted tablets bearing the portraits of some worshippers who prayed, then survived a subsequent shipwreck. The implication was that praying protects you from drowning. Diagoras asked, "Where were the pictures of those who prayed, then drowned?"

The drowned worshippers, being dead, would have a lot of trouble advertising their experiences from the bottom of the sea. This can fool the casual observer into believing in miracles.

We call this the problem of silent evidence. The idea is simple, yet potent and universal. While most thinkers try to put to shame those who came *before* them, Cicero puts to shame almost all empirical thinkers who came *after* him, until very recently.

Later on, both my hero of heroes, the essayist Michel de Montaigne and the empirical Francis Bacon, mentioned the point in their works, applying it to the formation of false beliefs. "And such is the way of all superstition, whether in astrology, dreams, omens, divine judgments, or the like," wrote Bacon in his *Novum Organum*. The problem, of course, is that unless they are drilled into us systematically, or integrated into our way of thinking, these great observations are rapidly forgotten.

Silent evidence pervades everything connected to the notion of *history*. By history, I don't just mean those learned-but-dull books in the history section (with Renaissance paintings on their cover to attract buyers). History, I will repeat, is *any succession of events* seen with the effect of *posteriority*.

This bias extends to the ascription of factors in the success of ideas and religions, to the illusion of skill in many professions, to success in artistic occupations, to the nature versus nurture debate, to mistakes in using evidence in the court of law, to illusions about the "logic" of history—and of course, most severely, in our perception of the nature of extreme events.

You are in a classroom listening to someone self-important, dignified, and ponderous (but dull), wearing a tweed jacket (white shirt, polka-dot tie), pontificating for two hours on the theories of history. You are too paralyzed by boredom to understand what on earth he is talking about, but you hear the names of big guns: Hegel, Fichte, Marx, Proudhon, Plato, Herodotus, Ibn Khaldoun, Toynbee, Spengler, Michelet, Carr, Bloch, Fukuyama, Schmukuyama, Trukuyama. He seems deep and knowledgeable, making sure that no attention lapse will make you forget that his approach is "post-Marxist," "postdialectical," or post-something, whatever that means. Then you realize that a large part of what he is saying reposes on a simple optical illusion! But this will not make a difference: he is so invested in it that if you questioned his method he would react by throwing even more names at you.

It is so easy to avoid looking at the cemetery while concocting historical theories. But this is not just a problem with history. It is a problem with the way we construct samples and gather evidence *in every domain*. We shall call this distortion a bias, i.e., the difference between what you see and what is there. By *bias* I mean a systematic error consistently showing a more positive, or negative, effect from the phenomenon, like a scale that unfailingly shows you a few pounds heavier or lighter than your true weight, or a video camera that adds a few sizes to your waistline. This bias has been rediscovered here and there throughout the past century across disciplines, often to be rapidly forgotten (like Cicero's insight). As drowned worshippers do not write histories of their experiences (it is better to be alive for that), so it is with the losers in history, whether people or ideas. Remarkably, historians and other scholars in the humanities who need to understand silent evidence the most do not seem to have a name for it (and I looked hard). As for journalists, fuhgedaboudit! They are industrial producers of the distortion.

The term *bias* also indicates the condition's potentially quantifiable nature: you may be able to calculate the distortion, and to correct for it by taking into account both the dead and the living, instead of only the living.

Silent evidence is what events use to conceal their own randomness, particularly the Black Swan type of randomness.

Sir Francis Bacon is an interesting and endearing fellow in many respects.

He harbored a deep-seated, skeptical, nonacademic, antidogmatic, and obsessively empirical nature, which, to someone skeptical, nonacademic, antidogmatic, and obsessively empirical, like this author, is a quality almost impossible to find in the thinking business. (Anyone can be skeptical; any scientist can be overly empirical—it is the rigor coming from the combination of skepticism and empiricism that's hard to come by.) The problem is that his empiricism wanted us to confirm, not disconfirm; thus he introduced the problem of confirmation, that beastly corroboration that generates the Black Swan.

THE CEMETERY OF LETTERS

The Phoenicians, we are often reminded, produced no literature, although they allegedly invented the alphabet. Commentators discuss their philistinism from the basis of this absence of a written legacy, asserting that by race or culture, they were more interested in commerce than in the arts. Accordingly, the Phoenician invention of the alphabet served the lower purpose of commercial record keeping rather than the more noble purpose of literary production. (I remember finding on the shelves of a country house I once rented a mildewed history book by Will and Ariel Durant describing the Phoenicians as the "merchant race." I was tempted to throw it in the fireplace.) Well, it now seems that the Phoenicians wrote quite a bit, but using a perishable brand of papyrus that did not stand the biodegradative assaults of time. Manuscripts had a high rate of extinction before copyists and authors switched to parchment in the second or third century. Those not copied during that period simply disappeared.

The neglect of silent evidence is endemic to the way we study comparative talent, particularly in activities that are plagued with winner-take-all attributes. We may enjoy what we see, but there is no point reading too much into success stories because we do not see the full picture.

Recall the *winner-take-all* effect from Chapter 3: notice the large number of people who call themselves writers but are (only "temporarily") operating the shiny cappuccino machines at Starbucks. The inequity in this field is larger than, say, medicine, since we rarely see medical doctors serving hamburgers. I can thus infer that I can largely gauge the performance of the latter profession's entire population from what sample is visible to me. Likewise with plumbers, taxi drivers, prostitutes, and those in professions devoid of superstar effects. Let us go beyond the discussion on Extremistan and Mediocristan in Chapter 3. The consequence of the superstar dynamic is that what we call "literary heritage" or "literary treasures" is a minute proportion of what has been produced cumulatively. This is the first point. How it invalidates the identification of talent can be derived immediately from it: say you attribute the success of the nineteenth-century novelist Honoré de Balzac to his superior "realism," "insights," "sensitivity," "treatment of characters," "ability to keep the reader riveted," and so on. These may be deemed "superior" qualities that lead to superior performance *if, and only if,* those who lack what we call talent also lack these qualities. But what if there are dozens of comparable literary masterpieces that happened to perish? And, following my logic, if there are indeed many perished manuscripts with similar attributes, then, I regret to say, your idol Balzac was just the beneficiary of disproportionate luck compared to his peers. Furthermore, you may be committing an injustice to others by favoring him.

My point, I will repeat, is not that Balzac is untalented, but that he is less *uniquely* talented than we think. Just consider the thousands of writers now completely vanished from consciousness: their record does not enter into analyses. We do not see the tons of rejected manuscripts because these writers have never been published. *The New Yorker* alone rejects close to a hundred manuscripts a day, so imagine the number of geniuses that we will never hear about. In a country like France, where more people write books while, sadly, fewer people read them, respectable literary publishers accept one in ten thousand manuscripts they receive from first-time authors. Consider the number of actors who have never passed an audition but would have done very well had they had that lucky break in life.

The next time you visit a Frenchman of comfortable means, you will likely spot the stern books from the collection *Bibliothèque de la Pléiade,* which their owner will never, almost never, read, mostly on account of their uncomfortable size and weight. Membership in the *Pléiade* means membership in the literary canon. The tomes are expensive; they have the distinctive smell of ultrathin India paper, compressing the equivalent of fifteen hundred pages into the size of a drugstore paperback. They are supposed to help you maximize the number of masterpieces per Parisian square foot. The publisher Gallimard has been extremely selective in electing writers into the *Pléiade* collection–only a few authors, such as the aesthete and adventurer André Malraux, have made it in while still alive. Dickens, Dostoyevsky, Hugo, and Stendhal are in, along with Mallarmé, Sartre, Camus, and . . . Balzac. Yet if you follow Balzac's own ideas, which I will examine next, you would accept that there is no ultimate justification for such an official corpus.

Balzac outlined the entire business of silent evidence in his novel *Lost Illusions*. Lucien de Rubempré (alias of Lucien Chardon), the penurious provincial genius, "goes up" to Paris to start a literary career. We are told that he is talented—actually *he* is told that he is talented by the semiaristocratic set in Angoulême. But it is difficult to figure out whether this is due to his good looks or to the literary quality of his works—or even whether literary quality is visible, or, as Balzac seems to wonder, if it has much to do with anything. Success is presented cynically, as the product of wile and promotion or the lucky surge of interest for reasons completely external to the works themselves. Lucien discovers the existence of the immense cemetery inhabited by what Balzac calls "nightingales."

> Lucien was told that this designation "nightingale" was given by bookstores to those works residing on the shelves in the solitary depths of their shops.

Balzac presents to us the sorry state of contemporary literature when Lucien's manuscript is rejected by a publisher who has never read it; later on, when Lucien's reputation has developed, the very same manuscript is accepted by another publisher who did not read it either! The work itself was a secondary consideration.

In another example of silent evidence, the book's characters keep bemoaning that things are no longer as they were *before*, implying that literary fairness prevailed in more ancient times—as if there was no cemetery before. They fail to take into account the nightingales among the ancients' work! Notice that close to two centuries ago people had an idealized opinion of their own past, just as we have an idealized opinion of today's past.

I mentioned earlier that to understand successes and analyze what *caused* them, we need to study the traits present in failures. It is to a more general version of this point that I turn next.

How to Become a Millionaire in Ten Steps

Numerous studies of millionaires aimed at figuring out the skills required for hotshotness follow the following methodology. They take a population of hotshots, those with big titles and big jobs, and study their attributes. They look at what those big guns have in common: courage, risk taking, optimism, and so on, and infer that these traits, most notably risk taking, help you to become successful. You would also probably get the same impression if you read CEOs' ghostwritten autobiographies or attended their presentations to fawning MBA students.

Now take a look at the cemetery. It is quite difficult to do so because people who fail do not seem to write memoirs, and, if they did, those business publishers I know would not even consider giving them the courtesy of a returned phone call (as to returned e-mail, fuhgedit). Readers would not pay $26.95 for a story of failure, even if you convinced them that it had more useful tricks than a story of success.* The entire notion of biography is grounded in the arbitrary ascription of a causal relation between specified traits and subsequent events. Now consider the cemetery. The graveyard of failed persons will be full of people who shared the following traits: courage, risk taking, optimism, et cetera. Just like the population of millionaires. There may be some differences in skills, but

* The best noncharlatanic finance book I know is called *What I Learned Losing a Million Dollars,* by D. Paul and B. Moynihan. The authors had to self-publish the book.

what truly separates the two is for the most part a single factor: luck. Plain luck.

You do not need a lot of empiricism to figure this out: a simple thought experiment suffices. The fund-management industry claims that some people are extremely skilled, since year after year they have outperformed the market. They will identify these "geniuses" and convince you of their abilities. My approach has been to manufacture cohorts of purely random investors and, by simple computer simulation, show how it would be impossible to not have these geniuses produced *just by luck*. Every year you fire the losers, leaving only the winners, and thus end up with long-term steady winners. Since you do not observe the cemetery of failed investors, you will think that it is a good business, and that some operators are considerably better than others. Of course an explanation will be readily provided for the success of the lucky survivors: "He eats tofu," "She works late; just the other day I called her office at eight P.M. . . ." Or of course, "She is naturally lazy. People with that type of laziness can see things clearly." By the mechanism of retrospective determinism we will find the "cause"—actually, we need to see the cause. I call these simulations of hypothetical cohorts, often done by computer, an engine of computational epistemology. Your thought experiments can be run on a computer. You just simulate an alternative world, plain random, and verify that it looks similar to the one in which we live. Not getting lucky billionaires in these experiments would be the exception.*

Recall the distinction between Mediocristan and Extremistan in Chapter 3. I said that taking a "scalable" profession is not a good idea, simply because there are far too few winners in these professions. Well, these professions produce a large cemetery: the pool of starving actors is larger than the one of starving accountants, even if you assume that, on average, they earn the same income.

* Doctors are rightfully and vigorously skeptical of anecdotal results, and require that studies of drug efficacy probe into the cemetery of silent evidence. However, the same doctors fall for the bias elsewhere! Where? In their personal lives, or in their investment activities. At the cost of being repetitive, I have to once again state my amazement at the aspect of human nature that allows us to mix the most rigorous skepticism and the most acute gullibility.

A HEALTH CLUB FOR RATS

The second, and more vicious, variety of the problem of silent evidence is as follows. When I was in my early twenties and still read the newspaper, and thought that steadily reading the newspapers was something useful to me, I came across an article discussing the mounting threat of the Russian Mafia in the United States and its displacement of the traditional Louie and Tony in some neighborhoods of Brooklyn. The article explained their toughness and brutality as a result of their being hardened by their Gulag experiences. The Gulag was a network of labor camps in Siberia where criminals and dissidents were routinely deported. Sending people to Siberia was one of the purification methods initially used by the czarist regimes and later continued and perfected by the Soviets. Many deportees did not survive these labor camps.

Hardened by the Gulag? The sentence jumped out at me as both profoundly flawed (and a reasonable inference). It took me a while to figure out the nonsense in it since it was protected by cosmetic wrapping; the following thought experiment will give the intuition. Assume that you're able to find a large, assorted population of rats: fat, thin, sickly, strong, well-proportioned, et cetera. (You can easily get them from the kitchens of fancy New York restaurants.) With these thousands of rats, you build a heterogeneous cohort, one that is well representative of the general New York rat population. You bring them to my laboratory on East Fifty-ninth Street in New York City and we put the entire collection in a large vat. We subject the rats to increasingly higher levels of radiation (since this is supposed to be a thought experiment, I am told that there is no cruelty in the process). At every level of radiation, those that are naturally stronger (and this is the key) will survive; the dead will drop out of your sample. We will progressively have a stronger and stronger collection of rats. Note the following central fact: every single rat, including the strong ones, will be *weaker* after the radiation than before.

An observer endowed with analytical abilities, who probably got excellent grades in college, would be led to believe that treatment in my laboratory is an excellent health-club replacement, and one that could be generalized to all mammals (think of the potential commercial success). His logic would run as follows: Hey, these rats are stronger than the rest of the rat population. What do they seem to have in common? They all came from that Black Swan guy Taleb's workshop. Not many people will have the temptation to go look at the dead rats.

Next we pull the following trick on *The New York Times:* we let these surviving rats loose in New York City and inform the chief rodent correspondent of the newsworthy disruption in the pecking order in the New York rat population. He will write a lengthy (and analytical) article on the social dynamics of New York rats that includes the following passage: "Those rats are now bullies in the rat population. They literally run the show. *Strengthened* by their experience in the laboratory of the reclusive (but friendly) statistician/philosopher/trader Dr. Taleb, they . . ."

Vicious Bias

There is a vicious attribute to the bias: it can hide best when its impact is largest. Owing to the invisibility of the dead rats, the more lethal the risks, the less visible they will be, since the severely victimized are likely to be eliminated from the evidence. The more injurious the treatment, the larger the difference between the surviving rats and the rest, and the more fooled you will be about the *strengthening* effect. One of the two following ingredients is necessary for this difference between the true effect (weakening) and the observed one (strengthening): a) a degree of inequality in strength, or diversity, in the base cohort, or b) unevenness, or diversity, somewhere in the treatment. Diversity here has to do with the degree of uncertainty inherent in the process.

More Hidden Applications

We can keep going with this argument; it has such universality that once we get the bug it is hard to look at reality with the same eyes again. Clearly it robs our observations of their realistic power. I will enumerate a few more cases to illustrate the weaknesses of our inferential machinery.

The stability of species. Take the number of species that we now consider extinct. For a long time scientists took the number of such species as that implied from an analysis of the extant fossils. But this number ignores the silent cemetery of species that came and left without leaving traces in the form of fossils; the fossils that we have managed to find correspond to a smaller proportion of all species that came and disappeared. This implies that our biodiversity was far greater than it seemed at first examination. A more worrisome consequence is that the rate of extinction of species may be far greater than we think—close to 99.5 percent of species that transited through earth are now extinct, a number of scientists have kept raising through time. Life is a great deal more fragile than we have allowed for. But this does not mean we (humans) should feel guilty for extinctions around us; nor does it mean that we should act to stop them—species were coming and going before we started messing up the environment. There is no need to feel moral responsibility for every endangered species.

Does crime pay? Newspapers report on the criminals who get caught. There is no section in *The New York Times* recording the stories of those who committed crimes but have not been caught. So it is with cases of tax evasion, government bribes, prostitution rings, poisoning of wealthy spouses (with substances that do not have a name and cannot be detected), and drug trafficking.

In addition, our representation of the standard criminal might be based on the properties of those less intelligent ones who were caught.

Once we seep ourselves into the notion of silent evidence, so many things around us that were previously hidden start manifesting themselves. Having spent a couple of decades in this mind-set, I am convinced (but cannot prove) that training and education can help us avoid its pitfalls.

The Evolution of the Swimmer's Body

What do the popular expressions "a swimmer's body" and "beginner's luck" have in common? What do they seem to share with the concept of history?

There is a belief among gamblers that beginners are almost always lucky. "It gets worse later, but gamblers are always lucky when they start out," you hear. This statement is actually empirically true: researchers confirm that gamblers have lucky beginnings (the same applies to stock market speculators). Does this mean that each one of us should become a gambler for a while, take advantage of lady luck's friendliness to beginners, then stop?

The answer is no. The same optical illusion prevails: those who start gambling will be either lucky or unlucky (given that the casino has the advantage, a slightly greater number will be unlucky). The lucky ones, with the feeling of having been selected by destiny, will continue gambling; the others, discouraged, will stop and will not show up in the sample. They will probably take up, depending on their temperaments, bird-watching, Scrabble, piracy, or other pastimes. Those who continue gambling will remember having been lucky as beginners. The dropouts, by definition, will no longer be part of the surviving gamblers' community. This explains beginner's luck.

There is an analogy with what is called in common parlance a "swimmer's body," which led to a mistake I shamefully made a few years ago (in spite of my specialty in this bias, I did not notice that I was being fooled). When asking around about the comparative physical elegance of athletes, I was often told that runners looked anorexic, cyclists bottom-heavy, and weight lifters insecure and a little primitive. I inferred that I should spend some time inhaling chlorine in the New York University pool to get those "elongated muscles." Now suspend the causality. Assume that a person's genetic variance allows for a certain type of body shape. Those born with a natural tendency to develop a swimmer's body become better swimmers. These are the ones you see in your sample splashing up and down at the pools. But they would have looked pretty much the same if they lifted weights. It is a fact that a given muscle grows exactly the same way whether you take steroids or climb walls at the local gym.

WHAT YOU SEE AND WHAT YOU DON'T SEE

Katrina, the devastating hurricane that hit New Orleans in 2005, got plenty of politicizing politicians on television. These legislators, moved by the images of devastation and the pictures of angry victims made homeless, made promises of "rebuilding." It was so noble on their part to do something humanitarian, to rise above our abject selfishness.

Did they promise to do so with their own money? No. It was with public money. Consider that such funds will be taken away from somewhere else, as in the saying "You take from Peter to give to Paul." That *somewhere else* will be less mediatized. It may be privately funded cancer research, or the next efforts to curb diabetes. Few seem to pay attention to the victims of cancer lying lonely in a state of untelevised depression. Not only do these cancer patients not vote (they will be dead by the next ballot), but they do not manifest themselves to our emotional system. More of them die every day than were killed by Hurricane Katrina; they are the ones who need us the most—not just our financial help, but our attention and kindness. And they may be the ones from whom the money will be taken—indirectly, perhaps even directly. Money (public or private) taken away from research might be responsible for killing them—in a crime that may remain silent.

A ramification of the idea concerns our decision making under a cloud of possibilities. We see the obvious and visible consequences, not the invisible and less obvious ones. Yet those unseen consequences can be—nay, generally are—more meaningful.

Frédéric Bastiat was a nineteenth-century humanist of a strange variety, one of those rare independent thinkers—independent to the point of being unknown in his own country, France, since his ideas ran counter to French political orthodoxy (he joins another of my favorite thinkers, Pierre Bayle, in being unknown at home and in his own language). But he has a large number of followers in America.

In his essay "What We See and What We Don't See," Bastiat offered the following idea: we can see what governments do, and therefore sing their praises—but we do not see the alternative. But there is an alternative; it is less obvious and remains unseen.

Recall the confirmation fallacy: governments are great at telling you what they did, but not what they did not do. In fact, they engage in what could be labeled as phony "philanthropy," the activity of helping people in a visible and sensational way without taking into account the unseen cemetery of invisible consequences. Bastiat inspired libertarians by attacking the usual arguments that showed the benefits of governments. But his ideas can be generalized to apply to both the Right and the Left.

Bastiat goes a bit deeper. If both the positive and the negative consequences of an action fell on its author, our learning would be fast. But often an action's positive consequences benefit only its author, since they are visible, while the negative consequences, being invisible, apply to others, with a net cost to society. Consider job-protection measures: you notice those whose jobs are made safe and ascribe social benefits to such protections. You do not notice the effect on those who cannot find a job as a result, since the measure will reduce job openings. In some cases, as with the cancer patients who may be punished by Katrina, the positive consequences of an action will immediately benefit the politicians and phony humanitarians, while the negative ones take a long time to appear—they may never become noticeable. One can even blame the press for directing charitable contributions toward those who may need them the least.

Let us apply this reasoning to September 11, 2001. Around twenty-five hundred people were directly killed by bin Laden's group in the Twin Towers of the World Trade Center. Their families benefited from the support of all manner of agencies and charities, as they should. But, according to researchers, during the remaining three months of the year, close to one thousand people died as silent victims of the terrorists. How? Those who were afraid of flying and switched to driving ran an increased risk of death. There was evidence of an increase of casualties on the road during that period; the road is considerably more lethal than the skies. These families got no support—they did not even know that their loved ones were also the victims of bin Laden.

In addition to Bastiat, I have a weakness for Ralph Nader (the activist and consumer advocate, certainly not the politician and political thinker). He may be the American citizen who saved the highest number of lives by exposing the safety record of car companies. But, in his political campaign a few years ago, even he forgot to trumpet the tens of thousands of lives saved by his seat belt laws. It is much easier to sell "Look what I did for you" than "Look what I avoided for you."

Recall from the Prologue the story of the hypothetical legislator whose actions might have avoided the attack of September 11. How many such people are walking the street without the upright gait of the phony hero?

Have the guts to consider the silent consequences when standing in front of the next snake-oil humanitarian.

Doctors

Our neglect of silent evidence kills people daily. Assume that a drug saves many people from a potentially dangerous ailment, but runs the risk of killing a few, with a net benefit to society. Would a doctor prescribe it? He has no incentive to do so. The lawyers of the person hurt by the side effects will go after the doctor like attack dogs, while the lives saved by the drug might not be accounted for anywhere.

A life saved is a statistic; a person hurt is an anecdote. Statistics are invisible; anecdotes are salient. Likewise, the risk of a Black Swan is invisible.

THE TEFLON-STYLE PROTECTION OF GIACOMO CASANOVA

This brings us to gravest of all manifestations of silent evidence, the illusion of stability. The bias lowers our perception of the risks we incurred in the past, particularly for those of us who were lucky to have survived them. Your life came under a serious threat but, having survived it, you retrospectively underestimate how risky the situation actually was.

The adventurer Giacomo Casanova, later self-styled Jacques, Chevalier de Seingalt, the wannabe intellectual and legendary seducer of women,

Giacomo Casanova, a.k.a. Jacques, Chevalier de Seingalt. Some readers might be surprised that the legendary seducer did not look quite like James Bond.

seems to have had a Teflon-style trait that would cause envy on the part of the most resilient of Mafia dons: misfortune did not stick to him. Casanova, while known for his seductions, viewed himself as some sort of a scholar. He aimed at literary fame with his twelve-volume *History of My Life,* written in bad (charmingly bad) French. In addition to the extremely useful lessons on how to become a seducer, the *History* provides an engrossing account of a succession of reversals of fortune. Casanova felt that every time he got into difficulties, his lucky star, his *étoile,* would pull him out of trouble. After things got bad for him, they somehow recovered by some invisible hand, and he was led to believe that it was his intrinsic property to recover from hardships by running every time into a new opportunity. He would somehow meet someone in extremis who offered him a financial transaction, a new patron that he had not betrayed in the past, or someone generous enough and with a weak enough memory to forget past betrayals. Could Casanova have been selected by destiny to bounce back from all hardships?

Not necessarily. Consider the following: of all the colorful adventurers who have lived on our planet, many were occasionally crushed, and a few did bounce back repeatedly. It is those who survive who will tend to believe that they are indestructible; they will have a long and interesting enough experience to write books about it. Until, of course . . .

Actually, adventurers who feel singled out by destiny abound, simply because there are plenty of adventurers, and we do not hear the stories of those down on their luck. As I started writing this chapter, I recalled a conversation with a woman about her flamboyant fiancé, the son of a civil servant, who managed through a few financial transactions to catapult himself into the life of a character in a novel, with handmade shoes, Cuban cigars, collectible cars, and so on. The French have a word for this, *flambeur*, which means a mixture of extravagant bon vivant, wild speculator, and risk taker, all the while bearing considerable personal charm; a word that does not seem to be available in Anglo-Saxon cultures. The fiancé was spending his money very quickly, and as we were having the conversation about his fate (she was going to marry him, after all), she explained to me that he was undergoing slightly difficult times, but that there was no need to worry since he always came back with a vengeance. That was a few years ago. Out of curiosity, I have just tracked him down (trying to do so tactfully): he has not recovered (yet) from his latest blow of fortune. He also dropped out of the scene and is no longer to be found among other *flambeurs*.

How does this relate to the dynamics of history? Consider what is generally called the resilience of New York City. For seemingly transcendental reasons, every time it gets close to the brink of disaster, the city manages to pull back and recover. Some people truly believe that this is an internal property of New York City. The following quote is from a *New York Times* article:

> Which is why New York still needs Samuel M. E. An economist who turns 77 today, Mr. E. studied New York City through half a century of booms and busts. . . . "We have a record of going through tough times and coming back stronger than ever," he said.

Now run the idea in reverse: think of cities as little Giacomo Casanovas, or as rats in my laboratory. As we put the thousands of rats through a very dangerous process, let's put a collection of cities in a simulator of history: Rome, Athens, Carthage, Byzantium, Tyre, Catal Hyuk (located in modern-day Turkey, it is one of the first known human settlements), Jericho, Peoria, and, of course, New York City. Some cities will survive the harsh conditions of the simulator. As to others, we know that history might not be too kind. I am sure that Carthage, Tyre, and Jericho had their local, no less eloquent, Samuel M. E., saying, "Our enemies have tried to destroy us many times; but we always came back more resilient than before. We are now invincible."

This bias causes the survivor to be an unqualified witness of the process. Unsettling? The fact that you survived is a condition that may weaken your interpretation of the properties of the survival, including the shallow notion of "cause."

You can do a lot with the above statement. Replace the retired economist Samuel E. with a CEO discussing his corporation's ability to recover from past problems. How about the taunted "resilience of the financial system"? How about a general who has had a good run?

The reader can now see why I use Casanova's unfailing luck as a generalized framework for the analysis of history, all histories. I generate artificial histories featuring, say, millions of Giacomo Casanovas, and observe the difference between the attributes of the successful Casanovas (because you generate them, you know their exact properties) and those an observer of the result would obtain. From that perspective, it is not a good idea to be a Casanova.

"I Am a Risk Taker"

Consider the restaurant business in a competitive place like New York City. One has indeed to be foolish to open one, owing to the enormous risks involved and the harrying quantity of work to get anywhere in the business, not counting the finicky fashion-minded clients. The cemetery of failed restaurants is very silent: walk around Midtown Manhattan and you will see these warm patron-filled restaurants with limos waiting outside for the diners to come out with their second, trophy, spouses. The owner is overworked but happy to have all these important people patronize his eatery. Does this mean that it makes sense to open a restaurant in such a competitive neighborhood? Certainly not, yet people do it out of the foolish risk-taking trait that pushes us to jump into such adventures blinded by the outcome.

Clearly there is an element of the surviving Casanovas in us, that of the risk-taking genes, which encourages us to take blind risks, unaware of the variability in the possible outcomes. We inherited the taste for uncalculated risk taking. Should we encourage such behavior?

In fact, economic growth comes from such risk taking. But some fool might argue the following: if someone followed reasoning such as mine, we would not have had the spectacular growth we experienced in the past. This is exactly like someone playing Russian roulette and finding it a good idea because he survived and pocketed the money.

We are often told that we humans have an optimistic bent, and that *it is supposed to be good for us*. This argument appears to justify general risk taking as a positive enterprise, and one that is glorified in the common culture. Hey, look, our ancestors took the challenges—while you, NNT, are encouraging us to do nothing (I am not).

We have enough evidence to confirm that, indeed, we humans are an extremely lucky species, and that we got the genes of the risk takers. The foolish risk takers, that is. In fact, the Casanovas who survived.

Once again, I am not dismissing the idea of risk taking, having been involved in it myself. I am only critical of the encouragement of *uninformed* risk taking. The überpsychologist Danny Kahneman has given us evidence that we generally take risks not out of bravado but out of ignorance and blindness to probability! The next few chapters will show in more depth how we tend to dismiss outliers and adverse outcomes when projecting the future. But I insist on the following: *that we got here by accident does not mean that we should continue to take the same risks*. We are mature enough a race to realize this point, enjoy our blessings, and try to preserve, by becoming more conservative, what we got by luck. We have been playing Russian roulette; now let's stop and get a real job.

I have two further points to make on this subject. First, justification of overoptimism on grounds that "it brought us here" arises from a far more serious mistake about human nature: the belief that we are built to understand nature and our own nature and that our decisions are, and have been, the result of our own choices. I beg to disagree. So many instincts drive us.

Second, a little more worrisome than the first point: evolutionary fitness is something that is continuously touted and aggrandized by the crowd who takes it as gospel. The more unfamiliar someone is with the wild Black Swan–generating randomness, the more he or she believes in the optimal working of evolution. Silent evidence is not present in their theories. Evolution is a series of flukes, some good, many bad. You only see the good. But, in the short term, it is not obvious which traits are really good for you, particularly if you are in the Black Swan–generating environment of Extremistan. This is like looking at rich gamblers coming out of the casino and claiming that a taste for gambling is good for the species because gambling makes you rich! Risk taking made many species head for extinction!

This idea that we are here, that this is the best of all possible worlds, and *that evolution did a great job* seems rather bogus in the light of the silent-evidence effect. The fools, the Casanovas, and the blind risk takers are often the ones who win in the short term. Worse, in a Black Swan environment, where one single but rare event can come shake up a species after a very long run of "fitness," the foolish risk takers can also win in the long term! I will revisit this idea in Part Three, where I show how Extremistan worsens the silent-evidence effect.

But there is another manifestation that merits a mention.

I AM A BLACK SWAN: THE ANTHROPIC BIAS

I want to stay closer to earth and avoid bringing higher-up metaphysical or cosmological arguments into this discussion—there are so many significant dangers to worry about down here on planet earth and it would be a good idea to postpone the metaphysical philosophizing for later. But it would be useful to take a peek (not more) at what is called the anthropic cosmological argument, as it points out the gravity of our misunderstanding of historical stability.

A recent wave of philosophers and physicists (and people combining the two categories) has been examining *the self-sampling assumption,* which is a generalization of the principle of the Casanova bias to our own existence.

Consider our own fates. Some people reason that the odds of any of us being in existence are so low that our being here cannot be attributed to an accident of fate. Think of the odds of the parameters being exactly where they need to be to induce our existence (any deviation from the optimal calibration would have made our world explode, collapse, or simply not come into existence). It is often said that the world seems to have been built to the specifications that would make our existence possible. According to such an argument, it could not come from luck.

However, *our presence in the sample* completely vitiates the computation of the odds. Again, the story of Casanova can make the point quite simple—much simpler than in its usual formulation. Think again of all the possible worlds as little Casanovas following their own fates. The one who is still kicking (by accident) will feel that, given that he cannot be so lucky, there had to be some transcendental force guiding him and supervising his destiny: "Hey, otherwise the odds would be too low to get here just by luck." For someone who observes *all* adventurers, the odds of finding a Casanova are not low at all: there so many adventurers, and someone is bound to win the lottery ticket.

The problem here with the universe and the human race is that *we are the surviving Casanovas*. When you start with many adventurous Casanovas, there is bound to be a survivor, and guess what: if you are here talking about it, you are likely to be that particular one (notice the "condition": you survived to talk about it). So we can no longer naïvely compute odds without considering that the condition that we are in existence imposes restrictions on the process that led us here.

Assume that history delivers either "bleak" (i.e., unfavorable) or "rosy" (i.e., favorable) scenarios. The bleak scenarios lead to extinction. Clearly, if I am now writing these lines, it is certainly because history delivered a "rosy" scenario, one that allowed me to be here, a historical route in which my forebears avoided massacre by the many invaders who roamed the Levant. Add to that beneficial scenarios free of meteorite collisions, nuclear war, and other large-scale terminal epidemics. But I do not have to look at humanity as a whole. Whenever I probe into my own biography I am alarmed at how tenuous my life has been so far. Once when I returned to Lebanon during the war, at the age of eighteen, I felt episodes of extraordinary fatigue and cold chills in spite of the summer heat. It was typhoid fever. Had it not been for the discovery of antibiotics, only a few decades earlier, I would not be here today. I was also later "cured" of another severe disease that would have left me for dead, thanks to a treatment that depends on another recent medical technology. As a human being alive here in the age of the Internet, capable of writing and reaching an audience, I have also benefited from society's luck and the remarkable absence of recent large-scale war. In addition, I am the result of the rise of the human race, itself an accidental event.

My being here is a consequential low-probability occurrence, and I tend to forget it.

Let us return to the touted recipes for becoming a millionaire in ten steps. A successful person will try to convince you that his achievements could not possibly be accidental, just as a gambler who wins at roulette seven times in a row will explain to you that the odds against such a streak are one in several million, so you either have to believe some transcendental intervention is in play or accept his skills and insight in picking the winning numbers. But if you take into account the quantity of gamblers out there, and the number of gambling sessions (several million episodes in total), then it becomes obvious that such strokes of luck are bound to happen. And if you are talking about them, they have happened to you.

The *reference point argument* is as follows: do not compute odds from the vantage point of the winning gambler (or the lucky Casanova, or the endlessly bouncing back New York City, or the invincible Carthage), but from all those who started in the cohort. Consider once again the example of the gambler. If you look at the population of beginning gamblers taken as a whole, you can be close to certain that one of them (but you do not know in advance which one) will show stellar results just by luck. So, from the *reference point* of the beginning cohort, this is not a big deal. But from the reference point of the winner (and, who does not, and this is key, take the losers into account), a long string of wins will appear to be too extraordinary an occurrence to be explained by luck. Note that a "history" is just a series of numbers through time. The numbers can represent degrees of wealth, fitness, weight, anything.

The Cosmetic Because

This in itself greatly weakens the notion of "because" that is often propounded by scientists, and almost always misused by historians. We have to accept the fuzziness of the familiar "because" no matter how queasy it makes us feel (and it does makes us queasy to remove the analgesic illusion of causality). I repeat that we are explanation-seeking animals who tend to think that everything has an identifiable cause and grab the most apparent one as *the* explanation. Yet there may not be a visible *because;* to the contrary, frequently there is nothing, not even a spectrum of possible explanations. But silent evidence masks this fact. Whenever our survival is in play, the very notion of *because* is severely weakened. The condition of survival drowns all possible explanations. The Aristotelian "because" is not there to account for a solid link between two items, but rather, as we saw in Chapter 6, to cater to our hidden weakness for imparting explanations.

Apply this reasoning to the following question: Why didn't the bubonic plague kill more people? People will supply quantities of cosmetic explanations involving theories about the intensity of the plague and "scientific models" of epidemics. Now, try the weakened causality argument that I have just emphasized in this chapter: had the bubonic plague killed more people, the observers (us) would not be here to observe. So it may not necessarily be the property of diseases to spare us humans. Whenever your survival is in play, don't immediately look for causes and effects. The main identifiable reason for our survival of such diseases might simply be inaccessible to us: we are here since, Casanova-style, the "rosy" scenario played out, and if it seems too hard to understand it is because we are too brainwashed by notions of causality and we think that it is smarter to say *because* than to accept randomness.

My biggest problem with the educational system lies precisely in that it forces students to squeeze explanations out of subject matters and *shames* them for withholding judgment, for uttering the "I don't know." Why did the Cold War end? Why did the Persians lose the battle of Salamis? Why did Hannibal get his behind kicked? Why did Casanova bounce back from hardship? In each of these examples, we are taking a condition, survival, and looking for the explanations, instead of flipping the argument on its head and stating that *conditional on such survival,* one cannot read *that* much into the process, and should learn instead to invoke some measure of randomness (randomness is what we don't know; to invoke randomness is to plead ignorance). It is not just your college professor who gives you bad habits. I showed in Chapter 6 how newspapers need to stuff their texts with causal links to make you enjoy the narratives. But have the integrity to deliver your "because" very sparingly; try to limit it to situations where the "because" is derived from experiments, not backward-looking history.

Note here that I am not saying causes do not exist; do not use this argument to avoid trying to learn from history. All I am saying is that it is *not so simple;* be suspicious of the "because" and handle it with care—particularly in situations where you suspect silent evidence.

We have seen several varieties of the silent evidence that cause deformations in our perception of empirical reality, making it appear more explainable (and more stable) than it actually is. In addition to the confirmation error and the narrative fallacy, the manifestations of silent evidence further distort the role and importance of Black Swans. In fact, they cause a gross overestimation at times (say, with literary success), and underestimation at others (the stability of history; the stability of our human species).

I said earlier that our perceptual system may not react to what does not lie in front of our eyes, or what does not arouse our emotional attention. We are made to be superficial, to heed what we see and not heed what does not vividly come to mind. We wage a double war against silent evidence. The unconscious part of our inferential mechanism (and there is one) will ignore the cemetery, even if we are intellectually aware of the need to take it into account. Out of sight, out of mind: we harbor a natural, even physical, scorn of the abstract.

This will be further illustrated in the next chapter.

Chapter Nine

THE LUDIC FALLACY, OR THE UNCERTAINTY OF THE NERD

Lunch at Lake Como (west)—The military as philosophers—Plato's randomness

---

FAT TONY

"Fat Tony" is one of Nero's friends who irritates Yevgenia Krasnova beyond measure. We should perhaps more thoughtfully style him "Horizontally-challenged Tony," since he is not as objectively overweight as his nickname indicates; it is just that his body shape makes whatever he wears seem ill-fitted. He wears only tailored suits, many of them cut for him in Rome, but they look as if he bought them from a Web catalog. He has thick hands, hairy fingers, wears a gold wrist chain, and reeks of licorice candies that he devours in industrial quantities as a substitute for an old smoking habit. He doesn't usually mind people calling him Fat Tony, but he much prefers to be called just Tony. Nero calls him, more politely, "Brooklyn Tony," because of his accent and his Brooklyn way of thinking, though Tony is one of the prosperous Brooklyn people who moved to New Jersey twenty years ago.

Tony is a successful nonnerd with a happy disposition. He leads a gregarious existence. His sole visible problem seems to be his weight and the corresponding nagging by his family, remote cousins, and friends, who keep warning him about that premature heart attack. Nothing seems to work; Tony often goes to a fat farm in Arizona to *not* eat, lose a few pounds, then gain almost all of them back in his first-class seat on the flight back. It is remarkable how his self-control and personal discipline, otherwise admirable, fail to apply to his waistline.

He started as a clerk in the back office of a New York bank in the early 1980s, in the letter-of-credit department. He pushed papers and did some grunt work. Later he grew into giving small business loans and figured out the game of how you can get financing from the monster banks, how their bureaucracies operate, and what they like to see on paper. All the while an employee, he started acquiring property in bankruptcy proceedings, buying it from financial institutions. His big insight is that bank employees who sell you a house that's not theirs just don't care as much as the owners; Tony knew very rapidly how to talk to them and maneuver. Later, he also learned to buy and sell gas stations with money borrowed from small neighborhood bankers.

Tony has this remarkable habit of trying to make a buck effortlessly, just for entertainment, without straining, without office work, without meeting, just by melding his deals into his private life. Tony's motto is "Finding who the sucker is." Obviously, they are often the banks: "The clerks don't care about nothing." Finding these suckers is second nature to him. If you took walks around the block with Tony you would feel considerably more informed about the texture of the world just "tawking" to him.

Tony is remarkably gifted at getting unlisted phone numbers, first-class seats on airlines for no additional money, or your car in a garage that is officially full, either through connections or his forceful charm.

Non-Brooklyn John

I found the perfect non-Brooklyn in someone I will call Dr. John. He is a former engineer currently working as an actuary for an insurance company. He is thin, wiry, and wears glasses and a dark suit. He lives in New Jersey not far from Fat Tony but certainly they rarely run into each other. Tony never takes the train, and, actually, never commutes (he drives a Cadillac, and sometimes his wife's Italian convertible, and jokes that he is more visible than the rest of the car). Dr. John is a master of the schedule; he is as predictable as a clock. He quietly and efficiently reads the newspaper on the train to Manhattan, then neatly folds it for the lunchtime continuation. While Tony makes restaurant owners rich (they beam when they see him coming and exchange noisy hugs with him), John meticulously packs his sandwich every morning, fruit salad in a plastic container. As for his clothing, he also wears a suit that looks like it came from a Web catalog, except that it is quite likely that it actually did.

Dr. John is a painstaking, reasoned, and gentle fellow. He takes his work seriously, so seriously that, unlike Tony, you can see a line in the sand between his working time and his leisure activities. He has a PhD in electrical engineering from the University of Texas at Austin. Since he knows both computers and statistics, he was hired by an insurance company to do computer simulations; he enjoys the business. Much of what he does consists of running computer programs for "risk management."

I know that it is rare for Fat Tony and Dr. John to breathe the same air, let alone find themselves at the same bar, so consider this a pure thought exercise. I will ask each of them a question and compare their answers.

NNT (that is, me): Assume that a coin is fair, i.e., has an equal probability of coming up heads or tails when flipped. I flip it ninety-nine times and get heads each time. What are the odds of my getting tails on my next throw?

Dr. John: Trivial question. One half, of course, since you are assuming 50 percent odds for each and independence between draws.

NNT: What do you say, Tony?

Fat Tony: I'd say no more than 1 percent, of course.

NNT: Why so? I gave you the initial assumption of a fair coin, meaning that it was 50 percent either way.

Fat Tony: You are either full of crap or a pure sucker to buy that "50 pehcent" business. The coin gotta be loaded. It can't be a fair game. (Translation: It is far more likely that your assumptions about the fairness are wrong than the coin delivering ninety-nine heads in ninety-nine throws.)

NNT: But Dr. John said 50 percent.

Fat Tony (whispering in my ear): I know these guys with the nerd examples from the bank days. They think way too slow. And they are too commoditized. You can take them for a ride.

Now, of the two of them, which would you favor for the position of mayor of New York City (or Ulan Bator, Mongolia)? Dr. John thinks entirely within the box, the box that was given to him; Fat Tony, almost entirely outside the box.

To set the terminology straight, what I call "a nerd" here doesn't have to look sloppy, unaesthetic, and sallow, and wear glasses and a portable computer on his belt as if it were an ostensible weapon. A nerd is simply someone who thinks exceedingly inside the box.

Have you ever wondered why so many of these straight-A students end up going nowhere in life while someone who lagged behind is now getting the shekels, buying the diamonds, and getting his phone calls returned? Or even getting the Nobel Prize in a real discipline (say, medicine)? Some of this may have something to do with luck in outcomes, but there is this sterile and obscurantist quality that is often associated with classroom knowledge that may get in the way of understanding what's going on in real life. In an IQ test, as well as in any academic setting (including sports), Dr. John would vastly outperform Fat Tony. But Fat Tony would outperform Dr. John in any other possible ecological, real-life situation. In fact, Tony, in spite of his lack of culture, has an enormous curiosity about the texture of reality, and his own erudition—to me, he is more scientific in the literal, though not in the social, sense than Dr. John.

We will get deep, very deep, into the difference between the answers of Fat Tony and Dr. John; this is probably the most vexing problem I know about the connections between two varieties of knowledge, what we dub Platonic and a-Platonic. Simply, people like Dr. John can cause Black Swans outside Mediocristan—their minds are closed. While the problem is very general, one of its nastiest illusions is what I call the ludic fallacy—the attributes of the uncertainty we face in real life have little connection to the sterilized ones we encounter in exams and games.

So I close Part One with the following story.

LUNCH AT LAKE COMO

One spring day a few years ago, I was surprised to receive an invitation from a think tank sponsored by the United States Defense Department to a brainstorming session on risk that was to take place in Las Vegas the following fall. The person who invited me announced on the phone, "We'll have lunch on a terrace overlooking Lake Como," which put me in a state of severe distress. Las Vegas (along with its sibling the emirate of Dubai) is perhaps one place I'd never wish to visit before I die. Lunch at "fake Como" would be torture. But I'm glad I went.

The think tank had gathered a nonpolitical collection of people they called doers and scholars (and practitioners like me who do not accept the distinction) involved in uncertainty in a variety of disciplines. And they symbolically picked a major casino as a venue.

The symposium was a closed-doors, synod-style assembly of people who would never have mixed otherwise. My first surprise was to discover that the military people there thought, behaved, and acted like philosophers—far more so than the philosophers we will see splitting hairs in their weekly colloquium in Part Three. They thought out of the box, like traders, except much better and without fear of introspection. An assistant secretary of defense was among us, but had I not known his profession I would have thought he was a practitioner of skeptical empiricism. Even an engineering investigator who had examined the cause of a space shuttle explosion was thoughtful and open-minded. I came out of the meeting realizing that only military people deal with randomness with genuine, introspective intellectual honesty—unlike academics and corporate executives using other people's money. This does not show in war movies, where they are usually portrayed as war-hungry autocrats. The people in front of me were not the people who initiate wars. Indeed, for many, the successful defense policy is the one that manages to eliminate potential dangers without war, such as the strategy of bankrupting the Russians through the escalation in defense spending. When I expressed my amazement to Laurence, another finance person who was sitting next to me, he told me that the military collected more genuine intellects and risk thinkers than most if not all other professions. Defense people wanted to understand the epistemology of risk.

In the group was a gentleman who ran a group of professional gamblers and who was banned from most casinos. He had come to share his wisdom with us. He sat not far from a stuffy professor of political science, dry like a bone and, as is characteristic of "big names," careful about his reputation, who said nothing out of the box, and who did not smile once. During the sessions, I tried to imagine the hotshot with a rat dropped down his back, putting him in a state of wriggling panic. He was perhaps good at writing Platonic models of something called game theory, but when Laurence and I went after him on his improper use of financial metaphors, he lost all his arrogance.

Now, when you think of the major risks casinos face, gambling situations come to mind. In a casino, one would think, the risks include lucky gamblers blowing up the house with a series of large wins and cheaters taking away money through devious methods. It is not just the general public that would believe so, but the casino management as well. Consequently, the casino had a high-tech surveillance system tracking cheaters, card counters, and other people who try to derive an advantage over them.

Each of the participants gave his presentation and listened to those of the others. I came to discuss Black Swans, and I intended to tell them that the only thing I know is that we know precious little about them, but that it was their property to sneak up on us, and that attempts at Platonifying them led to additional misunderstandings. Military people can understand such things, and the idea became recently prevalent in military circles with the expression *unknown unknown* (as opposed to the *known unknown*). But I had prepared my talk (on five restaurant napkins, some stained) and was ready to discuss a new phrase I coined for the occasion: the *ludic fallacy*. I intended to tell them that I should not be speaking at a casino because it had nothing to do with uncertainty.

The Uncertainty of the Nerd

What is the *ludic* fallacy? *Ludic* comes from *ludus,* Latin for games.

I was hoping that the representatives of the casino would speak before me so I could start harassing them by showing (politely) that a casino was precisely the venue *not* to pick for such a discussion, since the class of risks casinos encounter are very insignificant *outside* of the building, and their study not readily transferable. My idea is that gambling was *sterilized* and domesticated uncertainty. In the casino you know the rules, you can calculate the odds, and the type of uncertainty we encounter there, we will see later, is *mild,* belonging to Mediocristan. My prepared statement was this: "The casino is the only human venture I know where the probabilities are known, Gaussian (i.e., bell-curve), and almost computable." You cannot expect the casino to pay out a million times your bet, or to change the rules abruptly on you during the game—there are never days in which "36 black" is designed to pop up 95 percent of the time.*

In real life you do not know the odds; you need to discover them, and the sources of uncertainty are not defined. Economists, who do not con-

* My colleague Mark Spitznagel found a martial version of the ludic fallacy: organized competitive fighting trains the athlete to focus on the game and, in order not to dissipate his concentration, to ignore the possibility of what is not specifically allowed by the rules, such as kicks to the groin, a surprise knife, et cetera. So those who win the gold medal might be precisely those who will be most vulnerable in real life. Likewise, you see people with huge muscles (in black T-shirts) who can impress you in the artificial environment of the gym but are unable to lift a stone.

sider what was discovered by noneconomists worthwhile, draw an artificial distinction between Knightian risks (which you can compute) and Knightian uncertainty (which you cannot compute), after one Frank Knight, who rediscovered the notion of unkown uncertainty and did a lot of thinking but perhaps never took risks, or perhaps lived in the vicinity of a casino. Had he taken economic or financial risks he would have realized that these "computable" risks are largely absent from real life! They are laboratory contraptions!

Yet we automatically, spontaneously associate chance with these Platonified games. I find it infuriating to listen to people who, upon being informed that I specialize in problems of chance, immediately shower me with references to dice. Two illustrators for a paperback edition of one of my books spontaneously and independently added a die to the cover and below every chapter, throwing me into a state of rage. The editor, familiar with my thinking, warned them to "avoid the ludic fallacy," as if it were a well-known intellectual violation. Amusingly, they both reacted with an "Ah, sorry, we didn't know."

Those who spend too much time with their noses glued to maps will tend to mistake the map for the territory. Go buy a recent history of probability and probabilistic thinking; you will be showered with names of alleged "probability thinkers" who all base their ideas on these sterilized constructs. I recently looked at what college students are taught under the subject of chance and came out horrified; they were brainwashed with this ludic fallacy and the outlandish bell curve. The same is true of people doing PhD's in the field of probability theory. I'm reminded of a recent book by a thoughtful mathematician, Amir Aczel, called *Chance*. Excellent book perhaps, but like all other modern books it is grounded in the ludic fallacy. Furthermore, assuming chance has anything to do with mathematics, what little mathematization we can do in the real world does not assume the mild randomness represented by the bell curve, but rather scalable wild randomness. What can be mathematized is usually not Gaussian, but Mandelbrotian.

Now, go read any of the classical thinkers who had something practical to say about the subject of chance, such as Cicero, and you find something different: a notion of probability that remains fuzzy throughout, as it needs to be, since such fuzziness is the very nature of uncertainty. Probability is a liberal art; it is a child of skepticism, not a tool for people with calculators on their belts to satisfy their desire to produce fancy calculations and certainties. Before Western thinking drowned in its "scientific" mentality, what is arrogantly called the Enlightenment, people prompted their brain to think—not compute. In a beautiful treatise now vanished from our consciousness, *Dissertation on the Search for Truth*, published in 1673, the polemist Simon Foucher exposed our psychological predilection for certainties. He teaches us the art of doubting, how to position ourselves between doubting and believing. He writes: "One needs to exit doubt in order to produce science—but few people heed the importance of not exiting from it prematurely. . . . It is a fact that one usually exits doubt without realizing it." He warns us further: "We are dogma-prone from our mother's wombs."

By the confirmation error discussed in Chapter 5, we use the example of games, which probability theory was successful at tracking, and claim that this is a general case. Furthermore, just as we tend to underestimate the role of luck in life in general, we tend to *overestimate* it in games of chance.

"This building is inside the Platonic fold; life stands outside of it," I wanted to shout.

Gambling with the Wrong Dice

I was in for quite a surprise when I learned that the building too was outside the Platonic fold.

The casino's risk management, aside from setting its gambling policies, was geared toward reducing the losses resulting from cheaters. One does not need heavy training in probability theory to understand that the casino was sufficiently diversified across the different tables to not have to worry about taking a hit from an extremely lucky gambler (the diversification argument that leads to the bell curve, as we will see in Chapter 15). All they had to do was control the "whales," the high rollers flown in at the casino's expense from Manila or Hong Kong; whales can swing several million dollars in a gambling bout. Absent cheating, the performance of most individual gamblers would be the equivalent of a drop in the bucket, making the aggregate very stable.

I promised not to discuss any of the details of the casino's sophisticated surveillance system; all I am allowed to say is that I felt transported into a James Bond movie—I wondered if the casino was an imitation of the movies or if it was the other way around. Yet, in spite of such sophistication, their risks had nothing to do with what can be anticipated knowing that the business is a casino. For it turned out that the four largest losses incurred or narrowly avoided by the casino fell completely outside their sophisticated models.

First, they lost around $100 million when an irreplaceable performer in their main show was maimed by a tiger (the show, *Siegfried and Roy,* had been a major Las Vegas attraction). The tiger had been reared by the performer and even slept in his bedroom; until then, nobody suspected that the powerful animal would turn against its master. In scenario analyses, the casino had even conceived of the animal jumping into the crowd, but nobody came near to the idea of insuring against what happened.

Second, a disgruntled contractor was hurt during the construction of a hotel annex. He was so offended by the settlement offered him that he made an attempt to dynamite the casino. His plan was to put explosives around the pillars in the basement. The attempt was, of course, thwarted (otherwise, to use the arguments in Chapter 8, we would not have been there), but I shivered at the thought of possibly sitting above a pile of dynamite.

Third, casinos must file a special form with the Internal Revenue Service documenting a gambler's profit if it exceeds a given amount. The employee who was supposed to mail the forms hid them, instead, for completely unexplainable reasons, in boxes under his desk. This went on for years without anyone noticing that something was wrong. The employee's refraining from sending the documents was truly impossible to predict. Tax violations (and negligence) being serious offences, the casino faced the near loss of a gambling license or the onerous financial costs of a suspension. Clearly they ended up paying a monstrous fine (an undisclosed amount), which was the luckiest way out of the problem.

Fourth, there was a spate of other dangerous scenes, such as the kidnapping of the casino owner's daughter, which caused him, in order to secure cash for the ransom, to violate gambling laws by dipping into the casino coffers.

Conclusion: A back-of-the-envelope calculation shows that the dollar value of these Black Swans, the off-model hits and potential hits I've just outlined, swamp the on-model risks by a factor of close to 1,000 to 1. The casino spent hundreds of millions of dollars on gambling theory and high-tech surveillance while the bulk of their risks came from outside their models.

All this, and yet the rest of the world still learns about uncertainty and probability from gambling examples.

WRAPPING UP PART ONE

The Cosmetic Rises to the Surface

All of the topics in Part One are actually only one. You can think about a subject for a long time, to the point of being possessed by it. Somehow you have a lot of ideas, but they do not seem explicitly connected; the logic linking them remains concealed from you. Yet you know deep down that all these are *the same* idea. Meanwhile, what Nietzsche calls *bildungsphilisters,** or learned philistines, blue collars of the thinking business, tell you that you are spread out between fields; you reply that these disciplines are artificial and arbitrary, to no avail. Then you tell them that you are a limousine driver, and they leave you alone—you feel better because you do not identify with them, and thus you no longer need to be amputated to fit into the Procrustean bed of the disciplines. Finally, a little push and you see that it was all one single problem.

One evening I found myself at a cocktail party in Munich at the apartment of a former art historian who had more art books in its library than I thought existed. I stood drinking excellent Riesling in the spontaneously formed English-speaking corner of the apartment, in the hope of getting to a state where I would be able to start speaking my brand of fake German. One of the most insightful thinkers I know, the computer entrepreneur Yossi Vardi, prompted me to summarize "my idea" while standing on one leg. It was not too convenient to stand on one leg after a few glasses of perfumed Riesling, so I failed in my improvisation. The next day I experienced staircase wit. I jumped out of bed with the following idea: *the cosmetic and the Platonic rise naturally to the surface*. This is a simple extension of the problem of knowledge. It is simply that one side of Eco's library, the one we never see, has the property of being ignored. This is also the problem of silent evidence. It is why we do not see Black Swans: we worry about those that happened, not those that may happen but did not. It is why we Platonify, liking known schemas and well-organized knowledge—to the point of blindness to reality. It is why we fall for the problem of induction, why we *confirm*. It is why those who "study" and fare well in school have a tendency to be suckers for the ludic fallacy.

* What Nietzsche means by this term are the dogma-prone newspaper readers and opera lovers who have cosmetic exposure to culture and shallow depth. I extend the term here to the philistine hiding in academia who lacks in erudition out of lack of curiosity and is closely centered on his ideas.

And it is why we have Black Swans and never learn from their occurrence, because the ones that did not happen were too abstract. Thanks to Vardi, I now belonged to the club of single-idea people.

We love the tangible, the confirmation, the palpable, the real, the visible, the concrete, the known, the seen, the vivid, the visual, the social, the embedded, the emotionally laden, the salient, the stereotypical, the moving, the theatrical, the romanced, the cosmetic, the official, the scholarly-sounding verbiage (b******t), the pompous Gaussian economist, the mathematicized crap, the pomp, the Académie Française, Harvard Business School, the Nobel Prize, dark business suits with white shirts and Ferragamo ties, the moving discourse, and the lurid. Most of all we favor *the narrated*.

Alas, we are not manufactured, in our current edition of the human race, to understand abstract matters—we need context. Randomness and uncertainty are abstractions. We respect what has happened, ignoring what *could have* happened. In other words, we are naturally shallow and superficial—and we do not know it. This is not a psychological problem; it comes from the main property of information. The dark side of the moon is harder to see; beaming light on it costs energy. In the same way, beaming light on the unseen is costly in both computational and mental effort.

Distance from Primates

There have been in history many distinctions between higher and lower forms of humans. For the Greeks, there were the Greeks and the barbarians, those people of the north who uttered amorphous sentences similar, to the Attic ear, to an animal's shrieks. For the English, a higher form of life was the gentleman's—contrary to today's definition, a gentleman's life was practiced through idleness and a code of behavior that included, along with a set of manners, the avoidance of work beyond the necessities of comfortable subsistence. For New Yorkers, there are those with a Manhattan zip code and those with such a thing as a Brooklyn or, worse, Queens address. For the earlier Nietzsche, there was the Apollonian compared to the Dionysian; for the better-known Nietzsche, there was the Übermensch, something his readers interpret however it suits them. For a modern stoic, a higher individual subscribes to a dignified system of virtue that determines elegance in one's behavior and the ability to separate results from efforts. All of these distinctions aim at lengthening the distance between us and our relatives among other primates. (I keep insisting that, when it comes to decision making, the distance between us and these hairy cousins is far shorter than we think.)

I propose that if you want a simple step to a higher form of life, as distant from the animal as you can get, then you may have to denarrate, that is, shut down the television set, minimize time spent reading newspapers, ignore the blogs. Train your reasoning abilities to control your decisions; nudge System 1 (the heuristic or experiential system) out of the important ones. Train yourself to spot *the difference between the sensational and the empirical.* This insulation from the toxicity of the world will have an additional benefit: it will improve your well-being. Also, bear in mind how shallow we are with probability, the mother of all abstract notions. You do not have to do much more in order to gain a deeper understanding of things around you. Above all, learn to avoid "tunneling."

A bridge here to what is to come. The Platonic blindness I illustrated with the casino story has another manifestation: focusing. To be able to focus is a great virtue if you are a watch repairman, a brain surgeon, or a chess player. But the last thing you need to do when you deal with uncertainty is to "focus" (you should tell uncertainty to focus, not us). This "focus" makes you a sucker; it translates into prediction problems, as we will see in the next section. Prediction, not narration, is the real test of our understanding of the world.

Part 2

WE JUST CAN'T PREDICT

When I ask people to name three recently implemented technologies that most impact our world today, they usually propose the computer, the Internet, and the laser. All three were unplanned, unpredicted, and unappreciated upon their discovery, and remained unappreciated well after their initial use. They were consequential. They were Black Swans. Of course, we have this retrospective illusion of their partaking in some master plan. You can create your own lists with similar results, whether you use political events, wars, or intellectual epidemics.

You would expect our record of prediction to be horrible: the world is far, far more complicated than we think, which is not a problem, except when most of us don't know it. We tend to "tunnel" while looking into the future, making it business as usual, Black Swan–free, when in fact there is nothing usual about the future. It is not a Platonic category!

We have seen how good we are at narrating backward, at inventing stories that convince us that we understand the past. For many people, knowledge has the remarkable power of producing confidence instead of measurable aptitude. Another problem: the focus on the (inconsequential) regular, the Platonification that makes the forecasting "inside the box."

I find it scandalous that in spite of the empirical record we continue to project into the future as if we were good at it, using tools and methods that exclude rare events. Prediction is firmly institutionalized in our world. We are suckers for those who help us navigate uncertainty, whether the fortune-teller or the "well-published" (dull) academics or civil servants using phony mathematics.

From Yogi Berra to Henri Poincaré

The great baseball coach Yogi Berra has a saying, "It is tough to make predictions, especially about the future." While he did not produce the writings that would allow him to be considered a philosopher, in spite of his wisdom and intellectual abilities, Berra can claim to know something about randomness. He was a practitioner of uncertainty, and, as a baseball player and coach, regularly faced random outcomes, and had to face their results deep into his bones.

In fact, Yogi Berra is not the only thinker who thought about how much of the future lies beyond our abilities. Many less popular, less pithy, but not less competent thinkers than he have examined our inherent limitations in this regard, from the philosophers Jacques Hadamard and Henri Poincaré (commonly described as mathematicians), to the philosopher Friedrich von Hayek (commonly described, alas, as an economist), to the philosopher Karl Popper (commonly known as a philosopher). We can safely call this the Berra-Hadamard-Poincaré-Hayek-Popper conjecture, which puts structural, built-in limits to the enterprise of predicting.

"The future ain't what it used to be," Berra later said.* He seems to have been right: the gains in our ability to model (and predict) the world may be dwarfed by the increases in its complexity—implying a greater and greater role for the unpredicted. The larger the role of the Black Swan, the harder it will be for us to predict. Sorry.

Before going into the limits of prediction, we will discuss our track record in forecasting and the relation between gains in knowledge and the offsetting gains in confidence.

* Note that these sayings attributed to Yogi Berra might be apocryphal—it was the physicist Niels Bohr who came up with the first one, and plenty of others came up with the second. These sayings remain, however, quintessential Berraisms.

Chapter Ten

THE SCANDAL OF PREDICTION

Welcome to Sydney—How many lovers did she have?—How to be an economist, wear a nice suit, and make friends—Not right, just "almost" right—Shallow rivers can have deep spots

---

One March evening, a few men and women were standing on the esplanade overlooking the bay outside the Sydney Opera House. It was close to the end of the summer in Sydney, but the men were wearing jackets despite the warm weather. The women were more thermally comfortable than the men, but they had to suffer the impaired mobility of high heels.

They all had come to pay the price of sophistication. Soon they would listen for several hours to a collection of oversize men and women singing endlessly in Russian. Many of the opera-bound people looked like they worked for the local office of J. P. Morgan, or some other financial institution where employees experience differential wealth from the rest of the local population, with concomitant pressures on them to live by a sophisticated script (wine and opera). But I was not there to take a peek at the neosophisticates. I had come to look at the Sydney Opera House, a building that adorns every Australian tourist brochure. Indeed, it is striking, though it looks like the sort of building architects create in order to impress other architects.

That evening walk in the very pleasant part of Sydney called the Rocks was a pilgrimage. While Australians were under the illusion that they had built a monument to distinguish their skyline, what they had really done was to construct a monument to our failure to predict, to plan, and to come to grips with our *unknowledge* of the future—our systematic underestimation of what the future has in store.

The Australians had actually built a symbol of the epistemic arrogance of the human race. The story is as follows. The Sydney Opera House was supposed to open in early 1963 at a cost of AU$ 7 million. It finally opened its doors more than ten years later, and, although it was a less ambitious version than initially envisioned, it ended up costing around AU$ 104 million. While there are far worse cases of planning failures (namely the Soviet Union), or failures to forecast (all important historical events), the Sydney Opera House provides an aesthetic (at least in principle) illustration of the difficulties. This opera-house story is the mildest of all the distortions we will discuss in this section (it was only money, and it did not cause the spilling of innocent blood). But it is nevertheless emblematic.

This chapter has two topics. First, we are demonstrably arrogant about what we think we know. We certainly know a lot, but we have a built-in tendency to think that we know a little bit more than we actually do, enough of *that little bit* to occasionally get into serious trouble. We shall see how you can verify, even measure, such arrogance in your own living room.

Second, we will look at the implications of this arrogance for all the activities involving prediction.

Why on earth do we predict so much? Worse, even, and more interesting: Why don't we talk about our record in predicting? Why don't we see how we (almost) always miss the big events? I call this the scandal of prediction.

ON THE VAGUENESS OF CATHERINE'S LOVER COUNT

Let us examine what I call *epistemic arrogance,* literally, our hubris concerning the limits of our knowledge. *Epistēmē* is a Greek word that refers to knowledge; giving a Greek name to an abstract concept makes it sound important. True, our knowledge does grow, but it is threatened by greater increases in confidence, which make our increase in knowledge at the same time an increase in confusion, ignorance, and conceit.

Take a room full of people. Randomly pick a number. The number could correspond to anything: the proportion of psychopathic stockbrokers in western Ukraine, the sales of this book during the months with *r* in them, the average IQ of business-book editors (or business writers), the number of lovers of Catherine II of Russia, et cetera. Ask each person in the room to independently estimate a range of possible values for that number set in such a way that they believe that they have a 98 percent chance of being right, and less than 2 percent chance of being wrong. In other words, whatever they are guessing has about a 2 percent chance to fall outside their range. For example:

"I am 98 percent confident that the population of Rajastan is between 15 and 23 million."

"I am 98 percent confident that Catherine II of Russia had between 34 and 63 lovers."

You can make inferences about human nature by counting how many people in your sample guessed wrong; it is not expected to be too much higher than two out of a hundred participants. Note that the subjects (your victims) are free to set their range as wide as they want: you are not trying to gauge their knowledge but rather *their evaluation of their own knowledge.*

Now, the results. Like many things in life, the discovery was unplanned, serendipitous, surprising, and took a while to digest. Legend has it that Albert and Raiffa, the researchers who noticed it, were actually looking for something quite different, and more boring: how humans figure out probabilities in their decision making when uncertainty is involved (what the learned call *calibrating*). The researchers came out befuddled. The 2 percent error rate turned out to be close to 45 percent in the population being tested! It is quite telling that the first sample consisted of Harvard Business School students, a breed not particularly renowned for their humility or introspective orientation. MBAs are particularly nasty in this regard, which might explain their business success. Later studies document more humility, or rather a smaller degree of arrogance, in other populations. Janitors and cabdrivers are rather humble. Politicians and corporate executives, alas . . . I'll leave them for later.

Are we twenty-two times too comfortable with what we know? It seems so.

This experiment has been replicated dozens of times, across populations, professions, and cultures, and just about every empirical psychologist and decision theorist has tried it on his class to show his students the big problem of humankind: we are simply not wise enough to be trusted with knowledge. The intended 2 percent error rate usually turns out to be between 15 percent and 30 percent, depending on the population and the subject matter.

I have tested myself and, sure enough, failed, even while consciously trying to be humble by carefully setting a wide range—and yet such underestimation happens to be, as we will see, the core of my professional activities. This bias seems present in all cultures, even those that favor humility—there may be no consequential difference between downtown Kuala Lumpur and the ancient settlement of Amioun, (currently) Lebanon. Yesterday afternoon, I gave a workshop in London, and had been mentally writing on my way to the venue because the cabdriver had an above-average ability to "find traffic." I decided to make a quick experiment during my talk.

I asked the participants to take a stab at a range for the number of books in Umberto Eco's library, which, as we know from the introduction to Part One, contains 30,000 volumes. Of the sixty attendees, not a single one made the range wide enough to include the actual number (the 2 percent error rate became 100 percent). This case may be an aberration, but the distortion is exacerbated with quantities that are out of the ordinary. Interestingly, the crowd erred on the very high and the very low sides: some set their ranges at 2,000 to 4,000; others at 300,000 to 600,000.

True, someone warned about the nature of the test can play it safe and set the range between zero and infinity; but this would no longer be "calibrating"—that person would not be conveying any information, and could not produce an informed decision in such a manner. In this case it is more honorable to just say, "I don't want to play the game; I have no clue."

It is not uncommon to find counterexamples, people who overshoot in the opposite direction and actually overestimate their error rate: you may have a cousin particularly careful in what he says, or you may remember that college biology professor who exhibited pathological humility; the tendency that I am discussing here applies to the average of the population, not to every single individual. There are sufficient variations around the average to warrant occasional counterexamples. Such people are in the minority—and, sadly, since they do not easily achieve prominence, they do not seem to play too influential a role in society.

Epistemic arrogance bears a double effect: we overestimate what we know, and underestimate uncertainty, by compressing the range of possible uncertain states (i.e., by reducing the space of the unknown).

The applications of this distortion extend beyond the mere pursuit of knowledge: just look into the lives of the people around you. Literally any decision pertaining to the future is likely to be infected by it. Our human race is affected by a chronic underestimation of the possibility of the future straying from the course initially envisioned (in addition to other biases that sometimes exert a compounding effect). To take an obvious example, think about how many people divorce. Almost all of them are acquainted with the statistic that between one-third and one-half of all marriages fail, something the parties involved did not forecast while tying the knot. Of course, "not us," because "we get along so well" (as if others tying the knot got along poorly).

I remind the reader that I am not testing how much people know, but assessing *the difference between what people actually know and how much they think they know*. I am reminded of a measure my mother concocted, as a joke, when I decided to become a businessman. Being ironic about my (perceived) confidence, though not necessarily unconvinced of my abilities, she found a way for me to make a killing. How? Someone who could figure out how to buy me at the price I am truly worth and sell me at what I think I am worth would be able to pocket a huge difference. Though I keep trying to convince her of my internal humility and insecurity concealed under a confident exterior; though I keep telling her that I am an introspector—she remains skeptical. Introspector shmintrospector, she still jokes at the time of this writing that I am a little ahead of myself.

BLACK SWAN BLINDNESS REDUX

The simple test above suggests the presence of an ingrained tendency in humans to underestimate outliers—or Black Swans. Left to our own devices, we tend to think that what happens every decade in fact only happens once every century, and, furthermore, that we know what's going on.

This miscalculation problem is a little more subtle. In truth, outliers are not as sensitive to underestimation since they are fragile to estimation errors, which can go in both directions. As we saw in Chapter 6, there are conditions under which people overestimate the unusual or some specific unusual event (say when sensational images come to their minds)—which, we have seen, is how insurance companies thrive. So my general point is that these events are very fragile to *miscalculation,* with a general severe underestimation mixed with an occasional severe overestimation.

The errors get worse with the degree of remoteness to the event. So far, we have only considered a 2 percent error rate in the game we saw earlier, but if you look at, say, situations where the odds are one in a hundred, one in a thousand, or one in a million, then the errors become monstrous. The longer the odds, the larger the epistemic arrogance.

Note here one particularity of our intuitive judgment: even if we lived in Mediocristan, in which large events are rare, we would still underestimate extremes—we would think that they are even rarer. We underestimate our error rate even with Gaussian variables. Our intuitions are sub-Mediocristani. But we do not live in Mediocristan. The numbers we are likely to estimate on a daily basis belong largely in Extremistan, i.e., they are run by concentration and subjected to Black Swans.

Guessing and Predicting

There is no effective difference between my guessing a variable that is not random, but for which my information is partial or deficient, such as the number of lovers who transited through the bed of Catherine II of Russia, and predicting a random one, like tomorrow's unemployment rate or next year's stock market. In this sense, guessing (what I don't know, but what someone else may know) and predicting (what has not taken place yet) are the same thing.

To further appreciate the connection between guessing and predicting, assume that instead of trying to gauge the number of lovers of Catherine of Russia, you are estimating the less interesting but, for some, more important question of the population growth for the next century, the stock-market returns, the social-security deficit, the price of oil, the results of your great-uncle's estate sale, or the environmental conditions of Brazil two decades from now. Or, if you are the publisher of Yevgenia Krasnova's book, you may need to produce an estimate of the possible future sales. We are now getting into dangerous waters: just consider that most professionals who make forecasts are also afflicted with the mental impediment discussed above. Furthermore, people who make forecasts professionally are often *more* affected by such impediments than those who don't.

INFORMATION IS BAD FOR KNOWLEDGE

You may wonder how learning, education, and experience affect epistemic arrogance—how educated people might score on the above test, as compared with the rest of the population (using Mikhail the cabdriver as a benchmark). You will be surprised by the answer: it depends on the profession. I will first look at the advantages of the "informed" over the rest of us in the humbling business of prediction.

I recall visiting a friend at a New York investment bank and seeing a frenetic hotshot "master of the universe" type walking around with a set of wireless headphones wrapped around his ears and a microphone jutting out of the right side that prevented me from focusing on his lips during my twenty-second conversation with him. I asked my friend the purpose of that contraption. "He likes to keep in touch with London," I was told. When you are employed, hence dependent on other people's judgment, looking busy can help you claim responsibility for the results in a random environment. The appearance of busyness reinforces the perception of causality, of the link between results and one's role in them. This of course applies even more to the CEOs of large companies who need to trumpet a link between their "presence" and "leadership" and the results of the company. I am not aware of any studies that probe the usefulness of their time being invested in conversations and the absorption of small-time information—nor have too many writers had the guts to question how large the CEO's role is in a corporation's success.

Let us discuss one main effect of information: impediment to knowledge.

Aristotle Onassis, perhaps the first mediatized tycoon, was principally famous for being rich—and for exhibiting it. An ethnic Greek refugee from southern Turkey, he went to Argentina, made a lump of cash by importing Turkish tobacco, then became a shipping magnate. He was reviled when he married Jacqueline Kennedy, the widow of the American president John F. Kennedy, which drove the heartbroken opera singer Maria Callas to immure herself in a Paris apartment to await death.

If you study Onassis's life, which I spent part of my early adulthood doing, you would notice an interesting regularity: "work," in the conventional sense, was not his thing. He did not even bother to have a desk, let alone an office. He was not just a dealmaker, which does not necessitate having an office, but he also ran a shipping empire, which requires day-to-day monitoring. Yet his main tool was a notebook, which contained all the information he needed. Onassis spent his life trying to socialize with the rich and famous, and to pursue (and collect) women. He generally woke up at noon. If he needed legal advice, he would summon his lawyers to some nightclub in Paris at two A.M. He was said to have an irresistible charm, which helped him take advantage of people.

Let us go beyond the anecdote. There may be a "fooled by randomness" effect here, of making a causal link between Onassis's success and his modus operandi. I may never know if Onassis was skilled or lucky, though I am convinced that his charm opened doors for him, but I can subject his modus to a rigorous examination by looking at empirical research on the link between information and understanding. So this statement, *additional knowledge of the minutiae of daily business can be useless, even actually toxic,* is indirectly but quite effectively testable.

Show two groups of people a blurry image of a fire hydrant, blurry enough for them not to recognize what it is. For one group, increase the resolution slowly, in ten steps. For the second, do it faster, in five steps. Stop at a point where both groups have been presented an identical image and ask each of them to identify what they see. The members of the group that saw fewer intermediate steps are likely to recognize the hydrant much faster. Moral? The more information you give someone, the more hypotheses they will formulate along the way, and the worse off they will be. They see more random noise and mistake it for information.

The problem is that our ideas are sticky: once we produce a theory, we are not likely to change our minds—so those who delay developing their theories are better off. When you develop your opinions on the basis of weak evidence, you will have difficulty interpreting subsequent information that contradicts these opinions, even if this new information is obviously more accurate. Two mechanisms are at play here: the confirmation bias that we saw in Chapter 5, and belief perseverance, the tendency not to reverse opinions you already have. Remember that we treat ideas like possessions, and it will be hard for us to part with them.

The fire hydrant experiment was first done in the sixties, and replicated several times since. I have also studied this effect using the mathematics of information: the more detailed knowledge one gets of empirical reality, the more one will see the noise (i.e., the anecdote) and mistake it for actual information. Remember that we are swayed by the sensational. Listening to the news on the radio every hour is far worse for you than reading a weekly magazine, because the longer interval allows information to be filtered a bit.

In 1965, Stuart Oskamp supplied clinical psychologists with successive files, each containing an increasing amount of information about patients; the psychologists' diagnostic abilities did not grow with the additional supply of information. They just got more confident in their original diagnosis. Granted, one may not expect too much of psychologists of the 1965 variety, but these findings seem to hold across disciplines.

Finally, in another telling experiment, the psychologist Paul Slovic asked bookmakers to select from eighty-eight variables in past horse races those that they found useful in computing the odds. These variables included all manner of statistical information about past performances. The bookmakers were given the ten most useful variables, then asked to predict the outcome of races. Then they were given ten more and asked to predict again. The increase in the information set did not lead to an increase in their accuracy; their confidence in their choices, on the other hand, went up markedly. Information proved to be toxic. I've struggled much of my life with the common middlebrow belief that "more is better"—more is sometimes, but not always, better. This toxicity of knowledge will show in our investigation of the so-called expert.

THE EXPERT PROBLEM, OR THE TRAGEDY OF THE EMPTY SUIT

So far we have not questioned the authority of the professionals involved but rather their ability to gauge the boundaries of their own knowledge. Epistemic arrogance does not preclude skills. A plumber will almost always know more about plumbing than a stubborn essayist and mathematical trader. A hernia surgeon will rarely know less about hernias than a belly dancer. But their probabilities, on the other hand, will be off—and, this is the disturbing point, you may know much more on that score than the expert. No matter what anyone tells you, it is a good idea to question *the error rate* of an expert's procedure. Do not question his procedure, only his confidence. (As someone who was burned by the medical establishment, I learned to be cautious, and I urge everyone to be: if you walk into a doctor's office with a symptom, do not listen to his odds of its *not* being cancer.)

I will separate the two cases as follows. The mild case: *arrogance in the presence of (some) competence,* and the severe case: *arrogance mixed with incompetence (the empty suit)*. There are some professions in which you know more than the experts, who are, alas, people for whose opinions you are paying—instead of them paying you to listen to them. Which ones?

What Moves and What Does Not Move

There is a very rich literature on the so-called expert problem, running empirical testing on experts to verify their record. But it seems to be confusing at first. On one hand, we are shown by a class of expert-busting researchers such as Paul Meehl and Robyn Dawes that the "expert" is the closest thing to a fraud, performing no better than a computer using a single metric, their intuition getting in the way and blinding them. (As an example of a computer using a single metric, the ratio of liquid assets to debt fares better than the majority of credit analysts.) On the other hand, there is abundant literature showing that many people can beat computers thanks to their intuition. Which one is correct?

There must be some disciplines with true experts. Let us ask the following questions: Would you rather have your upcoming brain surgery performed by a newspaper's science reporter or by a certified brain surgeon? On the other hand, would you prefer to listen to an economic forecast by someone with a PhD in finance from some "prominent" institution such as the Wharton School, or by a newspaper's business writer? While the answer to the first question is empirically obvious, the answer to the second one isn't at all. We can already see the difference between "know-how" and "know-what." The Greeks made a distinction between *technē* and *epistēmē*. The empirical school of medicine of Menodotus of Nicomedia and Heraclites of Tarentum wanted its practitioners to stay closest to *technē* (i.e., "craft"), and away from *epistēmē* (i.e., "knowledge," "science").

The psychologist James Shanteau undertook the task of finding out which disciplines have experts and which have none. Note the confirmation problem here: if you want to prove that there are no experts, then you will be able to find *a* profession in which experts are useless. And you can prove the opposite just as well. But there is a regularity: there are professions where experts play a role, and others where there is no evidence of skills. Which are which?

Experts who tend to be experts: livestock judges, astronomers, test pilots, soil judges, chess masters, physicists, mathematicians (when they deal with mathematical problems, not empirical ones), accountants, grain inspectors, photo interpreters, insurance analysts (dealing with bell curve–style statistics).

Experts who tend to be . . . not experts: stockbrokers, clinical psychologists, psychiatrists, college admissions officers, court judges, councilors, personnel selectors, intelligence analysts (the CIA's record, in spite of its costs, is pitiful). I would add these results from my own examination of the literature: economists, financial forecasters, finance professors, political scientists, "risk experts," Bank for International Settlements staff, august members of the International Association of Financial Engineers, and personal financial advisers.

Simply, *things that move,* and therefore require knowledge, do not usually have experts, while things that don't move seem to have some experts. In other words, professions that deal with the future and base their studies on the nonrepeatable past have an expert problem (with the exception of the weather and businesses involving short-term physical processes, not socioeconomic ones). I am not saying that no one who deals with the future provides any valuable information (as I pointed out earlier, newspapers can predict theater opening hours rather well), but rather that those who provide no tangible added value are generally dealing with the future.

Another way to see it is that things that move are often Black Swan–prone. Experts are narrowly focused persons who need to "tunnel." In situations where tunneling is safe, because Black Swans are not consequential, the expert will do well.

Robert Trivers, an evolutionary psychologist and a man of supernormal insights, has another answer (he became one of the most influential evolutionary thinkers since Darwin with ideas he developed while trying to go to law school). He links it to self-deception. In fields where we have ancestral traditions, such as pillaging, we are very good at predicting outcomes by gauging the balance of power. Humans and chimps can immediately sense which side has the upper hand, and make a cost-benefit analysis about whether to attack and take the goods and the mates. Once you start raiding, you put yourself into a delusional mind-set that makes you ignore additional information—it is best to avoid wavering during battle. On the other hand, unlike raids, large-scale wars are not something present in human heritage—we are new to them—so we tend to misestimate their duration and overestimate our relative power. Recall the underestimation of the duration of the Lebanese war. Those who fought in the Great War thought it would be a mere cakewalk. So it was with the Vietnam conflict, so it is with the Iraq war, and just about every modern conflict.

You cannot ignore self-delusion. The problem with experts is that they do not know what they do not know. Lack of knowledge and delusion about the quality of your knowledge come together—the same process that makes you know less also makes you satisfied with your knowledge.

Next, instead of the range of forecasts, we will concern ourselves with the accuracy of forecasts, i.e., the ability to predict the number itself.

How to Have the Last Laugh

We can also learn about prediction errors from trading activities. We quants have ample data about economic and financial forecasts—from general data about large economic variables to the forecasts and market calls of the television "experts" or "authorities." The abundance of such data and the ability to process it on a computer make the subject invaluable for an empiricist. If I had been a journalist, or, God forbid, a historian, I would have had a far more difficult time testing the predictive effectiveness of these verbal discussions. You cannot process verbal commentaries with a computer—at least not so easily. Furthermore, many economists naïvely make the mistake of producing a lot of forecasts concerning many variables, giving us a database of economists and variables, which enables us to see whether some economists are better than others (there is no consequential difference) or if there are certain variables for which they are more competent (alas, none that are meaningful).

I was in a seat to observe from very close our ability to predict. In my full-time trader days, a couple of times a week, at 8:30 A.M., my screen would flash some economic number released by the Department of Commerce, or Treasury, or Trade, or some such honorable institution. I never had a clue about what these numbers meant and never saw any need to invest energy in finding out. So I would not have cared the least about them except that people got all excited and talked quite a bit about what these figures were going to mean, pouring verbal sauce around the forecasts. Among such numbers you have the Consumer Price Index (CPI), Nonfarm Payrolls (changes in the number of employed individuals), the Index of Leading Economic Indicators, Sales of Durable Goods (dubbed "doable girls" by traders), the Gross Domestic Product (the most important one), and many more that generate different levels of excitement depending on their presence in the discourse.

The data vendors allow you to take a peek at forecasts by "leading economists," people (in suits) who work for the venerable institutions, such as J. P. Morgan Chase or Morgan Stanley. You can watch these economists talk, theorizing eloquently and convincingly. Most of them earn seven figures and they rank as stars, with teams of researchers crunching numbers and projections. But the stars are foolish enough to publish their projected numbers, right there, for posterity to observe and assess their degree of competence.

Worse yet, many financial institutions produce booklets every year-end called "Outlook for 200X," reading into the following year. Of course they do not check how their previous forecasts fared *after* they were formulated. The public might have been even more foolish in buying the arguments without requiring the following simple tests—easy though they are, very few of them have been done. One elementary empirical test is to compare these star economists to a hypothetical cabdriver (the equivalent of Mikhail from Chapter 1): you create a synthetic agent, someone who takes the most recent number as the best predictor of the next, while assuming that he does not know anything. Then all you have to do is compare the error rates of the hotshot economists and your synthetic agent. The problem is that when you are swayed by stories you forget about the necessity of such testing.

Events Are Outlandish

The problem with prediction is a little more subtle. It comes mainly from the fact that we are living in Extremistan, not Mediocristan. Our predictors may be good at predicting the ordinary, but not the irregular, and this is where they ultimately fail. All you need to do is miss one interest-rates move, from 6 percent to 1 percent in a longer-term projection (what happened between 2000 and 2001) to have all your subsequent forecasts rendered completely ineffectual in correcting your cumulative track record. What matters is not how often you are right, but how large your cumulative errors are.

And these cumulative errors depend largely on the big surprises, the big opportunities. Not only do economic, financial, and political predictors miss them, but they are quite ashamed to say anything outlandish to their clients—and yet *events, it turns out, are almost always outlandish.* Furthermore, as we will see in the next section, economic forecasters tend to fall closer to one another than to the resulting outcome. Nobody wants to be off the wall.

Since my testing has been informal, for commercial and entertainment purposes, for my own consumption and not formatted for publishing, I will use the more formal results of other researchers who did the dog work of dealing with the tedium of the publishing process. I am surprised that so little introspection has been done to check on the usefulness of these professions. There are a few—but not many—formal tests in three domains: security analysis, political science, and economics. We will no doubt have more in a few years. Or perhaps not—the authors of such papers might become stigmatized by his colleagues. Out of close to a million papers published in politics, finance, and economics, there have been only a small number of checks on the predictive quality of such knowledge.

Herding Like Cattle

A few researchers have examined the work and attitude of security analysts, with amazing results, particularly when one considers the epistemic arrogance of these operators. In a study comparing them with weather forecasters, Tadeusz Tyszka and Piotr Zielonka document that the analysts are worse at predicting, while having a greater faith in their own skills. Somehow, the analysts' self-evaluation did not decrease their error margin after their failures to forecast.

Last June I bemoaned the dearth of such published studies to Jean-Philippe Bouchaud, whom I was visiting in Paris. He is a boyish man who looks half my age though he is only slightly younger than I, a matter that I half jokingly attribute to the beauty of physics. Actually he is not exactly a physicist but one of those quantitative scientists who apply methods of statistical physics to economic variables, a field that was started by Benoît Mandelbrot in the late 1950s. This community does not use Mediocristan mathematics, so they seem to care about the truth. They are completely outside the economics and business-school finance establishment, and survive in physics and mathematics departments or, very often, in trading houses (traders rarely hire economists for their own consumption, but rather to provide stories for their less sophisticated clients). Some of them also operate in sociology with the same hostility on the part of the "natives." Unlike economists who wear suits and spin theories, they use empirical methods to observe the data and do not use the bell curve.

He surprised me with a research paper that a summer intern had just finished under his supervision and that had just been accepted for publication; it scrutinized two thousand predictions by security analysts. What it showed was that these brokerage-house analysts predicted *nothing*—a naïve forecast made by someone who takes the figures from one period as predictors of the next would not do markedly worse. Yet analysts are informed about companies' orders, forthcoming contracts, and planned expenditures, so this advanced knowledge *should* help them do considerably better than a naïve forecaster looking at the past data without further information. Worse yet, the forecasters' errors were significantly larger than the average difference between individual forecasts, which indicates herding. Normally, forecasts should be as far from one another as they are from the predicted number. But to understand how they manage to stay in business, and why they don't develop severe nervous breakdowns (with weight loss, erratic behavior, or acute alcoholism), we must look at the work of the psychologist Philip Tetlock.

I Was "Almost" Right

Tetlock studied the business of political and economic "experts." He asked various specialists to judge the likelihood of a number of political, economic, and military events occurring within a specified time frame (about five years ahead). The outcomes represented a total number of around twenty-seven thousand predictions, involving close to three hundred specialists. Economists represented about a quarter of his sample. The study revealed that experts' error rates were clearly many times what they had estimated. His study exposed an expert problem: there was no difference in results whether one had a PhD or an undergraduate degree. Well-published professors had no advantage over journalists. The only regularity Tetlock found was the negative effect of reputation on prediction: those who had a big reputation were worse predictors than those who had none.

But Tetlock's focus was not so much to show the real competence of experts (although the study was quite convincing with respect to that) as to investigate why the experts did not realize that they were not so good at their own business, in other words, how they spun their stories. There seemed to be a logic to such incompetence, mostly in the form of belief defense, or the protection of self-esteem. He therefore dug further into the mechanisms by which his subjects generated ex post explanations.

I will leave aside how one's ideological commitments influence one's perception and address the more general aspects of this blind spot toward one's own predictions.

You tell yourself that you were playing a different game. Let's say you failed to predict the weakening and precipitous fall of the Soviet Union (which no social scientist saw coming). It is easy to claim that you were excellent at understanding the political workings of the Soviet Union, but that these Russians, being exceedingly Russian, were skilled at hiding from you crucial economic elements. Had you been in possession of such economic intelligence, you would certainly have been able to predict the demise of the Soviet regime. It is not your skills that are to blame. The same might apply to you if you had forecast the landslide victory for Al Gore over George W. Bush. You were not aware that the economy was in such dire straits; indeed, this fact seemed to be concealed from everyone. Hey, you are not an economist, and the game turned out to be about economics.

You invoke the outlier. Something happened that was outside the system, outside the scope of your science. Given that it was not predictable, you are not to blame. It was a Black Swan and you are not supposed to predict Black Swans. Black Swans, NNT tells us, are fundamentally unpredictable (but then I think that NNT would ask you, Why rely on predictions?). Such events are "exogenous," coming from outside your science. Or maybe it was an event of very, very low probability, a thousand-year flood, and we were unlucky to be exposed to it. But next time, it will not happen. This focus on the narrow game and linking one's performance to a given script is how the nerds explain the failures of mathematical methods in society. The model was right, it worked well, but the game turned out to be a different one than anticipated.

The "almost right" defense. Retrospectively, with the benefit of a revision of values and an informational framework, it is easy to feel that it was a close call. Tetlock writes, "Observers of the former Soviet Union who, in 1988, thought the Communist Party could not be driven from power by 1993 or 1998 were especially likely to believe that Kremlin hardliners almost overthrew Gorbachev in the 1991 coup attempt, and they would have if the conspirators had been more resolute and less inebriated, or if key military officers had obeyed orders to kill civilians challenging martial law or if Yeltsin had not acted so bravely."

I will go now into more general defects uncovered by this example. These "experts" were lopsided: on the occasions when they were right, they attributed it to their own depth of understanding and expertise; when wrong, it was either the situation that was to blame, since it was unusual, or, worse, they did not recognize that they were wrong and spun stories around it. They found it difficult to accept that their grasp was a little short. But this attribute is universal to all our activities: there is something in us designed to protect our self-esteem.

We humans are the victims of an asymmetry in the perception of random events. We attribute our successes to our skills, and our failures to external events outside our control, namely to randomness. We feel responsible for the good stuff, but not for the bad. This causes us to think that we are better than others at whatever we do for a living. Ninety-four percent of Swedes believe that their driving skills put them in the top 50 percent of Swedish drivers; 84 percent of Frenchmen feel that their lovemaking abilities put them in the top half of French lovers.

The other effect of this asymmetry is that we feel a little unique, unlike others, for whom we do not perceive such an asymmetry. I have mentioned the unrealistic expectations about the future on the part of people in the process of tying the knot. Also consider the number of families who tunnel on their future, locking themselves into hard-to-flip real estate thinking they are going to live there permanently, not realizing that the general track record for sedentary living is dire. Don't they see those well-dressed real-estate agents driving around in fancy two-door German cars? We are very nomadic, far more than we plan to be, and forcibly so. Consider how many people who have abruptly lost their job deemed it likely to occur, even a few days before. Or consider how many drug addicts entered the game willing to stay in it so long.

There is another lesson from Tetlock's experiment. He found what I mentioned earlier, that many university stars, or "contributors to top journals," are no better than the average *New York Times* reader or journalist in detecting changes in the world around them. These sometimes overspecialized experts failed tests in their own specialties.

The hedgehog and the fox. Tetlock distinguishes between two types of predictors, the hedgehog and the fox, according to a distinction promoted by the essayist Isaiah Berlin. As in Aesop's fable, the hedgehog knows one thing, the fox knows many things—these are the adaptable types you need in daily life. Many of the prediction failures come from hedgehogs who are mentally married to a single big Black Swan event, a big bet that is not likely to play out. The hedgehog is someone focusing on a single, improbable, and consequential event, falling for the narrative fallacy that makes us so blinded by one single outcome that we cannot imagine others.

Hedgehogs, because of the narrative fallacy, are easier for us to understand—their ideas work in sound bites. Their category is overrepresented among famous people; ergo famous people are on average worse at forecasting than the rest of the predictors.

I have avoided the press for a long time because whenever journalists hear my Black Swan story, they ask me to give them a list of future impacting events. They want me to be *predictive* of these Black Swans. Strangely, my book *Fooled by Randomness,* published a week before September 11, 2001, had a discussion of the possibility of a plane crashing into my office building. So I was naturally asked to show "how I predicted the event." I didn't predict it—it was a chance occurrence. I am not playing oracle! I even recently got an e-mail asking me to list the next ten Black Swans. Most fail to get my point about the error of specificity, the narrative fallacy, and the idea of prediction. Contrary to what people might expect, I am not recommending that anyone become a hedgehog—rather, be a fox with an open mind. I know that history is going to be dominated by an improbable event, I just don't know what that event will be.

Reality? What For?

I found no formal, Tetlock-like comprehensive study in economics journals. But, suspiciously, I found no paper trumpeting economists' ability to produce reliable projections. So I reviewed what articles and working papers in economics I could find. They collectively show no convincing evidence that economists as a community have an ability to predict, and, if they have some ability, their predictions are at best just *slightly* better than random ones—not good enough to help with serious decisions.

The most interesting test of how academic methods fare in the real world was run by Spyros Makridakis, who spent part of his career managing competitions between forecasters who practice a "scientific method" called econometrics—an approach that combines economic theory with statistical measurements. Simply put, he made people forecast *in real life* and then he judged their accuracy. This led to the series of "M-Competitions" he ran, with assistance from Michele Hibon, of which M3 was the third and most recent one, completed in 1999. Makridakis and Hibon reached the sad conclusion that "statistically sophisticated or complex methods do not necessarily provide more accurate forecasts than simpler ones."

I had an identical experience in my quant days—the foreign scientist with the throaty accent spending his nights on a computer doing complicated mathematics rarely fares better than a cabdriver using the simplest methods within his reach. The problem is that we focus on the rare occasion when these methods work and almost never on their far more numerous failures. I kept begging anyone who would listen to me: "Hey, I am an uncomplicated, no-nonsense fellow from Amioun, Lebanon, and have trouble understanding why something is considered valuable if it requires running computers overnight but does not enable me to predict better than any other guy from Amioun." The only reactions I got from these colleagues were related to the geography and history of Amioun rather than a no-nonsense explanation of their business. Here again, you see the narrative fallacy at work, except that in place of journalistic stories you have the more dire situation of the "scientists" with a Russian accent looking in the rearview mirror, narrating with equations, and refusing to look ahead because he may get too dizzy. The econometrician Robert Engel, an otherwise charming gentleman, invented a very complicated statistical method called GARCH and got a Nobel for it. No one tested it to see if it has any validity in real life. Simpler, less sexy methods fare exceedingly better, but they do not take you to Stockholm. You have an expert problem in Stockholm, and I will discuss it in Chapter 17.

This unfitness of complicated methods seems to apply to all methods. Another study effectively tested practitioners of something called game theory, in which the most notorious player is John Nash, the schizophrenic mathematician made famous by the film *A Beautiful Mind*. Sadly, for all the intellectual appeal of these methods and all the media attention, its practitioners are no better at predicting than university students.

There is another problem, and it is a little more worrisome. Makridakis and Hibon were to find out that the strong empirical evidence of their studies has been ignored by theoretical statisticians. Furthermore, they encountered shocking hostility toward their empirical verifications. "Instead [statisticians] have concentrated their efforts in building more sophisticated models without regard to the ability of such models to more accurately predict real-life data," Makridakis and Hibon write.

Someone may counter with the following argument: Perhaps economists' forecasts create feedback that cancels their effect (this is called the Lucas critique, after the economist Robert Lucas). Let's say economists predict inflation; in response to these expectations the Federal Reserve acts and lowers inflation. So you cannot judge the forecast accuracy in economics as you would with other events. I agree with this point, but I do not believe that it is the cause of the economists' failure to predict. The world is far too complicated for their discipline.

When an economist fails to predict outliers he often invokes the issue of earthquakes or revolutions, claiming that he is not into geodesics, atmospheric sciences, or political science, instead of incorporating these fields into his studies and accepting that his field does not exist in isolation. Economics is the most insular of fields; it is the one that quotes least from outside itself! Economics is perhaps the subject that currently has the highest number of philistine scholars—scholarship without erudition and natural curiosity can close your mind and lead to the fragmentation of disciplines.

"OTHER THAN THAT," IT WAS OKAY

We have used the story of the Sydney Opera House as a springboard for our discussion of prediction. We will now address another constant in human nature: a systematic error made by project planners, coming from a mixture of human nature, the complexity of the world, or the structure of organizations. In order to survive, institutions may need to give themselves and others the appearance of having a "vision."

Plans fail because of what we have called tunneling, the neglect of sources of uncertainty outside the plan itself.

The typical scenario is as follows. Joe, a nonfiction writer, gets a book contract with a set final date for delivery two years from now. The topic is relatively easy: the authorized biography of the writer Salman Rushdie, for which Joe has compiled ample data. He has even tracked down Rushdie's former girlfriends and is thrilled at the prospect of pleasant interviews. Two years later, minus, say, three months, he calls to explain to the publisher that he will be *a little* delayed. The publisher has seen this coming; he is used to authors being late. The publishing house now has cold feet because the subject has *unexpectedly* faded from public attention—the firm projected that interest in Rushdie would remain high, but attention has faded, seemingly because the Iranians, for some reason, lost interest in killing him.

Let's look at the source of the biographer's underestimation of the time for completion. He projected his own schedule, but he tunneled, as he did not forecast that some "external" events would emerge to slow him down. Among these external events were the disasters on September 11, 2001, which set him back several months; trips to Minnesota to assist his ailing mother (who eventually recovered); and many more, like a broken engagement (though not with Rushdie's ex-girlfriend). "Other than that," it was all within his plan; his own work did not stray the least from schedule. He does not feel responsible for his failure.*

The unexpected has a one-sided effect with projects. Consider the

* The book you have in your hands is approximately and "unexpectedly" fifteen months late.

track records of builders, paper writers, and contractors. The unexpected almost always pushes in a single direction: higher costs and a longer time to completion. On very rare occasions, as with the Empire State Building, you get the opposite: shorter completion and lower costs—these occasions are truly exceptional.

We can run experiments and test for repeatability to verify if such errors in projection are part of human nature. Researchers have tested how students estimate the time needed to complete their projects. In one representative test, they broke a group into two varieties, optimistic and pessimistic. Optimistic students promised twenty-six days; the pessimistic ones forty-seven days. The average actual time to completion turned out to be fifty-six days.

The example of Joe the writer is not acute. I selected it because it concerns a repeatable, routine task—for such tasks our planning errors are milder. With projects of great novelty, such as a military invasion, an all-out war, or something entirely new, errors explode upward. In fact, the more routine the task, the better you learn to forecast. But there is always something nonroutine in our modern environment.

There may be incentives for people to promise shorter completion dates—in order to win the book contract or in order for the builder to get your down payment and use it for his upcoming trip to Antigua. But the planning problem exists even where there is no incentive to underestimate the duration (or the costs) of the task. As I said earlier, we are too narrow-minded a species to consider the possibility of events straying from our mental projections, but furthermore, we are too focused on matters internal to the project to take into account external uncertainty, the "unknown unknown," so to speak, the contents of the unread books.

There is also the nerd effect, which stems from the mental elimination of off-model risks, or *focusing* on what you know. You view the world from *within* a model. Consider that most delays and cost overruns arise from unexpected elements that did not enter into the plan—that is, they lay outside the model at hand—such as strikes, electricity shortages, accidents, bad weather, or rumors of Martian invasions. These small Black Swans that threaten to hamper our projects do not seem to be taken into account. They are too abstract—we don't know how they look and cannot talk about them intelligently.

We cannot truly plan, because we do not understand the future—but this is not necessarily bad news. We could plan *while bearing in mind such limitations*. It just takes guts.

The Beauty of Technology: Excel Spreadsheets

In the not too distant past, say the precomputer days, projections remained vague and qualitative, one had to make a mental effort to keep track of them, and it was a strain to push scenarios into the future. It took pencils, erasers, reams of paper, and huge wastebaskets to engage in the activity. Add to that an accountant's love for tedious, slow work. The activity of projecting, in short, was effortful, undesirable, and marred with self-doubt.

But things changed with the intrusion of the spreadsheet. When you put an Excel spreadsheet into computer-literate hands you get a "sales projection" effortlessly extending ad infinitum! Once on a page or on a computer screen, or, worse, in a PowerPoint presentation, the projection takes on a life of its own, losing its vagueness and abstraction and becoming what philosophers call reified, invested with concreteness; it takes on a new life as a tangible object.

My friend Brian Hinchcliffe suggested the following idea when we were both sweating at the local gym. Perhaps the ease with which one can project into the future by dragging cells in these spreadsheet programs is responsible for the armies of forecasters confidently producing longer-term forecasts (all the while tunneling on their assumptions). We have become worse planners than the Soviet Russians thanks to these potent computer programs given to those who are incapable of handling their knowledge. Like most commodity traders, Brian is a man of incisive and sometimes brutally painful realism.

A classical mental mechanism, called anchoring, seems to be at work here. You lower your anxiety about uncertainty by producing a number, then you "anchor" on it, like an object to hold on to in the middle of a vacuum. This anchoring mechanism was discovered by the fathers of the psychology of uncertainty, Danny Kahneman and Amos Tversky, early in their heuristics and biases project. It operates as follows. Kahneman and Tversky had their subjects spin a wheel of fortune. The subjects first looked at the number on the wheel, *which they knew was random,* then they were asked to estimate the number of African countries in the United Nations. Those who had a low number on the wheel estimated a low number of African nations; those with a high number produced a higher estimate.

Similarly, ask someone to provide you with the last four digits of his social security number. Then ask him to estimate the number of dentists in

Manhattan. You will find that by making him aware of the four-digit number, you elicit an estimate that is correlated with it.

We use reference points in our heads, say sales projections, and start building beliefs around them because less mental effort is needed to compare an idea to a reference point than to evaluate it in the absolute (*System 1* at work!). We cannot work without a point of reference.

So the introduction of a reference point in the forecaster's mind will work wonders. This is no different from a starting point in a bargaining episode: you open with high number ("I want a million for this house"); the bidder will answer "only eight-fifty"—the discussion will be determined by that initial level.

The Character of Prediction Errors

Like many biological variables, life expectancy is from Mediocristan, that is, it is subjected to mild randomness. It is not scalable, since the older we get, the less likely we are to live. In a developed country a newborn female is expected to die at around 79, according to insurance tables. When she reaches her 79th birthday, her life expectancy, assuming that she is in typical health, is another 10 years. At the age of 90, she should have another 4.7 years to go. At the age of 100, 2.5 years. At the age of 119, if she miraculously lives that long, she should have about nine months left. As she lives beyond the expected date of death, the number of additional years to go decreases. This illustrates the major property of random variables related to the bell curve. The conditional expectation of additional life drops as a person gets older.

With human projects and ventures we have another story. These are often scalable, as I said in Chapter 3. With scalable variables, the ones from Extremistan, you will witness the exact opposite effect. Let's say a project is expected to terminate in 79 days, the same expectation in days as the newborn female has in years. On the 79th day, if the project is not finished, it will be expected to take another 25 days to complete. But on the 90th day, if the project is still not completed, it should have about 58 days to go. On the 100th, it should have 89 days to go. On the 119th, it should have an extra 149 days. On day 600, if the project is not done, you will be expected to need an extra 1,590 days. As you see, *the longer you wait, the longer you will be expected to wait.*

Let's say you are a refugee waiting for the return to your homeland. Each day that passes you are getting farther from, not closer to, the day of triumphal return. The same applies to the completion date of your next opera house. If it was expected to take two years, and three years later you are asking questions, do not expect the project to be completed any time soon. If wars last on average six months, and your conflict has been going on for two years, expect another few years of problems. The Arab-Israeli conflict is sixty years old, and counting—yet it was considered "a simple problem" sixty years ago. (Always remember that, in a modern environment, wars last longer and kill more people than is typically planned.) Another example: Say that you send your favorite author a letter, knowing that he is busy and has a two-week turnaround. If three weeks later your mailbox is still empty, do not expect the letter to come tomorrow—it will take on average another three weeks. If three months later you still have nothing, you will have to expect to wait another year. Each day will bring you closer to your death but further from the receipt of the letter.

This subtle but extremely consequential property of scalable randomness is unusually counterintuitive. We misunderstand the logic of large deviations from the norm.

I will get deeper into these properties of scalable randomness in Part Three. But let us say for now that they are central to our misunderstanding of the business of prediction.

DON'T CROSS A RIVER IF IT IS (ON AVERAGE) FOUR FEET DEEP

Corporate and government projections have an additional easy-to-spot flaw: they do not attach a *possible error rate* to their scenarios. Even in the absence of Black Swans this omission would be a mistake.

I once gave a talk to policy wonks at the Woodrow Wilson Center in Washington, D.C., challenging them to be aware of our weaknesses in seeing ahead.

The attendees were tame and silent. What I was telling them was against everything they believed and stood for; I had gotten carried away with my aggressive message, but they looked thoughtful, compared to the testosterone-charged characters one encounters in business. I felt guilty for my aggressive stance. Few asked questions. The person who organized the talk and invited me must have been pulling a joke on his colleagues. I was like an aggressive atheist making his case in front of a synod of cardinals, while dispensing with the usual formulaic euphemisms.

Yet some members of the audience were sympathetic to the message. One anonymous person (he is employed by a governmental agency) explained to me privately after the talk that in January 2004 his department was forecasting the price of oil for twenty-five years later at $27 a barrel, slightly higher than what it was at the time. Six months later, around June 2004, after oil doubled in price, they had to revise their estimate to $54 (the price of oil is currently, as I am writing these lines, close to $79 a barrel). It did not dawn on them that it was ludicrous to forecast a second time given that their forecast was off so early and so markedly, that this business of forecasting had to be somehow questioned. And they were looking *twenty-five years* ahead! Nor did it hit them that there was something called an error rate to take into account.*

Forecasting without incorporating an error rate uncovers three fallacies, all arising from the same misconception about the nature of uncertainty.

The first fallacy: *variability matters.* The first error lies in taking a projection too seriously, without heeding its accuracy. Yet, for planning purposes, the accuracy in your forecast matters far more the forecast itself. I will explain it as follows.

Don't cross a river if it is four feet deep on average. You would take a different set of clothes on your trip to some remote destination if I told you that the temperature was expected to be seventy degrees Fahrenheit, with an expected error rate of forty degrees than if I told you that my margin of error was only five degrees. The policies we need to make decisions on should depend far more on the range of possible outcomes than on the expected final number. I have seen, while working for a bank, how people project cash flows for companies without wrapping them in the thinnest layer of uncertainty. Go to the stockbroker and check on what method they use to forecast sales ten years ahead to "calibrate" their valuation models. Go find out how analysts forecast government deficits. Go to a bank or security-analysis training program and see how they teach

* While forecast errors have always been entertaining, commodity prices have been a great trap for suckers. Consider this 1970 forecast by U.S. officials (signed by the U.S. Secretaries of the Treasury, State, Interior, and Defense): "the standard price of foreign crude oil by 1980 may well decline and will in any event not experience a substantial increase." Oil prices went up tenfold by 1980. I just wonder if current forecasters lack in intellectual curiosity or if they are intentionally ignoring forecast errors.

Also note this additional aberration: since high oil prices are marking up their inventories, oil companies are making record bucks and oil executives are getting huge bonuses because "they did a good job"—as if they brought profits by *causing* the rise of oil prices.

trainees to make assumptions; they do not teach you to build an error rate around those assumptions—but their error rate is so large that it is far more significant than the projection itself!

The second fallacy lies in failing to take into account forecast degradation as the projected period lengthens. We do not realize the full extent of the difference between near and far futures. Yet the degradation in such forecasting through time becomes evident through simple introspective examination—without even recourse to scientific papers, which on this topic are suspiciously rare. Consider forecasts, whether economic or technological, made in 1905 for the following quarter of a century. How close to the projections did 1925 turn out to be? For a convincing experience, go read George Orwell's *1984*. Or look at more recent forecasts made in 1975 about the prospects for the new millennium. Many events have taken place and new technologies have appeared that lay outside the forecasters' imaginations; many more that were expected to take place or appear did not do so. Our forecast errors have traditionally been enormous, and there may be no reasons for us to believe that we are suddenly in a more privileged position to see into the future compared to our blind predecessors. Forecasting by bureaucrats tends to be used for anxiety relief rather than for adequate policy making.

The third fallacy, and perhaps the gravest, concerns a misunderstanding of the random character of the variables being forecast. Owing to the Black Swan, these variables can accommodate far more optimistic—or far more pessimistic—scenarios than are currently expected. Recall from my experiment with Dan Goldstein testing the domain-specificity of our intuitions, how we tend to make no mistakes in Mediocristan, but make large ones in Extremistan as we do not realize the consequences of the rare event.

What is the implication here? Even if you agree with a given forecast, you have to worry about the real possibility of significant divergence from it. These divergences may be welcomed by a speculator who does not depend on steady income; a retiree, however, with set risk attributes cannot afford such gyrations. I would go even further and, using the argument about the depth of the river, state that it is the lower bound of estimates (i.e., the worst case) that matters when engaging in a policy—the worst case is far more consequential than the forecast itself. This is particularly true if the bad scenario is not acceptable. Yet the current phraseology makes no allowance for that. None.

It is often said that "is wise he who can see things coming." Perhaps the wise one is the one who knows that he cannot see things far away.

Get Another Job

The two typical replies I face when I question forecasters' business are: "What should he do? Do you have a better way for us to predict?" and "If you're so smart, show me your own prediction." In fact, the latter question, usually boastfully presented, aims to show the superiority of the practitioner and "doer" over the philosopher, and mostly comes from people who do not know that I was a trader. If there is one advantage of having been in the daily practice of uncertainty, it is that one does not have to take any crap from bureaucrats.

One of my clients asked for my predictions. When I told him I had none, he was offended and decided to dispense with my services. There is in fact a routine, unintrospective habit of making businesses answer questionnaires and fill out paragraphs showing their "outlooks." I have never had an outlook and have never made professional predictions—but at least *I know that I cannot forecast* and a small number of people (those I care about) take that as an asset.

There are those people who produce forecasts uncritically. When asked why they forecast, they answer, "Well, that's what we're paid to do here."

My suggestion: get another job.

This suggestion is not too demanding: unless you are a slave, I assume you have some amount of control over your job selection. Otherwise this becomes a problem of ethics, and a grave one at that. People who are trapped in their jobs who forecast simply because "that's my job," knowing pretty well that their forecast is ineffectual, are not what I would call ethical. What they do is no different from repeating lies simply because "it's my job."

Anyone who causes harm by forecasting should be treated as either a fool or a liar. Some forecasters cause more damage to society than criminals. Please, don't drive a school bus blindfolded.

At JFK

At New York's JFK airport you can find gigantic newsstands with walls full of magazines. They are usually manned by a very polite family from

Caravaggio's *The Fortune-Teller.* We have always been suckers for those who tell us about the future. In this picture the fortune-teller is stealing the victim's ring.

the Indian subcontinent (just the parents; the children are in medical school). These walls present you with the entire corpus of what an "informed" person needs in order "to know what's going on." I wonder how long it would take to read every single one of these magazines, excluding the fishing and motorcycle periodicals (but including the gossip magazines—you might as well have some fun). Half a lifetime? An entire lifetime?

Sadly, all this knowledge would not help the reader to forecast what is to happen tomorrow. Actually, it might decrease his ability to forecast.

There is another aspect to the problem of prediction: its inherent limitations, those that have little to do with human nature, but instead arise from the very nature of information itself. I have said that the Black Swan has three attributes: unpredictability, consequences, and retrospective explainability. Let us examine this unpredictability business.*

* I owe the reader an answer concerning Catherine's lover count. She had only twelve.

Chapter Eleven

HOW TO LOOK FOR BIRD POOP

Popper's prediction about the predictors—Poincaré plays with billiard balls—Von Hayek is allowed to be irreverent—Anticipation machines—Paul Samuelson wants you to be rational—Beware the philosopher—Demand some certainties.

We've seen that a) we tend to both tunnel and think "narrowly" (epistemic arrogance), and b) our prediction record is highly overestimated—many people who think they can predict actually can't.

We will now go deeper into the unadvertised structural limitations on our ability to predict. These limitations may arise not from us but from the nature of the activity itself—too complicated, not just for us, but for any tools we have or can conceivably obtain. Some Black Swans will remain elusive, enough to kill our forecasts.

HOW TO LOOK FOR BIRD POOP

In the summer of 1998 I worked at a European-owned financial institution. It wanted to distinguish itself by being rigorous and farsighted. The unit involved in trading had five managers, all serious-looking (always in dark blue suits, even on dress-down Fridays), who had to meet throughout the summer in order "to formulate the five-year plan." This was supposed to be a meaty document, a sort of user's manual for the firm. A five-year plan? To a fellow deeply skeptical of the central planner, the notion was ludicrous; growth within the firm had been organic and unpredictable, bottom-up not top-down. It was well known that the firm's most lucrative department was the product of a chance call from a customer asking for a specific but strange financial transaction. The firm accidentally realized that they could build a unit just to handle these transactions, since they were profitable, and it rapidly grew to dominate their activities.

The managers flew across the world in order to meet: Barcelona, Hong Kong, et cetera. A lot of miles for a lot of verbiage. Needless to say they were usually sleep-deprived. Being an executive does not require very developed frontal lobes, but rather a combination of charisma, a capacity to sustain boredom, and the ability to shallowly perform on harrying schedules. Add to these tasks the "duty" of attending opera performances.

The managers sat down to brainstorm during these meetings, about, of course, the medium-term future—they wanted to have "vision." But then an event occurred that was not in the previous five-year plan: the Black Swan of the Russian financial default of 1998 and the accompanying meltdown of the values of Latin American debt markets. It had such an effect on the firm that, although the institution had a sticky employment policy of retaining managers, none of the five was still employed there a month after the sketch of the 1998 five-year plan.

Yet I am confident that today their replacements are still meeting to work on the next "five-year plan." We never learn.

Inadvertent Discoveries

The discovery of human epistemic arrogance, as we saw in the previous chapter, was allegedly inadvertent. But so were many other discoveries as well. Many more than we think.

The classical model of discovery is as follows: you search for what you know (say, a new way to reach India) and find something you didn't know was there (America).

If you think that the inventions we see around us came from someone sitting in a cubicle and concocting them according to a timetable, think again: almost everything of the moment is the product of serendipity. The term *serendipity* was coined in a letter by the writer Hugh Walpole, who derived it from a fairy tale, "The Three Princes of Serendip." These princes "were always making discoveries by accident or sagacity, of things which they were not in quest of."

In other words, you find something you are not looking for and it changes the world, while wondering after its discovery why it "took so long" to arrive at something so obvious. No journalist was present when the wheel was invented, but I am ready to bet that people did not just embark on the project of inventing the wheel (that main engine of growth) and then complete it according to a timetable. Likewise with most inventions.

Sir Francis Bacon commented that the most important advances are the least predictable ones, those "lying out of the path of the imagination." Bacon was not the last intellectual to point this out. The idea keeps popping up, yet then rapidly dying out. Almost half a century ago, the bestselling novelist Arthur Koestler wrote an entire book about it, aptly called *The Sleepwalkers*. It describes discoverers as sleepwalkers stumbling upon results and not realizing what they have in their hands. We think that the import of Copernicus's discoveries concerning planetary motions was obvious to him and to others in his day; he had been dead seventy-five years before the authorities started getting offended. Likewise we think that Galileo was a victim in the name of science; in fact, the church didn't take him too seriously. It seems, rather, that Galileo caused the uproar himself by ruffling a few feathers. At the end of the year in which Darwin and Wallace presented their papers on evolution by natural selection that changed the way we view the world, the president of the Linnean society, where the papers were presented, announced that the society saw "no striking discovery," nothing in particular that could revolutionize science.

We forget about unpredictability when it is our turn to predict. This is why people can read this chapter and similar accounts, agree entirely with them, yet fail to heed their arguments when thinking about the future.

Take this dramatic example of a serendipitous discovery. Alexander Fleming was cleaning up his laboratory when he found that penicillium mold had contaminated one of his old experiments. He thus happened upon the antibacterial properties of penicillin, the reason many of us are alive today (including, as I said in Chapter 8, myself, for typhoid fever is often fatal when untreated). True, Fleming was looking for "something," but the actual discovery was simply serendipitous. Furthermore, while in hindsight the discovery appears momentous, it took a very long time for health officials to realize the importance of what they had on their hands. Even Fleming lost faith in the idea before it was subsequently revived.

In 1965 two radio astronomists at Bell Labs in New Jersey who were mounting a large antenna were bothered by a background noise, a hiss, like the static that you hear when you have bad reception. The noise could not be eradicated—even after they cleaned the bird excrement out of the dish, since they were convinced that bird poop was behind the noise. It took a while for them to figure out that what they were hearing was the trace of the birth of the universe, the cosmic background microwave radiation. This discovery revived the big bang theory, a languishing idea that was posited by earlier researchers. I found the following comments on Bell Labs' website commenting on how this "discovery" was one of the century's greatest advances:

> Dan Stanzione, then Bell Labs president and Lucent's chief operating officer when Penzias [one of the radio astronomers involved in the discovery] retired, said Penzias "embodies the creativity and technical excellence that are the hallmarks of Bell Labs." He called him a Renaissance figure who "extended our fragile understanding of creation, and advanced the frontiers of science in many important areas."

Renaissance shmenaissance. The two fellows were looking for bird poop! Not only were they not looking for anything remotely like the evidence of the big bang but, as usual in these cases, they did not immediately see the importance of their find. Sadly, the physicist Ralph Alpher, the person who initially conceived of the idea, in a paper coauthored with heavyweights George Gamow and Hans Bethe, was surprised to read about the discovery in *The New York Times*. In fact, in the languishing papers positing the birth of the universe, scientists were doubtful whether such radiation could ever be measured. As happens so often in discovery, those looking for evidence did not find it; those not looking for it found it and were hailed as discoverers.

We have a paradox. Not only have forecasters generally failed dismally to foresee the drastic changes brought about by unpredictable discoveries, but incremental change has turned out to be generally slower than forecasters expected. When a new technology emerges, we either grossly underestimate or severely overestimate its importance. Thomas Watson, the founder of IBM, once predicted that there would be no need for more than just a handful of computers.

That the reader of this book is probably reading these lines not on a screen but in the pages of that anachronistic device, the book, would seem quite an aberration to certain pundits of the "digital revolution." That you are reading them in archaic, messy, and inconsistent English, French, or Swahili, instead of in Esperanto, defies the predictions of half a century ago that the world would soon be communicating in a logical, unambiguous, and Platonically designed lingua franca. Likewise, we are not spending long weekends in space stations as was universally predicted three decades ago. In an example of corporate arrogance, after the first moon landing the now-defunct airline Pan Am took advance bookings for round-trips between earth and the moon. Nice prediction, except that the company failed to forsee that it would be out of business not long after.

A Solution Waiting for a Problem

Engineers tend to develop tools for the pleasure of developing tools, not to induce nature to yield its secrets. It so happens that *some* of these tools bring us more knowledge; because of the silent evidence effect, we forget to consider tools that accomplished nothing but keeping engineers off the streets. Tools lead to unexpected discoveries, which themselves lead to other unexpected discoveries. But rarely do our tools seem to work as intended; it is only the engineer's gusto and love for the building of toys and machines that contribute to the augmentation of our knowledge. Knowledge does not progress from tools designed to verify or help theories, but rather the opposite. The computer was not built to allow us to develop new, visual, geometric mathematics, but for some other purpose. It happened to allow us to discover mathematical objects that few cared to look for. Nor was the computer invented to let you chat with your friends in Siberia, but it has caused some long-distance relationships to bloom. As an essayist, I can attest that the Internet has helped me to spread my ideas by bypassing journalists. But this was not the stated purpose of its military designer.

The laser is a prime illustration of a tool made for a given purpose (actually no real purpose) that then found applications that were not even dreamed of at the time. It was a typical "solution looking for a problem." Among the early applications was the surgical stitching of detached retinas. Half a century later, *The Economist* asked Charles Townes, the alleged inventor of the laser, if he had had retinas on his mind. He had not. He was satisfying his desire to split light beams, and that was that. In fact,

Townes's colleagues teased him quite a bit about the irrelevance of his discovery. Yet just consider the effects of the laser in the world around you: compact disks, eyesight corrections, microsurgery, data storage and retrieval—all unforeseen applications of the technology.*

We build toys. Some of those toys change the world.

Keep Searching

In the summer of 2005 I was the guest of a biotech company in California that had found inordinate success. I was greeted with T-shirts and pins showing a bell-curve buster and the announcement of the formation of the Fat Tails Club ("fat tails" is a technical term for Black Swans). This was my first encounter with a firm that lived off Black Swans of the positive kind. I was told that a scientist managed the company and that he had the instinct, as a scientist, to just let scientists look wherever their instinct took them. Commercialization came later. My hosts, scientists at heart, understood that research involves a large element of serendipity, which can pay off big as long as one knows how serendipitous the business can be and structures it around that fact. Viagra, which changed the mental outlook and social mores of retired men, was meant to be a hypertension drug. Another hypertension drug led to a hair-growth medication. My friend Bruce Goldberg, who understands randomness, calls these unintended side applications "corners." While many worry about unintended consequences, technology adventurers thrive on them.

The biotech company seemed to follow implicitly, though not explicitly, Louis Pasteur's adage about creating luck by sheer exposure. "Luck favors the prepared," Pasteur said, and, like all great discoverers, he knew something about accidental discoveries. The best way to get maximal exposure is to keep researching. Collect opportunities—on that, later.

To predict the spread of a technology implies predicting a large element of fads and social contagion, which lie outside the objective utility of the technology itself (assuming there is such an animal as objective utility). How many wonderfully useful ideas have ended up in the cemetery, such as the Segway, an electric scooter that, it was prophesized, would change

* Most of the debate between creationists and evolutionary theorists (of which I do not partake) lies in the following: creationists believe that the world comes from some form of design while evolutionary theorists see the world as a result of random changes by an aimless process. But it is hard to look at a computer or a car and consider them the result of aimless process. Yet they are.

the morphology of cities, and many others. As I was mentally writing these lines I saw a *Time* magazine cover at an airport stand announcing the "meaningful inventions" of the year. These inventions seemed to be meaningful as of the issue date, or perhaps for a couple of weeks after. Journalists can teach us how to *not* learn.

HOW TO PREDICT YOUR PREDICTIONS!

This brings us to Sir Doktor Professor Karl Raimund Popper's attack on historicism. As I said in Chapter 5, this was his most significant insight, but it remains his least known. People who do not really know his work tend to focus on Popperian falsification, which addresses the verification or nonverification of claims. This focus obscures his central idea: he made skepticism a *method,* he made of a skeptic someone constructive.

Just as Karl Marx wrote, in great irritation, a diatribe called *The Misery of Philosophy* in response to Proudhon's *The Philosophy of Misery,* Popper, irritated by some of the philosophers of his time who believed in the scientific understanding of history, wrote, as a pun, *The Misery of Historicism* (which has been translated as *The Poverty of Historicism*).*

Popper's insight concerns the limitations in forecasting historical events and the need to downgrade "soft" areas such as history and social science to a level slightly above aesthetics and entertainment, like butterfly or coin collecting. (Popper, having received a classical Viennese education, didn't go quite that far; I do. I am from Amioun.) What we call here soft historical sciences are narrative dependent studies.

Popper's central argument is that in order to predict historical events you need to predict technological innovation, itself fundamentally unpredictable.

"Fundamentally" unpredictable? I will explain what he means using a modern framework. Consider the following property of knowledge: If you expect that you will know *tomorrow* with certainty that your boyfriend has been cheating on you all this time, then you know *today* with certainty that your boyfriend is cheating on you and will take action *today,* say, by grabbing a pair of scissors and angrily cutting all his Ferragamo ties in half. You won't tell yourself, This is what I will figure out tomorrow, but

* Recall from Chapter 4 how Algazel and Averroës traded insults through book titles. Perhaps one day I will be lucky enough to read an attack on this book in a diatribe called *The White Swan.*

today is different so I will ignore the information and have a pleasant dinner. This point can be generalized to all forms of knowledge. There is actually a law in statistics called the *law of iterated expectations,* which I outline here in its strong form: if I expect to expect something at some date in the future, then I already expect that something at present.

Consider the wheel again. If you are a Stone Age historical thinker called on to predict the future in a comprehensive report for your chief tribal planner, you must project the invention of the wheel or you will miss pretty much all of the action. Now, if you can prophesy the invention of the wheel, you already know what a wheel looks like, and thus you already *know how* to build a wheel, so you are already on your way. The Black Swan needs to be predicted!

But there is a weaker form of this law of iterated knowledge. It can be phrased as follows: *to understand the future to the point of being able to predict it, you need to incorporate elements from this future itself.* If you know about the discovery you are about to make in the future, then you have almost made it. Assume that you are a special scholar in Medieval University's Forecasting Department specializing in the projection of future history (for our purposes, the remote twentieth century). You would need to hit upon the inventions of the steam machine, electricity, the atomic bomb, and the Internet, as well as the institution of the airplane onboard massage and that strange activity called the business meeting, in which well-fed, but sedentary, men voluntarily restrict their blood circulation with an expensive device called a necktie.

This incapacity is not trivial. The mere knowledge that something has been invented often leads to a series of inventions of a similar nature, even though not a single detail of this invention has been disseminated—there is no need to find the spies and hang them publicly. In mathematics, once a proof of an arcane theorem has been announced, we frequently witness the proliferation of similar proofs coming out of nowhere, with occasional accusations of leakage and plagiarism. There may be no plagiarism: the information that the solution exists is itself a big piece of the solution.

By the same logic, we are not easily able to conceive of future inventions (if we were, they would have already been invented). On the day when we are able to foresee inventions we will be living in a state where everything conceivable has been invented. Our own condition brings to mind the apocryphal story from 1899 when the head of the U.S. patent office resigned because he deemed that there was nothing left to discover—except that on that day the resignation would be justified.*

Popper was not the first to go after the limits to our knowledge. In Germany, in the late nineteenth century, Emil du Bois-Reymond claimed that *ignoramus et ignorabimus*—we are ignorant and will remain so. Somehow his ideas went into oblivion. But not before causing a reaction: the mathematician David Hilbert set to defy him by drawing a list of problems that mathematicians would need to solve over the next century.

Even du Bois-Reymond was wrong. We are not even good at understanding the unknowable. Consider the statements we make about things that we will never come to know—we confidently underestimate what knowledge we may acquire in the future. Auguste Comte, the founder of the school of positivism, which is (unfairly) accused of aiming at the scientization of everything in sight, declared that mankind would forever remain ignorant of the chemical composition of the fixed stars. But, as Charles Sanders Peirce reported, "The ink was scarcely dry upon the printed page before the spectroscope was discovered and that which he had deemed absolutely unknowable was well on the way of getting ascertained." Ironically, Comte's other projections, concerning what we would come to learn about the workings of society, were grossly—and dangerously—overstated. He assumed that society was like a clock that would yield its secrets to us.

I'll summarize my argument here: Prediction requires knowing about technologies that will be discovered in the future. But that very knowledge would almost automatically allow us to start developing those technologies right away. Ergo, we do not know what we will know.

Some might say that the argument, as phrased, seems obvious, that we always think that we have reached definitive knowledge but don't notice that those past societies we laugh at also thought the same way. My argument is trivial, so why don't we take it into account? The answer lies in a pathology of human nature. Remember the psychological discussions on asymmetries in the perception of skills in the previous chapter? We see flaws in others and not in ourselves. Once again we seem to be wonderful at self-deceit machines.

* Such claims are not uncommon. For instance the physicist Albert Michelson imagined, toward the end of the nineteenth century, that what was left for us to discover in the sciences of nature was no more than fine-tuning our precisions by a few decimal places.

Monsieur le professeur Henri Poincaré. Somehow they stopped making this kind of thinker. *Courtesy of Université Nancy-2.*

THE NTH BILLIARD BALL

Henri Poincaré, in spite of his fame, is regularly considered to be an undervalued scientific thinker, given that it took close to a century for some of his ideas to be appreciated. He was perhaps the last great thinking mathematician (or possibly the reverse, a mathematical thinker). Every time I see a T-shirt bearing the picture of the modern icon Albert Einstein, I cannot help thinking of Poincaré—Einstein is worthy of our reverence, but he has displaced many others. There is so little room in our consciousness; it is winner-take-all up there.

Third Republic–Style Decorum

Again, Poincaré is in a class by himself. I recall my father recommending Poincaré's essays, not just for their scientific content, but for the quality of his French prose. The grand master wrote these wonders as serialized articles and composed them like extemporaneous speeches. As in every masterpiece, you see a mixture of repetitions, digressions, everything a "me too" editor with a prepackaged mind would condemn—but these make his text even more readable owing to an iron consistency of thought.

Poincaré became a prolific essayist in his thirties. He seemed in a hurry and died prematurely, at fifty-eight; he was in such a rush that he did not bother correcting typos and grammatical errors in his text, even after spotting them, since he found doing so a gross misuse of his time. They no longer make geniuses like that—or they no longer let them write in their own way.

Poincaré's reputation as a thinker waned rapidly after his death. His idea that concerns us took almost a century to resurface, but in another form. It was indeed a great mistake that I did not carefully read his essays as a child, for in his magisterial *La science et l'hypothèse,* I discovered later, he angrily disparages the use of the bell curve.

I will repeat that Poincaré was the true kind of philosopher of science: his philosophizing came from his witnessing the limits of the subject itself, which is what true philosophy is all about. I love to tick off French literary intellectuals by naming Poincaré as my favorite French philosopher. "*Him a philosophe? What do you mean, monsieur?*" It is always frustrating to explain to people that the thinkers they put on the pedestals, such as Henri Bergson or Jean-Paul Sartre, are largely the result of fashion production and can't come close to Poincaré in terms of sheer influence that will continue for centuries to come. In fact, there is a scandal of prediction going on here, since it is the French Ministry of National Education that decides who is a philosopher and which philosophers need to be studied.

I am looking at Poincaré's picture. He was a bearded, portly and imposing, well-educated patrician gentleman of the French Third Republic, a man who lived and breathed general science, looked deep into his subject, and had an astonishing breadth of knowledge. He was part of the class of mandarins that gained respectability in the late nineteenth century: upper middle class, powerful, but not exceedingly rich. His father was a doctor and professor of medicine, his uncle was a prominent scientist and administrator, and his cousin Raymond became a president of the republic of France. These were the days when the grandchildren of businessmen and wealthy landowners headed for the intellectual professions.

However, I can hardly imagine him on a T-shirt, or sticking out his tongue like in that famous picture of Einstein. There is something nonplayful about him, a Third Republic style of dignity.

In his day, Poincaré was thought to be the king of mathematics and science, except of course by a few narrow-minded mathematicians like Charles Hermite who considered him too intuitive, too intellectual, or too "hand-waving." When mathematicians say "hand-waving," disparagingly, about someone's work, it means that the person has: a) insight, b) realism, c) something to say, and it means that d) he is right because that's what critics say when they can't find anything more negative. A nod from Poincaré made or broke a career. Many claim that Poincaré figured out relativity before Einstein—and that Einstein got the idea from him—but that he did not make a big deal out of it. These claims are naturally made by the French, but there seems to be some validation from Einstein's friend and biographer Abraham Pais. Poincaré was too aristocratic in both background and demeanor to complain about the ownership of a result.

Poincaré is central to this chapter because he lived in an age when we had made extremely rapid intellectual progress in the fields of prediction—think of celestial mechanics. The scientific revolution made us feel that we were in possession of tools that would allow us to grasp the future. Uncertainty was gone. The universe was like a clock and, by studying the movements of the pieces, we could project into the future. It was only a matter of writing down the right models and having the engineers do the calculations. The future was a mere extension of our technological certainties.

The Three Body Problem

Poincaré was the first known big-gun mathematician to understand and explain that there are fundamental limits to our equations. He introduced nonlinearities, small effects that can lead to severe consequences, an idea that later became popular, perhaps a bit too popular, as chaos theory. What's so poisonous about this popularity? Because Poincaré's entire point is about the limits that nonlinearities put on forecasting; they are not an invitation to use mathematical techniques to make extended forecasts. Mathematics can show us its own limits rather clearly.

There is (as usual) an element of the unexpected in this story. Poincaré initially responded to a competition organized by the mathematician Gösta Mittag-Leffer to celebrate the sixtieth birthday of King Oscar of Sweden. Poincaré's memoir, which was about the stability of the solar system, won the prize that was then the highest scientific honor (as these were the happy days before the Nobel Prize). A problem arose, however, when a mathematical editor checking the memoir before publication realized that there was a calculation error, and that, after consideration, it led to the opposite conclusion—unpredictability, or, more technically, nonintegrability. The memoir was discreetly pulled and reissued about a year later.

Poincaré's reasoning was simple: as you project into the future you may need an increasing amount of precision about the dynamics of the process that you are modeling, since your error rate grows very rapidly. The problem is that near precision is not possible since the degradation of your forecast compounds abruptly—you would eventually need to figure

FIGURE 2: PRECISION AND FORECASTING

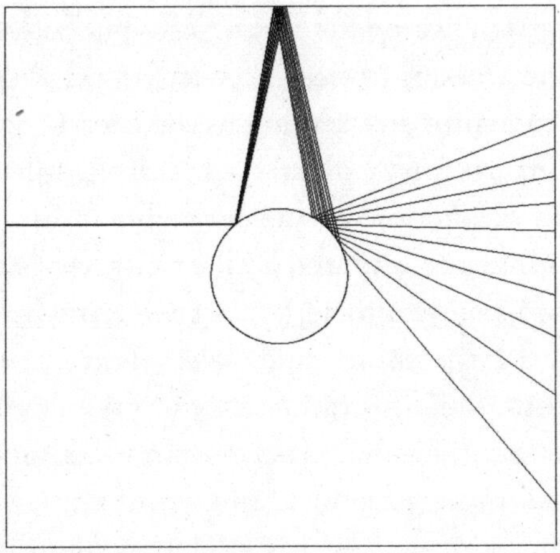

One of the readers of a draft of this book, David Cowan, gracefully drew this picture of scattering, which shows how, at the second bounce, variations in the initial conditions can lead to extremely divergent results. As the initial imprecision in the angle is multiplied, every additional bounce will be further magnified. This causes a severe multiplicative effect where the error grows out disproportionately.

out the past with infinite precision. Poincaré showed this in a very simple case, famously known as the "three body problem." If you have only two planets in a solar-style system, with nothing else affecting their course, then you may be able to indefinitely predict the behavior of these planets, no sweat. But add a third body, say a comet, ever so small, between the planets. Initially the third body will cause no drift, no impact; later, with time, its effects on the two other bodies may become explosive. Small differences in where this tiny body is located will eventually dictate the future of the behemoth planets.

Explosive forecasting difficulty comes from complicating the mechanics, ever so slightly. Our world, unfortunately, is far more complicated than the three body problem; it contains far more than three objects. We are dealing with what is now called a dynamical system—and the world, we will see, is a little too much of a dynamical system.

Think of the difficulty in forecasting in terms of branches growing out of a tree; at every fork we have a multiplication of new branches. To see how our intuitions about these nonlinear multiplicative effects are rather weak, consider this story about the chessboard. The inventor of the chessboard requested the following compensation: one grain of rice for the first square, two for the second, four for the third, eight, then sixteen, and so on, doubling every time, sixty-four times. The king granted this request, thinking that the inventor was asking for a pittance—but he soon realized that he was outsmarted. The amount of rice exceeded all possible grain reserves!

This multiplicative difficulty leading to the need for greater and greater precision in assumptions can be illustrated with the following simple exercise concerning the prediction of the movements of billiard balls on a table. I use the example as computed by the mathematician Michael Berry. If you know a set of basic parameters concerning the ball at rest, can compute the resistance of the table (quite elementary), and can gauge the strength of the impact, then it is rather easy to predict what would happen at the first hit. The second impact becomes more complicated, but possible; you need to be more careful about your knowledge of the initial states, and more precision is called for. The problem is that to correctly compute the ninth impact, you need to take into account the gravitational pull of someone standing next to the table (modestly, Berry's computations use a weight of less than 150 pounds). And to compute the fifty-sixth impact, every single elementary particle of the universe needs to be present in your assumptions! An electron at the edge of the universe, separated from us by 10 billion light-years, must figure in the calculations, since it exerts a meaningful effect on the outcome. Now, consider the additional burden of having to incorporate predictions about *where these variables will be in the future*. Forecasting the motion of a billiard ball on a pool table requires knowledge of the dynamics of the entire universe, down to every single atom! We can easily predict the movements of large objects like planets (though not too far into the future), but the smaller entities can be difficult to figure out—and there are so many more of them.

Note that this billiard-ball story assumes a plain and simple world; it does not even take into account these crazy social matters possibly endowed with free will. Billiard balls do not have a mind of their own. Nor does our example take into account relativity and quantum effects. Nor did we use the notion (often invoked by phonies) called the "uncertainty principle." We are not concerned with the limitations of the precision in measurements done at the subatomic level. We are just dealing with billiard balls!

In a dynamical system, where you are considering more than a ball on its own, where trajectories in a way depend on one another, the ability to project into the future is not just reduced, but is subjected to a fundamental limitation. Poincaré proposed that we can only work with qualitative matters—some property of systems can be *discussed,* but not computed. You can think rigorously, but you cannot use numbers. Poincaré even invented a field for this, analysis in situ, now part of topology. Prediction and forecasting are a more complicated business than is commonly accepted, but it takes someone who knows mathematics to understand that. To accept it takes both understanding and courage.

In the 1960s the MIT meteorologist Edward Lorenz rediscovered Poincaré's results on his own—once again, by accident. He was producing a computer model of weather dynamics, and he ran a simulation that projected a weather system a few days ahead. Later he tried to repeat the same simulation with the exact same model and what he thought were the same input parameters, but he got wildly different results. He initially attributed these differences to a computer bug or a calculation error. Computers then were heavier and slower machines that bore no resemblance to what we have today, so users were severely constrained by time. Lorenz subsequently realized that the consequential divergence in his results arose not from error, but from a small rounding in the input parameters. This became known as the butterfly effect, since a butterfly moving its wings in India could cause a hurricane in New York, two years later. Lorenz's findings generated interest in the field of chaos theory.

Naturally researchers found predecessors to Lorenz's discovery, not only in the work of Poincaré, but also in that of the insightful and intuitive Jacques Hadamard, who thought of the same point around 1898, and then went on to live for almost seven more decades—he died at the age of ninety-eight.*

They Still Ignore Hayek

Popper and Poincaré's findings limit our ability to see into the future, making it a very complicated reflection of the past—if it is a reflection of the past at all. A potent application in the social world comes from a friend of Sir Karl, the intuitive economist Friedrich Hayek. Hayek is one of the rare celebrated members of his "profession" (along with J. M. Keynes and G.L.S. Shackle) to focus on true uncertainty, on the limitations of knowledge, on the unread books in Eco's library.

In 1974 he received the Bank of Sweden Prize in Economic Sciences in

* There are more limits I haven't even attempted to discuss here. I am not even bringing up the class of incomputability people call NP completeness.

Memory of Alfred Nobel, but if you read his acceptance speech you will be in for a bit of a surprise. It was eloquently called "The Pretense of Knowledge," and he mostly railed about other economists and about the idea of the planner. He argued against the use of the tools of hard science in the social ones, and depressingly, right before the big boom for these methods in economics. Subsequently, the prevalent use of complicated equations made the environment for true empirical thinkers worse than it was before Hayek wrote his speech. Every year a paper or a book appears, bemoaning the fate of economics and complaining about its attempts to ape physics. The latest I've seen is about how economists should shoot for the role of lowly philosophers rather than that of high priests. Yet, in one ear and out the other.

For Hayek, a true forecast is done organically by a system, not by fiat. One single institution, say, the central planner, cannot *aggregate* knowledge; many important pieces of information will be missing. But society as a whole will be able to integrate into its functioning these multiple pieces of information. Society as a whole thinks outside the box. Hayek attacked socialism and managed economics as a product of what I have called *nerd knowledge, or Platonicity*—owing to the growth of scientific knowledge, we overestimate our ability to understand the subtle changes that constitute the world, and what weight needs to be imparted to each such change. He aptly called this "scientism."

This disease is severely ingrained in our institutions. It is why I fear governments and large corporations—it is hard to distinguish between them. Governments make forecasts; companies produce projections; every year various forecasters project the level of mortgage rates and the stock market at the end of the following year. Corporations survive not because they have made good forecasts, but because, like the CEOs visiting Wharton I mentioned earlier, they may have been the lucky ones. And, like a restaurant owner, they may be hurting themselves, not us—perhaps helping us and subsidizing our consumption by giving us goods in the process, like cheap telephone calls to the rest of the world funded by the overinvestment during the dotcom era. We consumers can let them forecast all they want if that's what is necessary for them to get into business. Let them go hang themselves if they wish.

As a matter of fact, as I mentioned in Chapter 8, we New Yorkers are all benefiting from the quixotic overconfidence of corporations and restaurant entrepreneurs. This is the benefit of capitalism that people discuss the least.

But corporations can go bust as often as they like, thus subsidizing us consumers by transferring their wealth into our pockets—the more bankruptcies, the better it is for us. Government is a more serious business and we need to make sure we do not pay the price for its folly. As individuals we should love free markets because operators in them can be as incompetent as they wish.

The only criticism one might have of Hayek is that he makes a hard and qualitative distinction between social sciences and physics. He shows that the methods of physics do not translate to its social science siblings, and he blames the engineering-oriented mentality for this. But he was writing at a time when physics, the queen of science, seemed to zoom in our world. It turns out that even the natural sciences are far more complicated than that. He was right about the social sciences, he is certainly right in trusting hard scientists more than social theorizers, but what he said about the weaknesses of social knowledge applies to all knowledge. All knowledge.

Why? Because of the confirmation problem, one can argue that we know very little about our natural world; we advertise the read books and forget about the unread ones. Physics has been successful, but it is a narrow field of hard science in which we have been successful, and people tend to generalize that success to all science. It would be preferable if we were better at understanding cancer or the (highly nonlinear) weather than the origin of the universe.

How Not to Be a Nerd

Let us dig deeper into the problem of knowledge and continue the comparison of Fat Tony and Dr. John in Chapter 9. Do nerds tunnel, meaning, do they focus on crisp categories and miss sources of uncertainty? Remember from the Prologue my presentation of Platonification as a top-down focus on a world composed of these crisp categories.*

Think of a bookworm picking up a new language. He will learn, say, Serbo-Croatian or !Kung by reading a grammar book cover to cover, and memorizing the rules. He will have the impression that some higher grammatical authority set the linguistic regulations so that nonlearned ordinary people could subsequently speak the language. In reality, languages grow

* This idea pops up here and there in history, under different names. Alfred North Whitehead called it the "fallacy of misplaced concreteness," e.g., the mistake of confusing a model with the physical entity that it means to describe.

organically; grammar is something people without anything more exciting to do in their lives codify into a book. While the scholastic-minded will memorize declensions, the a-Platonic nonnerd will acquire, say, Serbo-Croatian by picking up potential girlfriends in bars on the outskirts of Sarajevo, or talking to cabdrivers, then fitting (if needed) grammatical rules to the knowledge he already possesses.

Consider again the central planner. As with language, there is no grammatical authority codifying social and economic events; but try to convince a bureaucrat or social scientist that the world might not want to follow his "scientific" equations. In fact, thinkers of the Austrian school, to which Hayek belonged, used the designations *tacit* or *implicit* precisely for that part of knowledge that cannot be written down, but that we should avoid repressing. They made the distinction we saw earlier between "know-how" and "know-what"—the latter being more elusive and more prone to nerdification.

To clarify, Platonic is top-down, formulaic, closed-minded, self-serving, and commoditized; a-Platonic is bottom-up, open-minded, skeptical, and empirical.

The reason for my singling out the great Plato becomes apparent with the following example of the master's thinking: Plato believed that we should use both hands with equal dexterity. It would not "make sense" otherwise. He considered favoring one limb over the other a deformation caused by the "folly of mothers and nurses." Asymmetry bothered him, and he projected his ideas of elegance onto reality. We had to wait until Louis Pasteur to figure out that chemical molecules were either left- or right-handed and that this mattered considerably.

One can find similar ideas among several disconnected branches of thinking. The earliest were (as usual) the empirics, whose bottom-up, theory-free, "evidence-based" medical approach was mostly associated with Philnus of Cos, Serapion of Alexandria, and Glaucias of Tarentum, later made skeptical by Menodotus of Nicomedia, and currently well-known by its vocal practitioner, our friend the great skeptical philosopher Sextus Empiricus. Sextus who, we saw earlier, was perhaps the first to discuss the Black Swan. The empirics practiced the "medical art" without relying on reasoning; they wanted to benefit from chance observations by making guesses, and experimented and tinkered until they found something that worked. They did minimal theorizing.

Their methods are being revived today as evidence-based medicine, after two millennia of persuasion. Consider that before we knew of bacteria, and their role in diseases, doctors rejected the practice of hand washing because it *made no sense* to them, despite the evidence of a meaningful decrease in hospital deaths. Ignaz Semmelweis, the mid-nineteenth-century doctor who promoted the idea of hand washing, wasn't vindicated until decades after his death. Similarly it may not "make sense" that acupuncture works, but if pushing a needle in someone's toe systematically produces relief from pain (in properly conducted empirical tests), then it could be that there are functions too complicated for us to understand, so let's go with it for now while keeping our minds open.

Academic Libertarianism

To borrow from Warren Buffett, don't ask the barber if you need a haircut—and don't ask an academic if what he does is relevant. So I'll end this discussion of Hayek's libertarianism with the following observation. As I've said, the problem with organized knowledge is that there is an occasional divergence of interests between academic guilds and knowledge itself. So I cannot for the life of me understand why today's libertarians do not go after tenured faculty (except perhaps because many libertarians are academics). We saw that companies can go bust, while governments remain. But while governments remain, civil servants can be demoted and congressmen and senators can be eventually voted out of office. In academia a tenured faculty is permanent—the business of knowledge has permanent "owners." Simply, the charlatan is more the product of control than the result of freedom and lack of structure.

Prediction and Free Will

If you know all possible conditions of a physical system you can, in theory (though not, as we saw, in practice), project its behavior into the future. But this only concerns inanimate objects. We hit a stumbling block when social matters are involved. It is another matter to project a future when humans are involved, *if you consider them living beings and endowed with free will.*

If I can predict all of your actions, under given circumstances, then you may not be as free as you think you are. You are an automaton responding to environmental stimuli. You are a slave of destiny. And the illusion of free will could be reduced to an equation that describes the result of interactions among molecules. It would be like studying the mechanics of a clock: a genius with extensive knowledge of the initial conditions and the causal chains would be able to extend his knowledge to the future of *your* actions. Wouldn't that be stifling?

However, if you believe in free will you can't truly believe in social science and economic projection. You cannot predict how people will act. Except, of course, if there is a trick, and that trick is the cord on which neoclassical economics is suspended. You simply assume that individuals will be *rational* in the future and thus act predictably. There is a strong link between rationality, predictability, and mathematical tractability. A rational individual will perform a *unique* set of actions in specified circumstances. There is one and only one answer to the question of how "rational" people satisfying their best interests would act. Rational actors must be coherent: they cannot prefer apples to oranges, oranges to pears, then pears to apples. If they did, then it would be difficult to generalize their behavior. It would also be difficult to project their behavior in time.

In orthodox economics, rationality became a straitjacket. Platonified economists ignored the fact that people might prefer to do something other than maximize their economic interests. This led to mathematical techniques such as "maximization," or "optimization," on which Paul Samuelson built much of his work. Optimization consists in finding the mathematically optimal policy that an economic agent could pursue. For instance, what is the "optimal" quantity you should allocate to stocks? It involves complicated mathematics and thus raises a barrier to entry by non-mathematically trained scholars. I would not be the first to say that this optimization set back social science by reducing it from the intellectual and reflective discipline that it was becoming to an attempt at an "exact science." By "exact science," I mean a second-rate engineering problem for those who want to pretend that they are in the physics department—so-called physics envy. In other words, an intellectual fraud.

Optimization is a case of sterile modeling that we will discuss further in Chapter 17. It had no practical (or even theoretical) use, and so it became principally a competition for academic positions, a way to make people compete with mathematical muscle. It kept Platonified economists out of the bars, solving equations at night. The tragedy is that Paul Samuelson, a quick mind, is said to be one of the most intelligent scholars of his generation. This was clearly a case of very badly invested intelligence. Characteristically, Samuelson intimidated those who questioned his techniques with the statement "Those who can, do science, others do methodology." If you knew math, you could "do science." This is reminiscent of psychoanalysts who silence their critics by accusing them of having trouble with their fathers. Alas, it turns out that it was Samuelson and most of his followers who did not *know* much math, or did not know how to use what math they knew, how to apply it to reality. They only knew enough math to be blinded by it.

Tragically, before the proliferation of empirically blind idiot savants, interesting work had been begun by true thinkers, the likes of J. M. Keynes, Friedrich Hayek, and the great Benoît Mandelbrot, all of whom were displaced because they moved economics away from the precision of second-rate physics. Very sad. One great underestimated thinker is G.L.S. Shackle, now almost completely obscure, who introduced the notion of "unknowledge," that is, the unread books in Umberto Eco's library. It is unusual to see Shackle's work mentioned at all, and I had to buy his books from secondhand dealers in London.

Legions of empirical psychologists of the heuristics and biases school have shown that the model of rational behavior under uncertainty is not just grossly inaccurate but plain wrong as a description of reality. Their results also bother Platonified economists because they reveal that there are several ways to be irrational. Tolstoy said that happy families were all alike, while each unhappy one is unhappy in its own way. People have been shown to make errors equivalent to preferring apples to oranges, oranges to pears, and *pears to apples,* depending on how the relevant questions are presented to them. The sequence matters! Also, as we have seen with the anchoring example, subjects' estimates of the number of dentists in Manhattan are influenced by which random number they have just been presented with—the *anchor*. Given the randomness of the anchor, we will have randomness in the estimates. So if people make inconsistent choices and decisions, the central core of economic optimization fails. You can no longer produce a "general theory," and without one you cannot predict.

You have to learn to live without a general theory, for Pluto's sake!

THE GRUENESS OF EMERALD

Recall the turkey problem. You look at the past and derive some rule about the future. Well, the problems in projecting from the past can be even worse than what we have already learned, because the same past data can confirm a theory and also its exact opposite! If you survive until tomorrow, it could mean that either a) you are more likely to be immortal or b) that you are closer to death. Both conclusions rely on the exact same

FIGURE 3

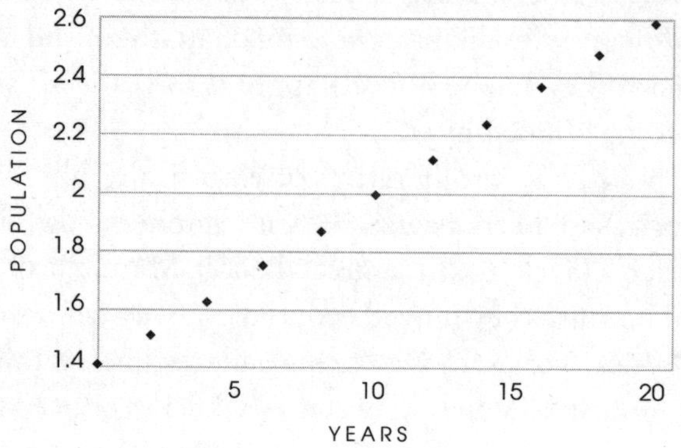

A series of a seemingly growing bacterial population (or of sales records, or of any variable observed through time—such as the total feeding of the turkey in Chapter 4).

FIGURE 4

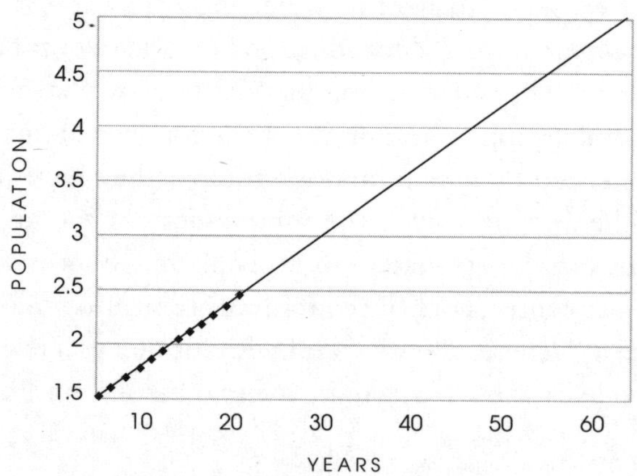

Easy to fit the trend—there is one and only one linear model that fits the data. You can project a continuation into the future

FIGURE 5

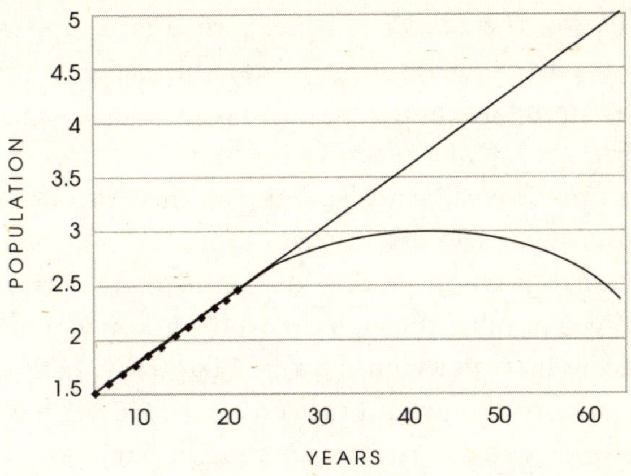

We look at a broader scale. Hey, other models also fit it rather well.

FIGURE 6

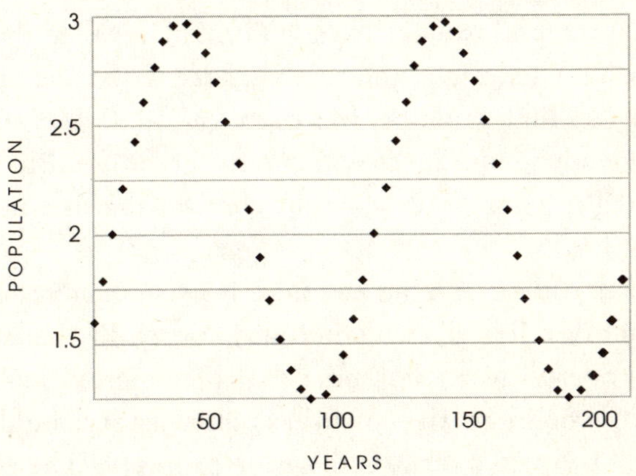

And the real "generating process" is extremely simple but it had nothing to do with a linear model! Some parts of it appear to be linear and we are fooled by extrapolating in a direct line.*

* These graphs also illustrate a statistical version of the narrative fallacy—you find a model that fits the past. "Linear regression" or "R-square" can ultimately fool you beyond measure, to the point where it is no longer funny. You can fit the linear part of the curve and claim a high R-square, meaning that your model fits the data very well and has high predictive powers. All that off hot air: you only fit the linear segment of the series. Always remember that "R-square" is unfit for Extremistan; it is only good for academic promotion.

data. If you are a turkey being fed for a long period of time, you can either naïvely assume that feeding *confirms your safety* or be shrewd and consider that it *confirms the danger* of being turned into supper. An acquaintance's unctuous past behavior may indicate his genuine affection for me and his concern for my welfare; it may also confirm his mercenary and calculating desire to get my business one day.

So not only can the past be misleading, but there are also many degrees of freedom in our interpretation of past events.

For the technical version of this idea, consider a series of dots on a page representing a number through time—the graph would resemble Figure 1 showing the first thousand days in Chapter 4. Let's say your high school teacher asks you to extend the series of dots. With a linear model, that is, using a ruler, you can run only a straight line, a *single* straight line from the past to the future. The linear model is unique. There is one and only one straight line that can project from a series of points. But it can get trickier. If you do not limit yourself to a straight line, you find that there is a huge family of curves that can do the job of connecting the dots. If you project from the past in a linear way, you continue a trend. But possible future deviations from the course of the past are infinite.

This is what the philosopher Nelson Goodman called the riddle of induction: We project a straight line only because we have a linear model in our head—the fact that a number has risen for 1,000 days straight should make you more confident that it will rise in the future. But if you have a nonlinear model in your head, it might confirm that the number should decline on day 1,001.

Let's say that you observe an emerald. It was green yesterday and the day before yesterday. It is green again today. Normally this would confirm the "green" property: we can assume that the emerald will be green tomorrow. But to Goodman, the emerald's color history could equally confirm the "grue" property. What is this grue property? The emerald's grue property is to be green until some specified date, say, December 31, 2006, and then blue thereafter.

The riddle of induction is another version of the narrative fallacy—you face an infinity of "stories" that explain what you have seen. The severity of Goodman's riddle of induction is as follows: if there is no longer even a single unique way to "generalize" from what you see, to make an inference about the unknown, then how should you operate? The answer, clearly, will be that you should employ "common sense," but your common sense may not be so well developed with respect to some Extremistan variables.

THAT GREAT ANTICIPATION MACHINE

The reader is entitled to wonder, So, NNT, why on earth do we plan? Some people do it for monetary gain, others because it's "their job." But we also do it without such intentions—spontaneously.

Why? The answer has to do with human nature. Planning may come with the package of what makes us human, namely, our consciousness.

There is supposed to be an evolutionary dimension to our need to project matters into the future, which I will rapidly summarize here, since it can be an excellent candidate explanation, an excellent conjecture, though, since it is linked to evolution, I would be cautious.

The idea, as promoted by the philosopher Daniel Dennett, is as follows: What is the most potent use of our brain? It is precisely the ability to project conjectures into the future and play the counterfactual game—"If I punch him in the nose, then he will punch me back right away, or, worse, call his lawyer in New York." One of the advantages of doing so is that we can let our conjectures die in our stead. Used correctly and in place of more visceral reactions, the ability to project effectively frees us from immediate, first-order natural selection—as opposed to more primitive organisms that were vulnerable to death and only grew by the improvement in the gene pool through the selection of the best. In a way, projecting allows us to cheat evolution: it now takes place in our head, as a series of projections and counterfactual scenarios.

This ability to mentally play with conjectures, even if it frees us from the laws of evolution, is itself supposed to be the product of evolution—it is as if evolution has put us on a long leash whereas other animals live on the very short leash of immediate dependence on their environment. For Dennett, our brains are "anticipation machines"; for him the human mind and consciousness are emerging properties, those properties necessary for our accelerated development.

Why do we listen to experts and their forecasts? A candidate explanation is that society reposes on specialization, effectively the division of knowledge. You do not go to medical school the minute you encounter a big health problem; it is less taxing (and certainly safer) for you to consult someone who has already done so. Doctors listen to car mechanics (not for health matters, just when it comes to problems with their cars); car mechanics listen to doctors. We have a natural tendency to listen to the expert, even in fields where there may be no experts.

Chapter Twelve

EPISTEMOCRACY, A DREAM

This is only an essay—Children and philosophers vs. adults and nonphilosophers—Science as an autistic enterprise—The past too has a past—Mispredict and live a long, happy life (if you survive)

Someone with a low degree of epistemic arrogance is not too visible, like a shy person at a cocktail party. We are not predisposed to respect humble people, those who try to suspend judgment. Now contemplate *epistemic humility*. Think of someone heavily introspective, tortured by the awareness of his own ignorance. He lacks the courage of the idiot, yet has the rare guts to say "I don't know." He does not mind looking like a fool or, worse, an ignoramus. He hesitates, he will not commit, and he agonizes over the consequences of being wrong. He introspects, introspects, and introspects until he reaches physical and nervous exhaustion.

This does not necessarily mean that he lacks confidence, only that he holds his own knowledge to be suspect. I will call such a person an *epistemocrat;* the province where the laws are structured with this kind of human fallibility in mind I will call an *epistemocracy.*

The major modern epistemocrat is Montaigne.

Monsieur de Montaigne, Epistemocrat

At the age of thirty-eight, Michel Eyquem de Montaigne retired to his estate, in the countryside of southwestern France. Montaigne, which means mountain in Old French, was the name of the estate. The area is known today for the Bordeaux wines, but in Montaigne's time not many people invested their mental energy and sophistication in wine. Montaigne had stoic tendencies and would not have been strongly drawn to such pursuits anyway. His idea was to write a modest collection of "attempts," that is, essays. The very word *essay* conveys the tentative, the speculative, and the nondefinitive. Montaigne was well grounded in the classics and wanted to meditate on life, death, education, knowledge, and some not uninteresting biological aspects of human nature (he wondered, for example, whether cripples had more vigorous libidos owing to the richer circulation of blood in their sexual organs).

The tower that became his study was inscribed with Greek and Latin sayings, almost all referring to the vulnerability of human knowledge. Its windows offered a wide vista of the surrounding hills.

Montaigne's subject, officially, was himself, but this was mostly as a means to facilitate the discussion; he was not like those corporate executives who write biographies to make a boastful display of their honors and accomplishments. He was mainly interested in *discovering* things about himself, making us discover things about himself, and presenting matters that could be generalized—generalized to the entire human race. Among the inscriptions in his study was a remark by the Latin poet Terence: *Homo sum, humani a me nil alienum puto*—I am a man, and nothing human is foreign to me.

Montaigne is quite refreshing to read after the strains of a modern education since he fully accepted human weaknesses and understood that no philosophy could be effective unless it took into account our deeply ingrained imperfections, the limitations of our rationality, the flaws that make us human. It is not that he was ahead of his time; it would be better said that later scholars (advocating rationality) were backward.

He was a thinking, ruminating fellow, and his ideas did not spring up in his tranquil study, but while on horseback. He went on long rides and came back with ideas. Montaigne was neither one of the academics of the Sorbonne nor a professional man of letters, and he was *not* these things on two planes. First, he was a *doer;* he had been a magistrate, a businessman, and the mayor of Bordeaux before he retired to mull over his life and, mostly, his own knowledge. Second, he was an antidogmatist: he was a skeptic with charm, a fallible, noncommittal, personal, introspective writer, and, primarily, someone who, in the great classical tradition, wanted to be a man. Had he been in a different period, he would have been an empirical skeptic—he had skeptical tendencies of the Pyrrhonian variety, the antidogmatic kind like Sextus Empiricus, particularly in his awareness of the need to suspend judgment.

Epistemocracy

Everyone has an idea of utopia. For many it means equality, universal justice, freedom from oppression, freedom from work (for some it may be the more modest, though no more attainable, society with commuter trains free of lawyers on cell phones). To me utopia is an epistemocracy, a society in which anyone of rank is an epistemocrat, and where epistemocrats manage to be elected. It would be a society governed from the basis of the awareness of ignorance, not knowledge.

Alas, one cannot assert authority by accepting one's own fallibility. Simply, people need to be blinded by knowledge—we are made to follow leaders who can gather people together because the advantages of being in groups trump the disadvantages of being alone. It has been more profitable for us to bind together in the wrong direction than to be alone in the right one. Those who have followed the assertive idiot rather than the introspective wise person have passed us some of their genes. This is apparent from a social pathology: psychopaths rally followers.

Once in a while you encounter members of the human species with so much intellectual superiority that they can change their minds effortlessly.

Note here the following Black Swan asymmetry. I believe that you can be dead certain about *some* things, and ought to be so. You can be more confident about disconfirmation than confirmation. Karl Popper was accused of promoting self-doubt while writing in an aggressive and confident tone (an accusation that is occasionally addressed to this author by people who don't follow my logic of skeptical empiricism). Fortunately, we have learned a lot since Montaigne about how to carry on the skeptical-empirical enterprise. The Black Swan asymmetry allows you to be confident *about what is wrong,* not about what you believe is right. Karl Popper was once asked whether one "could falsify falsification" (in other words, if one could be skeptical about skepticism). His answer was that he threw students out of his lectures for asking far more intelligent questions than that one. Quite tough, Sir Karl was.

THE PAST'S PAST, AND THE PAST'S FUTURE

Some truths only hit children—adults and nonphilosophers get sucked into the minutiae of practical life and need to worry about "serious matters," so they abandon these insights for seemingly more relevant questions. One of these truths concerns the larger difference in texture and quality between the past and the future. Thanks to my studying this distinction all my life, I understand it better than I did during my childhood, but I no longer envision it as vividly.

The only way you can imagine a future "similar" to the past is by assuming that it will be an *exact* projection of it, hence predictable. Just as you know with some precision when you were born, you would then know with equal precision when you will die. The notion of future mixed with *chance,* not a deterministic extension of your perception of the past, is a mental operation that our mind cannot perform. Chance is too fuzzy for us to be a category by itself. There is an asymmetry between past and future, and it is too subtle for us to understand naturally.

The first consequence of this asymmetry is that, in people's minds, the relationship between the past and the future does not learn from the relationship between the past and the past previous to it. There is a blind spot: when we think of tomorrow we do not frame it in terms of what we thought about yesterday or the day before yesterday. Because of this introspective defect we fail to learn about the difference between our past predictions and the subsequent outcomes. When we think of tomorrow, we just project it as another yesterday.

This small blind spot has other manifestations. Go to the primate section of the Bronx Zoo where you can see our close relatives in the happy primate family leading their own busy social lives. You can also see masses of tourists laughing at the caricature of humans that the lower primates represent. Now imagine being a member of a higher-level species (say a "real" philosopher, a truly wise person), far more sophisticated than the human primates. You would certainly laugh at the people laughing at the nonhuman primates. Clearly, to those people amused by the apes, the idea of a being who would look down on them the way they look down on the apes cannot immediately come to their minds—if it did, it would elicit self-pity. They would stop laughing.

Accordingly, an element in the mechanics of how the human mind learns from the past makes us believe in definitive solutions—yet not consider that those who preceded us thought that they too had definitive solutions. We laugh at others and we don't realize that someone will be just as justified in laughing at us on some not too remote day. Such a realization would entail the recursive, or second-order, thinking that I mentioned in the Prologue; we are not good at it.

This mental block about the future has not yet been investigated and labeled by psychologists, but it appears to resemble autism. Some autistic subjects can possess high levels of mathematical or technical intelligence. Their social skills are defective, but that is not the root of their problem. Autistic people cannot put themselves in the shoes of others, cannot view the world from their standpoint. They see others as inanimate objects, like machines, moved by explicit rules. They cannot perform such simple mental operations as "he knows that I don't know that I know," and it is this inability that impedes their social skills. (Interestingly, autistic subjects, regardless of their "intelligence," also exhibit an inability to comprehend uncertainty.)

Just as autism is called "mind blindness," this inability to think dynamically, to position oneself with respect to a future observer, we should call "future blindness."

Prediction, Misprediction, and Happiness

I searched the literature of cognitive science for any research on "future blindness" and found nothing. But in the literature on happiness I did find an examination of our chronic errors in prediction that will make us happy.

This prediction error works as follows. You are about to buy a new car. It is going to change your life, elevate your status, and make your commute a vacation. It is so quiet that you can hardly tell if the engine is on, so you can listen to Rachmaninoff's nocturnes on the highway. This new car will bring you to a permanently elevated plateau of contentment. People will think, Hey, he has a great car, every time they see you. Yet you forget that the last time you bought a car, you also had the same expectations. You do not anticipate that the effect of the new car will eventually wane and that you will revert to the initial condition, as you did last time. A few weeks after you drive your new car out of the showroom, it will become dull. If you had expected this, you probably would not have bought it.

You are about to commit a prediction error that you have already made. Yet it would cost so little to introspect!

Psychologists have studied this kind of misprediction with respect to both pleasant and unpleasant events. We overestimate the effects of both kinds of future events on our lives. We seem to be in a psychological predicament that makes us do so. This predicament is called "anticipated utility" by Danny Kahneman and "affective forecasting" by Dan Gilbert. The point is not so much that we tend to mispredict our future happiness, but rather that we do not learn recursively from past experiences. We have evidence of a mental block and distortions in the way we fail to learn from our past errors in projecting the future of our affective states.

We grossly overestimate the length of the effect of misfortune on our lives. You think that the loss of your fortune or current position will be devastating, but you are probably wrong. More likely, you will adapt to anything, as you probably did after past misfortunes. You may feel a sting, but it will not be as bad as you expect. This kind of misprediction may have a purpose: to motivate us to perform *important* acts (like buying new cars or getting rich) and to prevent us from taking certain unnecessary risks. And it is part of a more general problem: we humans are supposed to fool ourselves a little bit here and there. According to Trivers's theory of self-deception, this is supposed to orient us favorably toward the future. But self-deception is not a desirable feature outside of its natural domain. It prevents us from taking some unnecessary risks—but we saw in Chapter 6 how it does not as readily cover a spate of modern risks that we do not fear because they are not vivid, such as investment risks, environmental dangers, or long-term security.

Helenus and the Reverse Prophecies

If you are in the business of being a seer, describing the future to other less-privileged mortals, you are judged on the merits of your predictions.

Helenus, in *The Iliad*, was a different kind of seer. The son of Priam and Hecuba, he was the cleverest man in the Trojan army. It was he who, under torture, told the Achaeans how they would capture Troy (apparently he didn't predict that he himself would be captured). But this is not what distinguished him. Helenus, unlike other seers, was able to predict *the past* with great precision—without having been given any details of it. He predicted backward.

Our problem is not just that we do not know the future, we do not know much of the past either. We badly need someone like Helenus if we are to know history. Let us see how.

The Melting Ice Cube

Consider the following thought experiment borrowed from my friends Aaron Brown and Paul Wilmott:

Operation 1 (the melting ice cube): Imagine an ice cube and consider how it may melt over the next two hours while you play a few rounds of poker with your friends. Try to envision the shape of the resulting puddle.

Operation 2 (where did the water come from?): Consider a puddle of water on the floor. Now try to reconstruct in your mind's eye the shape of the ice cube it may once have been. Note that the puddle may not have necessarily originated from an ice cube.

The second operation is harder. Helenus indeed had to have skills.

The difference between these two processes resides in the following. If you have the right models (and some time on your hands, and nothing better to do) you can predict with great precision how the ice cube will melt—this is a specific engineering problem devoid of complexity, easier than the one involving billiard balls. However, from the pool of water you can build infinite possible ice cubes, if there was in fact an ice cube there at all. The first direction, from the ice cube to the puddle, is called the *forward process*. The second direction, the *backward process*, is much, much more complicated. The forward process is generally used in physics and engineering; the backward process in nonrepeatable, nonexperimental historical approaches.

In a way, the limitations that prevent us from unfrying an egg also prevent us from reverse engineering history.

Now, let me increase the complexity of the forward-backward problem just a bit by assuming nonlinearity. Take what is generally called the "butterfly in India" paradigm from the discussion of Lorenz's discovery in the previous chapter. As we have seen, a small input in a complex system can lead to nonrandom large results, depending on very special conditions. A single butterfly flapping its wings in New Delhi may be the certain *cause* of a hurricane in North Carolina, though the hurricane may take place a couple of years later. However, *given the observation of a hurricane in North Carolina,* it is dubious that you could figure out the causes with any precision: there are billions of billions of such small things as wing-flapping butterflies in Timbuktu or sneezing wild dogs in Australia that could have caused it. The process from the butterfly to the hurricane is greatly simpler than the reverse process *from* the hurricane *to* the potential butterfly.

Confusion between the two is disastrously widespread in common culture. This "butterfly in India" metaphor has fooled at least one filmmaker. For instance, *Happenstance* (a.k.a. *The Beating of a Butterfly's Wings*), a French-language film by one Laurent Firode, meant to encourage people to focus on small things that can change the course of their lives. Hey, since a small event (a petal falling on the ground and getting your attention) can lead to your choosing one person over another as a mate for life, you should focus on these very small details. Neither the filmmaker nor the critics realized that they were dealing with the backward process; there are trillions of such small things in the course of a simple day, and examining all of them lies outside of our reach.

Once Again, Incomplete Information

Take a personal computer. You can use a spreadsheet program to generate a random sequence, a succession of points we can call a history. How? The computer program responds to a very complicated equation of a nonlinear nature that produces numbers that seem random. The equation is very simple: if you know it, you can predict the sequence. It is almost impossible, however, for a human being to reverse engineer the equation and predict further sequences. I am talking about a simple one-line computer program (called the "tent map") generating a handful of data points, not about the billions of simultaneous events that constitute the real history of the world. In other words, even if history were a nonrandom series generated by some "equation of the world," as long as reverse engineering such an equation does not seem within human possibility, it should be deemed random and not bear the name "deterministic chaos." Historians should stay away from chaos theory and the difficulties of reverse engineering except to discuss general properties of the world and learn the limits of what they can't know.

This brings me to a greater problem with the historian's craft. I will state the fundamental problem of practice as follows: while in theory randomness is an intrinsic property, in practice, randomness is *incomplete information,* what I called *opacity* in Chapter 1.

Nonpractitioners of randomness do not understand the subtlety. Often, in conferences when they hear me talk about uncertainty and randomness, philosophers, and sometimes mathematicians, bug me about the least relevant point, namely whether the randomness I address is "true randomness" or "deterministic chaos" that masquerades as randomness. A true random system is in fact random and does not have predictable properties. A chaotic system has entirely predictable properties, but they are hard to know. So my answer to them is dual.

a) There is no functional difference in practice between the two since we will never get to make the distinction—the difference is mathematical, not practical. If I see a pregnant woman, the sex of her child is a purely random matter to me (a 50 percent chance for either sex)—but not to her doctor, who might have done an ultrasound. In practice, randomness is fundamentally incomplete information.

b) The mere fact that a person is talking about the difference implies that he has never made a meaningful decision under uncertainty—which is why he does not realize that they are indistinguishable in practice.

Randomness, in the end, is just unknowledge. The world is opaque and appearances fool us.

What They Call Knowledge

One final word on history.

History is like a museum where one can go to see the repository of the past, and taste the charm of olden days. It is a wonderful mirror in which we can see our own narratives. You can even track the past using DNA analyses. I am fond of literary history. Ancient history satisfies my desire to build my own self-narrative, my identity, to connect with my (complicated) Eastern Mediterranean roots. I even prefer the accounts of older, patently less accurate books to modern ones. Among the authors I've reread (the ultimate test of whether you like an author is if you've reread him) the following come to mind: Plutarch, Livy, Suetonius, Diodorus Siculus, Gibbon, Carlyle, Renan, and Michelet. These accounts are patently substandard, compared to today's works; they are largely anecdotal, and full of myths. But I know this.

History is useful for the thrill of knowing the past, and for the narrative (indeed), provided it remains a harmless narrative. One should learn under severe caution. History is certainly not a place to theorize or derive general knowledge, nor is it meant to help in the future, without some caution. We can get negative confirmation from history, which is invaluable, but we get plenty of illusions of knowledge along with it.

This brings me back once again to Menodotus and the treatment of the turkey problem and how to not be a sucker for the past. The empirical doctor's approach to the problem of induction was to *know* history without theorizing from it. Learn to read history, get all the knowledge you can, do not frown on the anecdote, but do not draw any causal links, do not try to reverse engineer too much—but if you do, do not make big scientific claims. Remember that the empirical skeptics had respect for custom: they used it as a default, a basis for action, but not for more than that. This clean approach to the past they called *epilogism.**

But most historians have another opinion. Consider the representative introspection *What Is History?* by Edward Hallett Carr. You will catch him explicitly pursuing causation as a central aspect of his job. You can even go higher up: Herodotus, deemed to be the father of the subject, defined his purpose in the opening of his work:

> To preserve a memory of the deeds of the Greeks and barbarians, "and in particular, beyond everything else, to give a *cause* [emphasis mine] to their fighting one another."

You see the same with all theoreticians of history, whether Ibn Khaldoun, Marx, or Hegel. The more we try to turn history into anything other than an enumeration of accounts to be enjoyed with minimal theorizing, the more we get into trouble. Are we so plagued with the narrative fallacy?†

* Yogi Berra might have a theory of epilogism with his saying, "You can observe a lot by just watching."

† While looking at the past it would be a good idea to resist naïve analogies. Many people have compared the United States today to Ancient Rome, both from a military standpoint (the destruction of Carthage was often invoked as an incentive for the destruction of enemy regimes) and from a social one (the endless platitudinous warnings of the upcoming decline and fall). Alas, we need to be extremely careful in transposing knowledge from a simple environment that is closer to type 1, like the one we had in antiquity, to today's type 2, complex system, with its intricate webs of casual links. Another error is to draw casual conclusions from the absence of nuclear war, since, invoking the Casanova argument of Chapter 8, I would repeat that we would not be here had a nuclear war taken place, and it is not a good idea for us to derive a "cause" when our survival is conditioned on that cause.

We may have to wait for a generation of skeptical-empiricist historians capable of understanding the difference between a forward process and a reverse one.

Just as Popper attacked the historicists in their making claims about the future, I have just presented the weakness of the historical approach in knowing the *past* itself.

After this discussion about future (and past) blindness, let us see what to do about it. Remarkably, there are extremely practical measures we can take. We will explore this next.

Chapter Thirteen

APPELLES THE PAINTER, OR WHAT DO YOU DO IF YOU CANNOT PREDICT?*

You should charge people for advice—My two cents here—Nobody knows anything, but, at least, he knows it—Go to parties

ADVICE IS CHEAP, VERY CHEAP

It is not a good habit to stuff one's text with quotations from prominent thinkers, except to make fun of them or provide a historical reference. They "make sense," but well-sounding maxims force themselves on our gullibility and do not always stand up to empirical tests. So I chose the following statement by the überphilosopher Bertrand Russell precisely because I disagree with it.

> The demand for certainty is one which is natural to man, but is nevertheless an intellectual vice. If you take your children for a picnic on a doubtful day, they will demand a dogmatic answer as to whether it will be fine or wet, and be disappointed in you when you cannot be sure. . . .

* This chapter provides a general conclusion for those who by now say, "Taleb, I get the point, but what should I do?" My answer is that if you got the point, you are pretty much there. But here is a nudge.

> But so long as men are not *trained* [emphasis mine] to withhold judgment in the absence of evidence, they will be led astray by cocksure prophets . . . For the learning of every virtue there is an appropriate discipline, and for the learning of suspended judgment the best discipline is philosophy.

The reader may be surprised that I disagree. It is hard to disagree that the demand for certainty is an intellectual vice. It is hard to disagree that we can be led astray by some cocksure prophet. Where I beg to differ with the great man is that I do not believe in the track record of advice-giving "philosophy" in helping us deal with the problem; nor do I believe that virtues can be *easily* taught; nor do I urge people to strain in order to avoid making a judgment. Why? Because we have to deal with humans as humans. We cannot *teach* people to withhold judgment; judgments are embedded in the way we view objects. I do not see a "tree"; I see a pleasant or an ugly tree. It is not possible without great, paralyzing effort to strip these small values we attach to matters. Likewise, it is not possible to hold a situation in one's head without some element of bias. Something in our dear human nature makes us want to believe; so what?

Philosophers since Aristotle have taught us that we are deep-thinking animals, and that we can learn by reasoning. It took a while to discover that we do effectively think, but that we more readily narrate backward in order to give ourselves the illusion of understanding, and give a cover to our past actions. The minute we forgot about this point, the "Enlightenment" came to drill it into our heads for a second time.

I'd rather degrade us humans to a level certainly above other known animals but not quite on a par with the ideal Olympian man who can absorb philosophical statements and act accordingly. Indeed, if philosophy were *that* effective, the self-help section of the local bookstore would be of some use in consoling souls experiencing pain—but it isn't. We forget to philosophize when under strain.

I'll end this section on prediction with the following two lessons, one very brief (for the small matters), one rather lengthy (for the large, important decisions).

Being a Fool in the Right Places

The lesson for the small is: *be human!* Accept that being human involves some amount of epistemic arrogance in running your affairs. Do not be ashamed of that. Do not try to always withhold judgment—opinions are the stuff of life. Do not try to avoid predicting—yes, after this diatribe about prediction I am *not* urging you to stop being a fool. Just be a fool in the right places.*

What you should avoid is unnecessary dependence on large-scale harmful predictions—those and only those. Avoid the big subjects that may hurt your future: be fooled in small matters, not in the large. Do not listen to economic forecasters or to predictors in social science (they are mere entertainers), but do make your own forecast for the picnic. By all means, demand certainty for the next picnic; but avoid government social-security forecasts for the year 2040.

Know how to rank beliefs not according to their plausibility but by the harm they may cause.

Be Prepared

The reader might feel queasy reading about these general failures to see the future and wonder what to do. But if you shed the idea of full predictability, there are plenty of things to do provided you remain conscious of their limits. Knowing that you cannot predict does not mean that you cannot benefit from unpredictability.

The bottom line: be prepared! Narrow-minded prediction has an analgesic or therapeutic effect. Be aware of the numbing effect of magic numbers. Be prepared for all relevant eventualities.

THE IDEA OF POSITIVE ACCIDENT

Recall the empirics, those members of the Greek school of empirical medicine. They considered that you should be open-minded in your medical diagnoses to let luck play a role. By luck, a patient might be cured, say, by

* Dan Gilbert showed in a famous paper, "How Mental Systems Believe," that we are not natural skeptics and that not believing required an expenditure of mental effort.

eating some food that accidentally turns out to be the cure for his disease, so that the treatment can then be used on subsequent patients. The *positive* accident (like hypertension medicine producing side benefits that led to Viagra) was the empirics' central method of medical discovery.

This same point can be generalized to life: maximize the serendipity around you.

Sextus Empiricus retold the story of Apelles the Painter, who, while doing a portrait of a horse, was attempting to depict the foam from the horse's mouth. After trying very hard and making a mess, he gave up and, in irritation, took the sponge he used for cleaning his brush and threw it at the picture. Where the sponge hit, it left a perfect representation of the foam.

Trial and error means trying a lot. In *The Blind Watchmaker,* Richard Dawkins brilliantly illustrates this notion of the world without grand design, moving by small incremental random changes. Note a slight disagreement on my part that does not change the story by much: the world, rather, moves by *large* incremental random changes.

Indeed, we have psychological and intellectual difficulties with trial and error, and with accepting that series of small failures are necessary in life. My colleague Mark Spitznagel understood that we humans have a mental hang-up about failures: "You need to love to lose" was his motto. In fact, the reason I felt immediately at home in America is precisely because American culture encourages the process of failure, unlike the cultures of Europe and Asia where failure is met with stigma and embarrassment. America's specialty is to take these small risks for the rest of the world, which explains this country's disproportionate share in innovations. Once established, an idea or a product is later "perfected" over there.

Volatility and Risk of Black Swan

People are often ashamed of losses, so they engage in strategies that produce very little volatility but contain the risk of a large loss—like collecting nickels in front of steamrollers. In Japanese culture, which is ill-adapted to randomness and badly equipped to understand that bad performance can come from bad luck, losses can severely tarnish someone's reputation. People hate volatility, thus engage in strategies exposed to blowups, leading to occasional suicides after a big loss.

Furthermore, this trade-off between volatility and risk can show up in careers that give the appearance of being stable, like jobs at IBM until the 1990s. When laid off, the employee faces a total void: he is no longer fit for anything else. The same holds for those in protected industries. On the other hand, consultants can have volatile earnings as their clients' earnings go up and down, but face a lower risk of starvation, since their skills match demand—*fluctuat nec mergitur* (fluctuates but doesn't sink). Likewise, dictatorships that do not appear volatile, like, say, Syria or Saudi Arabia, face a larger risk of chaos than, say, Italy, as the latter has been in a state of continual political turmoil since the second war. I learned about this problem from the finance industry, in which we see "conservative" bankers sitting on a pile of dynamite but fooling themselves because their operations seem dull and lacking in volatility.

Barbell Strategy

I am trying here to generalize to real life the notion of the "barbell" strategy I used as a trader, which is as follows. If you know that you are vulnerable to prediction errors, and if you accept that most "risk measures" are flawed, because of the Black Swan, then your strategy is to be as hyperconservative and hyperaggressive as you can be instead of being mildly aggressive or conservative. Instead of putting your money in "medium risk" investments (how do you know it is medium risk? by listening to tenure-seeking "experts"?), you need to put a portion, say 85 to 90 percent, in extremely safe instruments, like Treasury bills—as safe a class of instruments as you can manage to find on this planet. The remaining 10 to 15 percent you put in extremely speculative bets, as leveraged as possible (like options), preferably venture capital–style portfolios.* That way you do not depend on errors of risk management; no Black Swan can hurt you at all, beyond your "floor," the nest egg that you have in maximally safe investments. Or, equivalently, you can have a speculative portfolio and insure it (if possible) against losses of more than, say, 15 percent. You are "clipping" your incomputable risk , the one that is harmful to you. Instead

* Make sure that you have plenty of these small bets; avoid being blinded by the vividness of one single Black Swan. Have as many of these small bets as you can conceivably have. Even venture capital firms fall for the narrative fallacy with a few stories that "make sense" to them; they do not have as many bets as they should. If venture capital firms are profitable, it is not because of the stories they have in their heads, but because they are exposed to unplanned rare events.

of having medium risk, you have high risk on one side and no risk on the other. The average will be medium risk but constitutes a positive exposure to the Black Swan. More technically, this can be called a "convex" combination. Let us see how this can be implemented in all aspects of life.

"Nobody Knows Anything"

The legendary screenwriter William Goldman was said to have shouted "Nobody knows anything!" in relation to the prediction of movie sales. Now, the reader may wonder how someone as successful as Goldman can figure out what to do without making predictions. The answer stands perceived business logic on its head. He knew that he could not predict individual events, but he was well aware that the unpredictable, namely a movie turning into a blockbuster, would benefit him immensely.

So the second lesson is more aggressive: you can actually take advantage of the problem of prediction and epistemic arrogance! As a matter of fact, I suspect that the most successful businesses are precisely those that know how to work around inherent unpredictability and even exploit it.

Recall my discussion of the biotech company whose managers understood that the essence of research is in the unknown unknowns. Also, notice how they seized on the "corners," those free lottery tickets in the world.

Here are the (modest) tricks. But note that the more modest they are, the more effective they will be.

a. *First, make a distinction between* positive contingencies *and* negative ones. Learn to distinguish between those human undertakings in which the lack of predictability can be (or has been) extremely beneficial and those where the failure to understand the future caused harm. There are both positive and negative Black Swans. William Goldman was involved in the movies, a positive–Black Swan business. Uncertainty did occasionally pay off there.

   A negative–Black Swan business is one where the unexpected can hit hard and hurt severely. If you are in the military, in catastrophe insurance, or in homeland security, you face only downside. Likewise, as we saw in Chapter 7, if you are in banking and lending, surprise outcomes are likely to be negative for you. You lend, and in the best of circumstances you get your loan back—but you may lose all of your money if the borrower defaults. In the event that the borrower enjoys great financial success, he is not likely to offer you an additional dividend.

Aside from the movies, examples of positive–Black Swan businesses are: some segments of publishing, scientific research, and venture capital. In these businesses, you lose small to make big. You have little to lose per book and, for completely unexpected reasons, any given book might take off. The downside is small and easily controlled. The problem with publishers, of course, is that they regularly pay up for books, thus making their upside rather limited and their downside monstrous. (If you pay $10 million for a book, your Black Swan is it not being a bestseller.) Likewise, while technology can carry a great payoff, paying for the hyped-up story, as people did with the dot-com bubble, can make any upside limited and any downside huge. It is the venture capitalist who invested in a speculative company and sold his stake to unimaginative investors who is the beneficiary of the Black Swan, not the "me, too" investors.

In these businesses you are lucky if you don't know anything—particularly if others don't know anything either, but aren't aware of it. And you fare best if you know where your ignorance lies, if you are the only one looking at the unread books, so to speak. This dovetails into the "barbell" strategy of taking maximum exposure to the positive Black Swans while remaining paranoid about the negative ones. For your exposure to the positive Black Swan, you do not need to have any precise understanding of the structure of uncertainty. I find it hard to explain that when you have a very limited loss you need to get as aggressive, as speculative, and sometimes as "unreasonable" as you can be.

Middlebrow thinkers sometimes make the analogy of such strategy with that of collecting "lottery tickets." It is plain wrong. First, lottery tickets do not have a scalable payoff; there is a known upper limit to what they can deliver. The ludic fallacy applies here—the scalability of real-life payoffs compared to lottery ones makes the payoff unlimited or of unknown limit. Secondly, the lottery tickets have known rules and laboratory-style well-presented possibilities; here we do not know the rules and can benefit from this additional uncertainty, since it cannot hurt you and can only benefit you.*

b. *Don't look for the* precise *and the* local. Simply, do not be narrow-minded. The great discoverer Pasteur, who came up with the notion that chance favors the prepared, understood that you do not look for something particular every morning but work hard to let contingency enter your working life. As Yogi Berra, another great thinker, said, "You got to be very careful if you don't know where you're going, because you might not get there."

Likewise, do not try to predict precise Black Swans—it tends to make you more vulnerable to the ones you did not predict. My friends Andy Marshall and Andrew Mays at the Department of Defense face the same problem. The impulse on the part of the military is to devote resources to predicting the next problems. These thinkers advocate the opposite: invest in preparedness, not in prediction.

Remember that infinite vigilance is just not possible.

c. *Seize any opportunity, or anything that looks like opportunity.* They are rare, much rarer than you think. Remember that positive Black Swans have a necessary first step: you need to be exposed to them. Many people do not realize that they are getting a lucky break in life when they get it. If a big publisher (or a big art dealer or a movie executive or a hotshot banker or a big thinker) suggests

* There is a finer epistemological point. Remember that in a virtuous Black Swan business, what the past did not reveal is almost certainly going to be good for you. When you look at past biotech revenues, you do not see the superblockbuster in them, and owing to the potential for a cure for cancer (or headaches, or baldness, or bad sense of humor, etc.), there is a small probability that the sales in that industry may turn out to be monstrous, far larger than might be expected. On the other hand, consider negative Black Swan businesses. The track record you see is likely to overestimate the properties. Recall the 1982 blowup of banks: they appeared to the naïve observer to be more profitable than they seemed. Insurance companies are of two kinds: the regular diversifiable kind that belongs to Mediocristan (say, life insurance) and the more critical and explosive Black Swan–prone risks that are usually sold to reinsurers. According to the data, reinsurers have lost money on underwriting over the past couple of decades, but, unlike bankers, they are introspective enough to know that it actually could have been far worse, because the past twenty years did not have a big catastrophe, and all you need is one of those per century to kiss the business good-bye. Many finance academics doing "valuation" on insurance seem to have missed the point.

an appointment, cancel anything you have planned: you may never see such a window open up again. I am sometimes shocked at how little people realize that these opportunities do not grow on trees. Collect as many free nonlottery tickets (those with open-ended payoffs) as you can, and, once they start paying off, do not discard them. Work hard, not in grunt work, but in chasing such opportunities and maximizing exposure to them. This makes living in big cities invaluable because you increase the odds of serendipitous encounters—you gain exposure to the envelope of serendipity. The idea of settling in a rural area on grounds that one has good communications "in the age of the Internet" tunnels out of such sources of positive uncertainty. Diplomats understand that very well: casual chance discussions at cocktail parties usually lead to big breakthroughs—not dry correspondence or telephone conversations. Go to parties! If you're a scientist, you will chance upon a remark that might spark new research. And if you are autistic, send your associates to these events.

d. *Beware of precise plans by governments.* As discussed in Chapter 10, let governments predict (it makes officials feel better about themselves and justifies their existence) but do not set much store by what they say. Remember that the interest of these civil servants is to survive and self-perpetuate—not to get to the truth. It does not mean that governments are useless, only that you need to keep a vigilant eye on their side effects. For instance, regulators in the banking business are prone to a severe expert problem and they tend to condone reckless but (hidden) risk taking. Andy Marshall and Andy Mays asked me if the private sector could do better in predicting. Alas, no. Once again, recall the story of banks hiding explosive risks in their portfolios. It is not a good idea to trust corporations with matters such as rare events because the performance of these executives is not observable on a short-term basis, and they will game the system by showing good performance so they can get their yearly bonus. The Achilles' heel of capitalism is that if you make corporations compete, it is sometimes the one that is most exposed to the negative Black Swan that will appear to be the most fit for survival. Also recall from the footnote on Ferguson's discovery in Chapter 1 that markets are not good predictors of wars. No one in particular is a good predictor of anything. Sorry.

e. "There are some people who, if they don't already know, you can't tell 'em," as the great philosopher of uncertainty Yogi Berra once said. *Do not waste your time trying to fight forecasters, stock analysts, economists, and social scientists, except to play pranks on them.* They are considerably easy to make fun of, and many get angry quite readily. It is ineffective to moan about unpredictability: people will continue to predict foolishly, especially if they are paid for it, and you cannot put an end to institutionalized frauds. If you ever do have to heed a forecast, keep in mind that its accuracy degrades rapidly as you extend it through time.

   If you hear a "prominent" economist using the word *equilibrium,* or *normal distribution,* do not argue with him; just ignore him, or try to put a rat down his shirt.

The Great Asymmetry

All these recommendations have one point in common: asymmetry. Put yourself in situations where favorable consequences are much larger than unfavorable ones.

Indeed, the notion of *asymmetric outcomes* as the central idea of this book: I will never get to know the unknown since, by definition, it is unknown. However, I can always guess how it might affect me, and I should base my decisions around that.

This idea is often erroneously called Pascal's wager, after the philosopher and (thinking) mathematician Blaise Pascal. He presented it something like this: I do not know whether God exists, but I know that I have nothing to gain from being an atheist if he does not exist, whereas I have plenty to lose if he does. Hence, this justifies my belief in God.

Pascal's argument is severely flawed theologically: one has to be naïve enough to believe that God would not penalize us for false belief. Unless, of course, one is taking the quite restrictive view of a naïve God. (Bertrand Russell was reported to have claimed that God would need to have created fools for Pascal's argument to work.)

But the idea behind Pascal's wager has fundamental applications outside of theology. It stands the entire notion of knowledge on its head. It eliminates the need for us to understand the probabilities of a rare event (there are fundamental limits to our knowledge of these); rather, we can focus on the payoff and benefits of an event if it takes place. The probabilities of very rare events are not computable; the effect of an event on us is considerably easier to ascertain (the rarer the event, the fuzzier the odds). We can have a clear idea of the consequences of an event, even if we do not know how likely it is to occur. I don't know the odds of an earthquake, but I can imagine how San Francisco might be affected by one. This idea that in order to make a decision you need to focus on the consequences (which you can know) rather than the probability (which you can't know) is the *central idea of uncertainty*. Much of my life is based on it.

You can build an overall theory of decision making on this idea. All you have to do is mitigate the consequences. As I said, if my portfolio is exposed to a market crash, the odds of which I can't compute, all I have to do is buy insurance, or get out and invest the amounts I am not willing to ever lose in less risky securities.

Effectively, if free markets have been successful, it is precisely because they allow the trial-and-error process I call "stochastic tinkering" on the part of competing individual operators who fall for the narrative fallacy—but are effectively collectively partaking of a grand project. We are increasingly learning to practice stochastic tinkering without knowing it—thanks to overconfident entrepreneurs, naïve investors, greedy investment bankers, and aggressive venture capitalists brought together by the free-market system. The next chapter shows why I am optimistic that the academy is losing its power and ability to put knowledge in straitjackets and that more out-of-the-box knowledge will be generated Wiki-style.

In the end we are being driven by history, all the while thinking that we are doing the driving.

I'll sum up this long section on prediction by stating that we can easily narrow down the reasons we can't figure out what's going on. There are: a) epistemic arrogance and our corresponding future blindness; b) the Platonic notion of categories, or how people are fooled by reductions, particularly if they have an academic degree in an expert-free discipline; and, finally c) flawed tools of inference, particularly the Black Swan–free tools from Mediocristan.

In the next section we will go deeper, much deeper, into these tools from Mediocristan, into the "plumbing," so to speak. Some readers may see it as an appendix; others may consider it the heart of the book.

Part 3

THOSE GRAY SWANS OF EXTREMISTAN

It's time to deal in some depth with four final items that bear on our Black Swan.

Primo, I have said earlier that the world is moving deeper into Extremistan, that it is less and less governed by Mediocristan—in fact, this idea is more subtle than that. I will show how and present the various ideas we have about the formation of inequality. *Secondo,* I have been describing the Gaussian bell curve as a contagious and severe delusion, and it is time to get into that point in some depth. *Terso,* I will present what I call Mandelbrotian, or fractal, randomness. Remember that for an event to be a Black Swan, it does not just have to be rare, or just wild; it has to be unexpected, has to lie outside our tunnel of possibilities. You must be a sucker for it. As it happens, many rare events can yield their structure to us: it is not easy to compute their probability, but it is easy to get a *general* idea about the possibility of their occurrence. We can turn these Black Swans into Gray Swans, so to speak, reducing their surprise effect. A person aware of the possibility of such events can come to belong to the non-sucker variety.

Finally, I will present the ideas of those philosophers who focus on phony uncertainty. I organized this book in such a way that the more technical (though nonessential) sections are here; these can be skipped without any loss to the thoughtful reader, particularly Chapters 15, 17, and the second half of Chapter 16. I will alert the reader with footnotes. The reader less interested in the mechanics of deviations can then directly proceed to Part 4.

Chapter Fourteen

FROM MEDIOCRISTAN TO EXTREMISTAN, AND BACK

I prefer Horowitz—How to fall from favor—The long tail—Get ready for some surprises—It's not just money

---

Let us see how an increasingly man-made planet can evolve away from mild into wild randomness. First, I describe how we get to Extremistan. Then, I will take a look at its evolution.

The World Is Unfair

Is the world that unfair? I have spent my entire life studying randomness, practicing randomness, hating randomness. The more that time passes, the worse things seem to me, the more scared I get, the more disgusted I am with Mother Nature. The more I think about my subject, the more I see evidence that the world we have in our minds is different from the one playing outside. Every morning the world appears to me more random than it did the day before, and humans seem to be even more fooled by it than they were the previous day. It is becoming unbearable. I find writing these lines painful; I find the world revolting.

Two "soft" scientists propose intuitive models for the development of this inequity: one is a mainstream economist, the other a sociologist. Both simplify a little too much. I will present their ideas because they are easy to understand, not because of the scientific quality of their insights or any consequences in their discoveries; then I will show the story as seen from the vantage point of the natural scientists.

Let me start with the economist Sherwin Rosen. In the early eighties, he wrote papers about "the economics of superstars." In one of the papers he conveyed his sense of outrage that a basketball player could earn $1.2 million a year, or a television celebrity could make $2 million. To get an idea of how this concentration is increasing—i.e., of how we are moving away from Mediocristan—consider that television celebrities and sports stars (even in Europe) get contracts today, only two decades later, worth in the hundreds of millions of dollars! The extreme is about (so far) twenty times higher than it was two decades ago!

According to Rosen, this inequality comes from a tournament effect: someone who is marginally "better" can easily win the entire pot, leaving the others with nothing. Using an argument from Chapter 3, people prefer to pay $10.99 for a recording featuring Horowitz to $9.99 for a struggling pianist. Would you rather read Kundera for $13.99 or some unknown author for $1? So it looks like a tournament, where the winner grabs the whole thing—and he does not have to win by much.

But the role of luck is missing in Rosen's beautiful argument. The problem here is the notion of "better," this focus on skills as leading to success. Random outcomes, or an arbitrary situation, can also explain success, and provide the initial push that leads to a winner-take-all result. A person can get slightly ahead for entirely random reasons; because we like to imitate one another, we will flock to him. The world of contagion is so underestimated!

As I am writing these lines I am using a Macintosh, by Apple, after years of using Microsoft-based products. The Apple technology is vastly better, yet the inferior software won the day. How? Luck.

The Matthew Effect

More than a decade before Rosen, the sociologist of science Robert K. Merton presented his idea of the Matthew effect, by which people take from the poor to give to the rich.* He looked at the performance of scien-

* These scalable laws were already discussed in the scriptures: "For onto everyone that hath shall be given, and he shall have abundance; but from him that hath not shall be taken away even that which he hath." Matthew (Matthew 25:29, King James Version).

tists and showed how an initial advantage follows someone through life. Consider the following process.

Let's say someone writes an academic paper quoting fifty people who have worked on the subject and provided background materials for his study; assume, for the sake of simplicity, that all fifty are of equal merit. Another researcher working on the exact same subject will randomly cite three of those fifty in his bibliography. Merton showed that many academics cite references without having read the original work; rather, they'll read a paper and draw their own citations from among its sources. So a third researcher reading the second article selects three of the previously referenced authors for *his* citations. These three authors will receive cumulatively more and more attention as their names become associated more tightly with the subject at hand. The difference between the winning three and the other members of the original cohort is mostly luck: they were initially chosen not for their greater skill, but simply for the way their names appeared in the prior bibliography. Thanks to their reputations, these successful academics will go on writing papers and their work will be easily accepted for publication. Academic success is partly (but significantly) a lottery.*

It is easy to test the effect of reputation. One way would be to find papers that were written by famous scientists, had their authors' identities changed by mistake, and got rejected. You could verify how many of these rejections were subsequently overturned after the true identities of the authors were established. Note that scholars are judged mostly on how many times their work is referenced in other people's work, and thus cliques of people who quote one another are formed (it's an "I quote you, you quote me" type of business).

Eventually, authors who are not often cited will drop out of the game by, say, going to work for the government (if they are of a gentle nature), or for the Mafia, or for a Wall Street firm (if they have a high level of hormones). Those who got a good push in the beginning of their scholarly careers will keep getting persistent cumulative advantages throughout life. It is easier for the rich to get richer, for the famous to become more famous.

In sociology, Matthew effects bear the less literary name "cumulative

* Much of the perception of the importance of precocity in the career of researchers can be owed to the misunderstanding of the perverse role of this effect, especially when reinforced by bias. Enough counterexamples, even in fields like mathematics meant to be purely a "young man's game," illustrate the age fallacy: simply, it is necessary to be successful early, and even very early at that.

advantage." This theory can easily apply to companies, businessmen, actors, writers, and anyone else who benefits from past success. If you get published in *The New Yorker* because the color of your letterhead attracted the attention of the editor, who was daydreaming of daisies, the resultant reward can follow you for life. More significantly, it will follow *others* for life. Failure is also cumulative; losers are likely to also lose in the future, even if we don't take into account the mechanism of demoralization that might exacerbate it and cause additional failure.

Note that art, because of its dependence on word of mouth, is extremely prone to these cumulative-advantage effects. I mentioned clustering in Chapter 1, and how journalism helps perpetuate these clusters. Our opinions about artistic merit are the result of arbitrary contagion even more than our political ideas are. One person writes a book review; another person reads it and writes a commentary that uses the same arguments. Soon you have several hundred reviews that actually sum up in their contents to no more than two or three because there is so much overlap. For an anecdotal example read *Fire the Bastards!*, whose author, Jack Green, goes systematically through the reviews of William Gaddis's novel *The Recognitions*. Green shows clearly how book reviewers anchor on other reviews and reveals powerful mutual influence, even in their wording. This phenomenon is reminiscent of the herding of financial analysts I discussed in Chapter 10.

The advent of the modern media has accelerated these cumulative advantages. The sociologist Pierre Bourdieu noted a link between the increased concentration of success and the globalization of culture and economic life. But I am not trying to play sociologist here, only show that unpredictable elements can play a role in social outcomes.

Merton's cumulative-advantage idea has a more general precursor, "preferential attachment," which, reversing the chronology (though not the logic), I will present next. Merton was interested in the social aspect of knowledge, not in the dynamics of social randomness, so his studies were derived separately from research on the dynamics of randomness in more mathematical sciences.

Lingua Franca

The theory of preferential attachment is ubiquitous in its applications: it can explain why city size is from Extremistan, why vocabulary is concentrated among a small number of words, or why bacteria populations can vary hugely in size.

The scientists J. C. Willis and G. U. Yule published a landmark paper in *Nature* in 1922 called "Some Statistics of Evolution and Geographical Distribution in Plants and Animals, and Their Significance." Willis and Yule noted the presence in biology of the so-called power laws, atractable versions of the scalable randomness that I discussed in Chapter 3. These power laws (on which more technical information in the following chapters) had been noticed earlier by Vilfredo Pareto, who found that they applied to the distribution of income. Later, Yule presented a simple model showing how power laws can be generated. His point was as follows: Let's say species split in two at some constant rate, so that new species arise. The richer in species a genus is, the richer it will tend to get, with the same logic as the Mathew effect. Note the following caveat: in Yule's model the species never die out.

During the 1940s, a Harvard linguist, George Zipf, examined the properties of language and came up with an empirical regularity now known as Zipf's law, which, of course, is not a law (and if it were, it would not be Zipf's). It is just another way to think about the process of inequality. The mechanisms he described were as follows: the more you use a word, the less effortful you will find it to use that word again, so you borrow words from your private dictionary in proportion to their past use. This explains why out of the sixty thousand main words in English, only a few hundred constitute the bulk of what is used in writings, and even fewer appear regularly in conversation. Likewise, the more people aggregate in a particular city, the more likely a stranger will be to pick that city as his destination. The big get bigger and the small stay small, or get relatively smaller.

A great illustration of preferential attachment can be seen in the mushrooming use of English as a lingua franca—though not for its intrinsic qualities, but because people need to use one single language, or stick to one as much as possible, when they are having a conversation. So whatever language appears to have the upper hand will suddenly draw people in droves; its usage will spread like an epidemic, and other languages will be rapidly dislodged. I am often amazed to listen to conversations between people from two neighboring countries, say, between a Turk and an Iranian, or a Lebanese and a Cypriot, communicating in bad English, moving their hands for emphasis, searching for these words that come out of their throats at the cost of great physical effort. Even members of the Swiss Army use English (not French) as a lingua franca (it would be fun to listen). Consider that a very small minority of Americans of northern European descent is from England; traditionally the preponderant ethnic groups are of German, Irish, Dutch, French, and other northern European extraction. Yet because all these groups now use English as their main tongue, they have to study the roots of their adoptive tongue and develop a cultural association with parts of a particular wet island, along with its history, its traditions, and its customs!

Ideas and Contagions

The same model can be used for the contagions and concentration of ideas. But there are some restrictions on the nature of epidemics I must discuss here. Ideas do not spread without some form of structure. Recall the discussion in Chapter 4 about how we come prepared to make inferences. Just as we tend to generalize some matters but not others, so there seem to be "basins of attraction" directing us to certain beliefs. Some ideas will prove contagious, but not others; some forms of superstitions will spread, but not others; some types of religious beliefs will dominate, but not others. The anthropologist, cognitive scientist, and philosopher Dan Sperber has proposed the following idea on the epidemiology of representations. What people call "memes," ideas that spread and that compete with one another using people as carriers, are not truly like genes. Ideas spread because, alas, they have for carriers self-serving agents who are interested in them, and interested in distorting them in the replication process. You do not make a cake for the sake of merely replicating a recipe—you try to make *your* own cake, using ideas from others to improve it. We humans are not photocopiers. So contagious mental categories must be those in which we are prepared to believe, perhaps even programmed to believe. To be contagious, a mental category must agree with our nature.

NOBODY IS SAFE IN EXTREMISTAN

There is something extremely naïve about all these models of the dynamics of concentration I've presented so far, particularly the socioeconomic ones. For instance, although Merton's idea includes luck, it misses an additional layer of randomness. In all these models the winner stays a winner. Now, a loser might always remain a loser, but a winner could be unseated by someone new popping up out of nowhere. Nobody is safe.

Preferential-attachment theories are intuitively appealing, but they do not account for the possibility of being supplanted by newcomers—what every schoolchild knows as the decline of civilizations. Consider the logic of cities: How did Rome, with a population of 1.2 million in the first century A.D., end up with a population of twelve thousand in the third? How did Baltimore, once a principal American city, become a relic? And how did Philadelphia come to be overshadowed by New York?

A Brooklyn Frenchman

When I started trading foreign exchange, I befriended a fellow named Vincent who exactly resembled a Brooklyn trader, down to the mannerisms of Fat Tony, except that he spoke the French version of Brooklynese. Vincent taught me a few tricks. Among his sayings were "Trading may have princes, but nobody stays a king" and "The people you meet on the way up, you will meet again on the way down."

There were theories when I was a child about class warfare and struggles by innocent individuals against powerful monster-corporations capable of swallowing the world. Anyone with intellectual hunger was fed these theories, which were inherited from the Marxist belief that the tools of exploitation were self-feeding, that the powerful would grow more and more powerful, furthering the unfairness of the system. But one had only to look around to see that these large corporate monsters dropped like flies. Take a cross section of the dominant corporations at any particular time; many of them will be out of business a few decades later, while firms nobody ever heard of will have popped onto the scene from some garage in California or from some college dorm.

Consider the following sobering statistic. Of the five hundred largest U.S. companies in 1957, only seventy-four were still part of that select group, the Standard and Poor's 500, forty years later. Only a few had disappeared in mergers; the rest either shrank or went bust.

Interestingly, almost all these large corporations were located in the most capitalist country on earth, the United States. The more socialist a country's orientation, the easier it was for the large corporate monsters to stick around. Why did capitalism (and not socialism) destroy these ogres?

In other words, if you leave companies alone, they tend to get eaten up. Those in favor of economic freedom claim that beastly and greedy corporations pose no threat because competition keeps them in check. What I saw at the Wharton School convinced me that the real reason includes a large share of something else: chance.

But when people discuss chance (which they rarely do), they usually only look at their own luck. The luck *of others* counts greatly. Another corporation may luck out thanks to a blockbuster product and displace the current winners. Capitalism is, among other things, the revitalization of the world thanks to the opportunity to be lucky. Luck is the grand equalizer, because almost everyone can benefit from it. The socialist governments protected their monsters and, by doing so, killed potential newcomers in the womb.

Everything is transitory. Luck both made and unmade Carthage; it both made and unmade Rome.

I said earlier that randomness is bad, but it is not always so. Luck is far more egalitarian than even intelligence. If people were rewarded strictly according to their abilities, things would still be unfair—people don't choose their abilities. Randomness has the beneficial effect of reshuffling society's cards, knocking down the big guy.

In the arts, fads do the same job. A newcomer may benefit from a fad, as followers multiply thanks to a preferential attachment–style epidemic. Then, guess what? He too becomes history. It is quite interesting to look at the acclaimed authors of a particular era and see how many have dropped out of consciousness. It even happens in countries such as France where the government supports established reputations, just as it supports ailing large companies.

When I visit Beirut, I often spot in relatives' homes the remnants of a series of distinctively white-leather-bound "Nobel books." Some hyperactive salesman once managed to populate private libraries with these beautifully made volumes; many people buy books for decorative purposes and want a simple selection criterion. The criterion this series offered was one book by a Nobel winner in literature every year—a simple way to build the ultimate library. The series was supposed to be updated every year, but I presume the company went out of business in the eighties. I feel a pang every time I look at these volumes: Do you hear much today about Sully Prudhomme (the first recipient), Pearl Buck (an American woman), Romain Rolland, Anatole France (the last two were the most famous French authors of their generations), St. John Perse, Roger Martin du Gard, or Frédéric Mistral?

The Long Tail

I have said that nobody is safe in Extremistan. This has a converse: nobody is threatened with complete extinction either. Our current environment allows the little guy to bide his time in the antechamber of success—as long as there is life, there is hope.

This idea was recently revived by Chris Anderson, one of a very few who get the point that the dynamics of fractal concentration has another layer of randomness. He packaged it with his idea of the "long tail," about which in a moment. Anderson is lucky not to be a professional statistician (people who have had the misfortune of going through conventional statistical training think we live in Mediocristan). He was able to take a fresh look at the dynamics of the world.

True, the Web produces acute concentration. A large number of users visit just a few sites, such as Google, which, at the time of this writing, has total market dominance. At no time in history has a company grown so dominant so quickly—Google can service people from Nicaragua to southwestern Mongolia to the American West Coast, without having to worry about phone operators, shipping, delivery, and manufacturing. This is the ultimate winner-take-all case study.

People forget, though, that before Google, Alta Vista dominated the search-engine market. I am prepared to revise the Google metaphor by replacing it with a new name for future editions of this book.

What Anderson saw is that the Web causes something *in addition* to concentration. The Web enables the formation of a reservoir of proto-Googles waiting in the background. It also promotes the *inverse Google,* that is, it allows people with a technical specialty to find a small, stable audience.

Recall the role of the Web in Yevgenia Krasnova's success. Thanks to the Internet, she was able to bypass conventional publishers. Her publisher with the pink glasses would not even have been in business had it not been for the Web. Let's assume that Amazon.com does not exist, and that you have written a sophisticated book. Odds are that a very small bookstore that carries only 5,000 volumes will not be interested in letting your "beautifully crafted prose" occupy premium shelf space. And the megabookstore, such as the average American Barnes & Noble, might stock 130,000 volumes, which is still not sufficient to accommodate marginal titles. So your work is stillborn.

Not so with Web vendors. A Web bookstore can carry a near-infinite number of books since it need not have them physically in inventory. Actually, nobody needs to have them physically in inventory since they can remain in digital form until they are needed in print, an emerging business called print-on-demand.

So as the author of this little book, you can sit there, bide your time, be available in search engines, and perhaps benefit from an occasional epidemic. In fact, the quality of readership has improved markedly over the past few years thanks to the availability of these more sophisticated books. This is a fertile environment for diversity.*

Plenty of people have called me to discuss the idea of the long tail, which seems to be the exact opposite of the concentration implied by scalability. The long tail implies that the small guys, collectively, should control a large segment of culture and commerce, thanks to the niches and subspecialties that can now survive thanks to the Internet. But, strangely, it can also imply a large measure of inequality: a large base of small guys and a very small number of supergiants, together representing a share of the world's culture—with some of the small guys, on occasion, rising to knock out the winners. (This is thc "double tail": a large tail of the small guys, a small tail of the big guys.)

The role of the long tail is fundamental in changing the dynamics of success, destabilizing the well-seated winner, and bringing about another winner. In a snapshot this will always be Extremistan, always ruled by the concentration of type-2 randomness; but it will be an ever-changing Extremistan.

The long tail's contribution is not yet numerical; it is still confined to the Web and its small-scale online commerce. But consider how the long tail could affect the future of culture, information, and political life. It could free us from the dominant political parties, from the academic system, from the clusters of the press—anything that is currently in the hands of ossified, conceited, and self-serving authority. The long tail will help foster cognitive diversity. One highlight of the year 2006 was to find in my

* The Web's bottom-up feature is also making book reviewers more accountable. While writers were helpless and vulnerable to the arbitrariness of book reviews, which can distort their messages and, thanks to the confirmation bias, expose small irrelevant weak points in their text, they now have a much stronger hand. In place of the moaning letter to the editor, they can simply post their review of a review on the Web. If attacked ad hominem, they can reply ad hominem and go directly after the credibility of the reviewer, making sure that their statement shows rapidly in an Internet search or on Wikipedia, the bottom-up encyclopedia.

mailbox a draft manuscript of a book called *Cognitive Diversity: How Our Individual Differences Produce Collective Benefits,* by Scott Page. Page examines the effects of cognitive diversity on problem solving and shows how variability in views and methods acts like an engine for tinkering. It works like evolution. By subverting the big structures we also get rid of the Platonified *one way* of doing things—in the end, the bottom-up theory-free empiricist should prevail.

In sum, the long tail is a by-product of Extremistan that makes it somewhat less unfair: the world is made no less unfair for the little guy, but it now becomes extremely unfair for the big man. Nobody is truly established. The little guy is very subversive.

Naïve Globalization

We are gliding into disorder, but not necessarily bad disorder. This implies that we will see more periods of calm and stability, with most problems concentrated into a small number of Black Swans.

Consider the nature of past wars. The twentieth century was not the deadliest (in percentage of the total population), but it brought something new: the beginning of the Extremistan warfare—a small probability of a conflict degenerating into total decimation of the human race, a conflict from which nobody is safe anywhere.

A similar effect is taking place in economic life. I spoke about globalization in Chapter 3; it is here, but it is not all for the good: it creates interlocking fragility, while reducing volatility and giving the appearance of stability. In other words it creates devastating Black Swans. We have never lived before under the threat of a global collapse. Financial institutions have been merging into a smaller number of very large banks. Almost all banks are now interrelated. So the financial ecology is swelling into gigantic, incestuous, bureaucratic banks (often Gaussianized in their risk measurement)—when one falls, they all fall.* The increased concentration

* As if we did not have enough problems, banks are now more vulnerable to the Black Swan and the ludic fallacy than ever before with "scientists" among their staff taking care of exposures. The giant firm J. P. Morgan put the entire world at risk by introducing in the nineties RiskMetrics, a phony method aiming at managing people's risks, causing the generalized use of the ludic fallacy, and bringing Dr. Johns into power in place of the skeptical Fat Tonys. (A related method called "Value-at-Risk," which relies on the quantitative measurement of risk, has been spreading.) Likewise, the government-sponsored institution Fanny Mae, when I look at their risks, seems to be sitting on a barrel of dynamite, vulnerable to the

among banks seems to have the effect of making financial crisis less likely, but when they happen they are more global in scale and hit us very hard. We have moved from a diversified ecology of small banks, with varied lending policies, to a more homogeneous framework of firms that all resemble one another. True, we now have fewer failures, but when they occur . . . I shiver at the thought. I rephrase here: we will have fewer but more severe crises. The rarer the event, the less we know about its odds. It mean that we know less and less about the possibility of a crisis.

And we have some idea how such a crisis would happen. A network is an assemblage of elements called nodes that are somehow connected to one another by a link; the world's airports constitute a network, as does the World Wide Web, as do social connections and electricity grids. There is a branch of research called "network theory" that studies the organization of such networks and the links between their nodes, with such researchers as Duncan Watts, Steven Strogatz, Albert-Laszlo Barabasi, and many more. They all understand Extremistan mathematics and the inadequacy of the Gaussian bell curve. They have uncovered the following property of networks: there is a concentration among a few nodes that serve as central connections. Networks have a natural tendency to organize themselves around an extremely concentrated architecture: a few nodes are extremely connected; others barely so. The distribution of these connections has a scalable structure of the kind we will discuss in Chapters 15 and 16. Concentration of this kind is not limited to the Internet; it appears in social life (a small number of people are connected to others), in electricity grids, in communications networks. This seems to make networks more robust: random insults to most parts of the network will not be consequential since they are likely to hit a poorly connected spot. But it also makes networks more vulnerable to Black Swans. Just consider what would happen if there is a problem with a major node. The electricity blackout experienced in the northeastern United States during August 2003, with its consequential mayhem, is a perfect example of what could take place if one of the big banks went under today.

But banks are in a far worse situation than the Internet. The financial industry has no significant long tail! We would be far better off if there were a different ecology, in which financial institutions went bust on occasion and were rapidly replaced by new ones, thus mirroring the diversity

---

slightest hiccup. But not to worry: their large staff of scientists deemed these events "unlikely."

of Internet businesses and the resilience of the Internet economy. Or if there were a long tail of government officials and civil servants coming to reinvigorate bureaucracies.

REVERSALS AWAY FROM EXTREMISTAN

There is, inevitably, a mounting tension between our society, full of concentration, and our classical idea of aurea mediocritas, the golden mean, so it is conceivable that efforts may be made to reverse such concentration. We live in a society of one person, one vote, where progressive taxes have been enacted precisely to weaken the winners. Indeed, the rules of society can be easily rewritten by those at the bottom of the pyramid to prevent concentration from hurting them. But it does not require voting to do so—religion could soften the problem. Consider that before Christianity, in many societies the powerful had many wives, thus preventing those at the bottom from accessing wombs, a condition that is not too different from the reproductive exclusivity of alpha males in many species. But Christianity reversed this, thanks to the one man–one woman rule. Later, Islam came to limit the number of wives to four. Judaism, which had been polygenic, became monogamous in the Middle Ages. One can say that such a strategy has been successful—the institution of tightly monogamous marriage (with no official concubine, as in the Greco-Roman days), even when practiced the "French way," provides social stability since there is no pool of angry, sexually deprived men at the bottom fomenting a revolution just so they can have the chance to mate.

But I find the emphasis on economic inequality, at the expense of other types of inequality, extremely bothersome. Fairness is not exclusively an economic matter; it becomes less and less so when we are satisfying our basic material needs. It is pecking order that matters! The superstars will always be there. The Soviets may have flattened the economic structure, but they encouraged their own brand of übermensch. What is poorly understood, or denied (owing to its unsettling implications), is the absence of a role for the *average* in intellectual production. The disproportionate share of the very few in intellectual influence is even more unsettling than the unequal distribution of wealth—unsettling because, unlike the income gap, no social policy can eliminate it. Communism could conceal or compress income discrepancies, but it could not eliminate the superstar system in intellectual life.

It has even been shown, by Michael Marmot of the Whitehall Studies, that those at the top of the pecking order live longer, even when adjusting for disease. Marmot's impressive project shows how social rank alone can affect longevity. It was calculated that actors who win an Oscar tend to live on average about five years longer than their peers who don't. People live longer in societies that have flatter social gradients. Winners kill their peers as those in a steep social gradient live shorter lives, regardless of their economic condition.

I do not know how to remedy this (except through religious beliefs). Is insurance against your peers' demoralizing success possible? Should the Nobel Prize be banned? Granted the Nobel medal in economics has not been good for society or knowledge, but even those rewarded for *real* contributions in medicine and physics too rapidly displace others from our consciousness, and steal longevity away from them. Extremistan is here to stay, so we have to live with it, and find the tricks that make it more palatable.

Chapter Fifteen

THE BELL CURVE, THAT GREAT INTELLECTUAL FRAUD*

Not worth a pastis—Quételet's error—The average man is a monster—Let's deify it—Yes or no—Not so literary an experiment

Forget everything you heard in college statistics or probability theory. If you never took such a class, even better. Let us start from the very beginning.

THE GAUSSIAN AND THE MANDELBROTIAN

I was transiting through the Frankfurt airport in December 2001, on my way from Oslo to Zurich.

I had time to kill at the airport and it was a great opportunity for me to buy dark European chocolate, especially since I have managed to successfully convince myself that airport calories don't count. The cashier handed me, among other things, a ten deutschmark bill, an (illegal) scan of which can be seen on the next page. The deutschmark banknotes were going to be put out of circulation in a matter of days, since Europe was

* The nontechnical (or intuitive) reader can skip this chapter, as it goes into some details about the bell curve. Also, you can skip it if you belong to the category of fortunate people who do not know about the bell curve.

The last ten deutschmark bill, representing Gauss and, to his right, the bell curve of Mediocristan.

switching to the euro. I kept it as a valedictory. Before the arrival of the euro, Europe had plenty of national currencies, which was good for printers, money changers, and of course currency traders like this (more or less) humble author. As I was eating my dark European chocolate and wistfully looking at the bill, I almost choked. I suddenly noticed, for the first time, that there was something curious about it. The bill bore the portrait of Carl Friedrich Gauss and a picture of his Gaussian bell curve.

The striking irony here is that the last possible object that can be linked to the German currency is precisely such a curve: the reichsmark (as the currency was previously called) went from four per dollar to *four trillion* per dollar in the space of a few years during the 1920s, an outcome that tells you that the bell curve is meaningless as a description of the randomness in currency fluctuations. All you need to reject the bell curve is for such a movement to occur once, and only once—just consider the consequences. Yet there was the bell curve, and next to it Herr Professor Doktor Gauss, unprepossessing, a little stern, certainly not someone I'd want to spend time with lounging on a terrace, drinking pastis, and holding a conversation without a subject.

Shockingly, the bell curve is used as a risk-measurement tool by those regulators and central bankers who wear dark suits and talk in a boring way about currencies.

The Increase in the Decrease

The main point of the Gaussian, as I've said, is that most observations hover around the mediocre, the average; the odds of a deviation decline faster and faster (exponentially) as you move away from the average. If you must have only one single piece of information, this is the one: the dramatic increase in the speed of decline in the odds as you move away from the center, or the average. Look at the list below for an illustration of this. I am taking an example of a Gaussian quantity, such as height, and simplifying it a bit to make it more illustrative. Assume that the average height (men and women) is 1.67 meters, or 5 feet 7 inches. Consider what I call a *unit of deviation* here as 10 centimeters. Let us look at increments above 1.67 meters and consider the odds of someone being that tall.*

10 centimeters taller than the average (i.e., taller than 1.77 m, or 5 feet 10): 1 in 6.3

20 centimeters taller than the average (i.e., taller than 1.87 m, or 6 feet 2): 1 in 44

30 centimeters taller than the average (i.e., taller than 1.97 m, or 6 feet 6): 1 in 740

40 centimeters taller than the average (i.e., taller than 2.07 m, or 6 feet 9): 1 in 32,000

50 centimeters taller than the average (i.e., taller than 2.17 m, or 7 feet 1): 1 in 3,500,000

60 centimeters taller than the average (i.e., taller than 2.27 m, or 7 feet 5): 1 in 1,000,000,000

70 centimeters taller than the average (i.e., taller than 2.37 m, or 7 feet 9): 1 in 780,000,000,000

80 centimeters taller than the average (i.e., taller than 2.47 m, or 8 feet 1): 1 in 1,600,000,000,000,000

90 centimeters taller than the average (i.e., taller than 2.57 m, or 8 feet 5): 1 in 8,900,000,000,000,000,000

100 centimeters taller than the average (i.e., taller than 2.67 m, or 8 feet 9): 1 in 130,000,000,000,000,000,000,000

. . . and,

* I have fudged the numbers a bit for simplicity's sake.

110 centimeters taller than the average (i.e., taller than 2.77 m, or 9 feet 1): 1 in 36,000,000,000,000,000,000,000,000,000,000,000,000,000,000,000,000,000,000,000,000,000,000,000,000,000,000,000,000,000.

Note that soon after, I believe, 22 deviations, or 220 centimeters taller than the average, the odds reach a googol, which is 1 with 100 zeroes behind it.

The point of this list is to illustrate the acceleration. Look at the difference in odds between 60 and 70 centimeters taller than average: for a mere increase of four inches, we go from one in 1 billion people to one in 780 billion! As for the jump between 70 and 80 centimeters: an additional 4 inches above the average, we go from one in 780 billion to one in 1.6 million billion!*

This precipitous decline in the odds of encountering something is what allows you to ignore outliers. Only one curve can deliver this decline, and it is the bell curve (and its nonscalable siblings).

The Mandelbrotian

By comparison, look at the odds of being rich in Europe. Assume that wealth there is scalable, i.e., Mandelbrotian. (This is not an accurate description of wealth in Europe; it is simplified to emphasize the logic of scalable distribution.)†

Scalable Wealth Distribution

People with a net worth higher than €1 million: 1 in 62.5

Higher than €2 million: 1 in 250

Higher than €4 million: 1 in 1,000

* One of the most misunderstood aspects of a Gaussian is its fragility and vulnerability in the estimation of tail events. The odds of a 4 sigma move are twice that of a 4.15 sigma. The odds of a 20 sigma are a trillion times higher than those of a 21 sigma! It means that a small measurement error of the sigma will lead to a massive underestimation of the probability. We can be a trillion times wrong about some events.

† My main point, which I repeat in some form or another throughout Part Three, is as follows. Everything is made easy, conceptually, when you consider that there are two, and only two, possible paradigms: nonscalable (like the Gaussian) and *other* (such as Mandebrotian randomness). The rejection of the application of the nonscalable is sufficient, as we will see later, *to eliminate a certain vision of the world.* This is like negative empiricism: I know a lot by determining what is wrong.

Higher than €8 million: 1 in 4,000
Higher than €16 million: 1 in 16,000
Higher than €32 million: 1 in 64,000
Higher than €320 million: 1 in 6,400,000

The speed of the decrease here remains constant (or does not decline)! When you double the amount of money you cut the incidence by a factor of four, no matter the level, whether you are at €8 million or €16 million. This, in a nutshell, illustrates the difference between Mediocristan and Extremistan.

Recall the comparison between the scalable and the nonscalable in Chapter 3. Scalability means that there is no headwind to slow you down.

Of course, Mandelbrotian Extremistan can take many shapes. Consider wealth in an extremely concentrated version of Extremistan; there, if you double the wealth, you halve the incidence. The result is quantitatively different from the above example, but it obeys the same logic.

Fractal Wealth Distribution with Large Inequalities

People with a net worth higher than €1 million: 1 in 63
Higher than €2 million: 1 in 125
Higher than €4 million: 1 in 250
Higher than €8 million: 1 in 500
Higher than €16 million: 1 in 1,000
Higher than €32 million: 1 in 2,000
Higher than €320 million: 1 in 20,000
Higher than €640 million: 1 in 40,000

If wealth were Gaussian, we would observe the following divergence away from €1 million.

Wealth Distribution Assuming a Gaussian Law

People with a net worth higher than €1 million: 1 in 63
Higher than €2 million: 1 in 127,000
Higher than €3 million: 1 in 14,000,000,000
Higher than €4 million: 1 in 886,000,000,000,000,000
Higher than €8 million:
1 in 16,000,000,000,000,000,000,000,000,000,000,000
Higher than €16 million: 1 in . . . *none of my computers is capable of handling the computation.*

What I want to show with these lists is the qualitative difference in the paradigms. As I have said, the second paradigm is scalable; it has no headwind. Note that another term for the scalable is power laws.

Just knowing that we are in a power-law environment does not tell us much. Why? Because we have to measure the coefficients in real life, which is much harder than with a Gaussian framework. Only the Gaussian yields its properties rather rapidly. The method I propose is a general way of viewing the world rather than a precise solution.

What to Remember

Remember this: the Gaussian–bell curve variations face a headwind that makes probabilities drop at a faster and faster rate as you move away from the mean, while "scalables," or Mandelbrotian variations, do not have such a restriction. That's pretty much most of what you need to know.*

Inequality

Let us look more closely at the nature of inequality. In the Gaussian framework, inequality decreases as the deviations get larger—caused by the increase in the rate of decrease. Not so with the scalable: inequality stays the same throughout. The inequality among the superrich is the same as the inequality among the simply rich—it does not slow down.†

* Note that variables may not be infinitely scalable; there could be a very, very remote upper limit—but we do not know where it is so we treat a given situation as if it were infinitely scalable. Technically, you cannot sell more of one book than there are denizens of the planet—but that upper limit is large enough to be treated as if it didn't exist. Furthermore, who knows, by repackaging the book, you might be able to sell it to a person twice, or get that person to watch the same movie several times.

† As I was revising this draft, in August 2006, I stayed at a hotel in Dedham, Massachusetts, near one of my children's summer camps. There, I was a little intrigued by the abundance of weight-challenged people walking around the lobby and causing problems with elevator backups. It turned out that the annual convention of NAFA, the National Association for Fat Acceptance, was being held there. As most of the members were extremely overweight, I was not able to figure out which delegate was the heaviest: some form of equality prevailed among the very heavy (someone much heavier than the persons I saw would have been dead). I am sure that at the NARA convention, the National Association for Rich Acceptance, one person would dwarf the others, and, even among the superrich, a very small percentage would represent a large section of the total wealth.

Consider this effect. Take a random sample of any two people from the U.S. population who jointly earn $1 million per annum. What is the most likely breakdown of their respective incomes? In Mediocristan, the most likely combination is half a million each. In Extremistan, it would be $50,000 and $950,000.

The situation is even more lopsided with book sales. If I told you that two authors sold a total of a million copies of their books, the most likely combination is 993,000 copies sold for one and 7,000 for the other. This is far more likely than that the books each sold 500,000 copies. *For any large total, the breakdown will be more and more asymmetric.*

Why is this so? The height problem provides a comparison. If I told you that the total height of two people is fourteen feet, you would identify the most likely breakdown as seven feet each, not two feet and twelve feet; not even eight feet and six feet! Persons taller than eight feet are so rare that such a combination would be impossible.

Extremistan and the 80/20 Rule

Have you ever heard of the 80/20 rule? It is the common signature of a power law—actually it is how it all started, when Vilfredo Pareto made the observation that 80 percent of the land in Italy was owned by 20 percent of the people. Some use the rule to imply that 80 percent of the work is done by 20 percent of the people. Or that 80 percent worth of effort contributes to only 20 percent of results, and vice versa.

As far as axioms go, this one wasn't phrased to impress you the most: it could easily be called the 50/01 rule, that is, 50 percent of the work comes from 1 percent of the workers. This formulation makes the world look even more unfair, yet the two formulae are exactly the same. How? Well, if there is inequality, then those who constitute the 20 percent in the 80/20 rule also contribute unequally—only a few of them deliver the lion's share of the results. This trickles down to about one in a hundred contributing a little more than half the total.

The 80/20 rule is only metaphorical; it is not a rule, even less a rigid law. In the U.S. book business, the proportions are more like 97/20 (i.e., 97 percent of book sales are made by 20 percent of the authors); it's even worse if you focus on literary nonfiction (twenty books of close to eight thousand represent half the sales).

Note here that it is not all uncertainty. In some situations you may have a concentration, of the 80/20 type, with very predictable and tractable properties, which enables clear decision making, because you can identify *beforehand* where the meaningful 20 percent are. These situations are very easy to control. For instance, Malcolm Gladwell wrote in an article in *The New Yorker* that most abuse of prisoners is attributable to a very small number of vicious guards. Filter those guards out and your rate of prisoner abuse drops dramatically. (In publishing, on the other hand, you do not know beforehand which book will bring home the bacon. The same with wars, as you do not know beforehand which conflict will kill a portion of the planet's residents.)

Grass and Trees

I'll summarize here and repeat the arguments previously made throughout the book. Measures of uncertainty that are based on the bell curve simply disregard the possibility, and the impact, of sharp jumps or discontinuities and are, therefore, inapplicable in Extremistan. Using them is like focusing on the grass and missing out on the (gigantic) trees. Although unpredictable large deviations are rare, they cannot be dismissed as outliers because, cumulatively, their impact is so dramatic.

The traditional Gaussian way of looking at the world begins by focusing on the ordinary, and then deals with exceptions or so-called outliers as ancillaries. But there is a second way, which takes the exceptional as a starting point and treats the ordinary as subordinate.

I have emphasized that there are two varieties of randomness, qualitatively different, like air and water. One does not care about extremes; the other is severely impacted by them. One does not generate Black Swans; the other does. We cannot use the same techniques to discuss a gas as we would use with a liquid. And if we could, we wouldn't call the approach "an approximation." A gas does not "approximate" a liquid.

We can make good use of the Gaussian approach in variables for which there is a rational reason for the largest not to be too far away from the average. If there is gravity pulling numbers down, or if there are physical limitations preventing very large observations, we end up in Mediocristan. If there are strong forces of equilibrium bringing things back rather rapidly after conditions diverge from equilibrium, then again you can use the Gaussian approach. Otherwise, fuhgedaboudit. This is why much of economics is based on the notion of equilibrium: among other benefits, it allows you to treat economic phenomena as Gaussian.

Note that I am not telling you that the Mediocristan type of randomness does not allow for *some* extremes. But it tells you that they are so rare that they do not play a significant role in the total. The effect of such extremes is pitifully small and decreases as your population gets larger.

To be a little bit more technical here, if you have an assortment of giants and dwarfs, that is, observations several orders of magnitude apart, you could still be in Mediocristan. How? Assume you have a sample of one thousand people, with a large spectrum running from the dwarf to the giant. You are likely to see many giants in your sample, not a rare occasional one. Your average will not be impacted by the occasional additional giant because some of these giants are expected to be part of your sample, and your average is likely to be high. In other words, the largest observation cannot be too far away from the average. The average will always contain both kinds, giants and dwarves, so that neither should be too rare—unless you get a megagiant or a microdwarf on very rare occasion. This would be Mediocristan with a large unit of deviation.

Note once again the following principle: the rarer the event, the higher the error in our estimation of its probability—even when using the Gaussian.

Let me show you how the Gaussian bell curve sucks randomness out of life—which is why it is popular. We like it because it allows certainties! How? Through averaging, as I will discuss next.

How Coffee Drinking Can Be Safe

Recall from the Mediocristan discussion in Chapter 3 that no single observation will impact your total. This property will be more and more significant as your population increases in size. The averages will become more and more stable, to the point where all samples will look alike.

I've had plenty of cups of coffee in my life (it's my principal addiction). I have never seen a cup jump two feet from my desk, nor has coffee spilled spontaneously on this manuscript without intervention (even in Russia). Indeed, it will take more than a mild coffee addiction to witness such an event; it would require more lifetimes than is perhaps conceivable—the odds are so small, one in so many zeroes, that it would be impossible for me to write them down in my free time.

Yet physical reality makes it possible for my coffee cup to jump—very unlikely, but possible. Particles jump around all the time. How come the coffee cup, itself composed of jumping particles, does not? The reason is, simply, that for the cup to jump would require that all of the particles

FIGURE 7: How the Law of Large Numbers Works

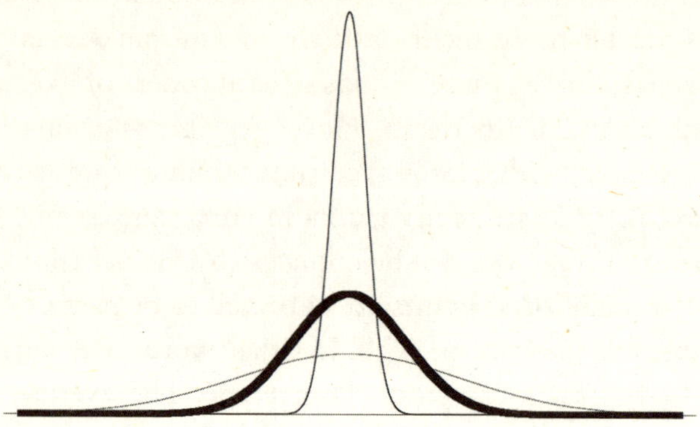

In Mediocristan, as your sample size increases, the observed average will present itself with less and less dispersion—as you can see, the distribution will be narrower and narrower. This, in a nutshell, is how everything in statistical theory works (or is supposed to work). Uncertainty in Mediocristan vanishes under averaging. This illustrates the hackneyed "law of large numbers."

jump in the *same* direction, and do so in lockstep several times in a row (with a compensating move of the table in the opposite direction). All several trillion particles in my coffee cup are not going to jump in the same direction; this is not going to happen in the lifetime of this universe. So I can safely put the coffee cup on the edge of my writing table and worry about more serious sources of uncertainty.

The safety of my coffee cup illustrates how the randomness of the Gaussian is tamable by averaging. If my cup were one large particle, or acted as one, then its jumping would be a problem. But my cup is the sum of trillions of very small particles.

Casino operators understand this well, which is why they never (if they do things right) lose money. They simply do not let one gambler make a massive bet, instead preferring to have plenty of gamblers make series of bets of limited size. Gamblers may bet a total of $20 million, but you needn't worry about the casino's health: the bets run, say, $20 on average; the casino caps the bets at a maximum that will allow the casino owners to sleep at night. So the variations in the casino's returns are going to be ridiculously small, no matter the total gambling activity. You will not see anyone leaving the casino with $1 billion—in the lifetime of this universe.

The above is an application of the supreme law of Mediocristan: when you have plenty of gamblers, no single gambler will impact the total more than minutely.

The consequence of this is that variations around the average of the Gaussian, also called "errors," are not truly worrisome. They are small and they wash out. They are domesticated fluctuations around the mean.

Love of Certainties

If you ever took a (dull) statistics class in college, did not understand much of what the professor was excited about, and wondered what "standard deviation" meant, there is nothing to worry about. The notion of standard deviation is meaningless outside of Mediocristan. Clearly it would have been more beneficial, and certainly more entertaining, to have taken classes in the neurobiology of aesthetics or postcolonial African dance, and this is easy to see empirically.

Standard deviations do not exist outside the Gaussian, or if they do exist they do not matter and do not explain much. But it gets worse. The Gaussian family (which includes various friends and relatives, such as the Poisson law) are the only class of distributions that the standard deviation (and the average) is sufficient to describe. You need nothing else. The bell curve satisfies the reductionism of the deluded.

There are other notions that have little or no significance outside of the Gaussian: *correlation* and, worse, *regression*. Yet they are deeply ingrained in our methods; it is hard to have a business conversation without hearing the word *correlation*.

To see how meaningless correlation can be outside of Mediocristan, take a historical series involving two variables that are patently from Extremistan, such as the bond and the stock markets, or two securities prices, or two variables like, say, changes in book sales of children's books in the United States, and fertilizer production in China; or real-estate prices in New York City and returns of the Mongolian stock market. Measure correlation between the pairs of variables in different subperiods, say, for 1994, 1995, 1996, etc. The correlation measure will be likely to exhibit severe instability; it will depend on the period for which it was computed. Yet people talk about correlation as if it were something real, making it tangible, investing it with a physical property, reifying it.

The same illusion of concreteness affects what we call "standard" deviations. Take any series of historical prices or values. Break it up into subsegments and measure its "standard" deviation. Surprised? Every sample will yield a different "standard" deviation. Then why do people talk about standard deviations? Go figure.

Note here that, as with the narrative fallacy, when you look at past data and compute one single correlation or standard deviation, you do not notice such instability.

How to Cause Catastrophes

If you use the term *statistically significant,* beware of the illusions of certainties. Odds are that someone has looked at his observation errors and assumed that they were Gaussian, which necessitates a Gaussian context, namely, Mediocristan, for it to be acceptable.

To show how endemic the problem of misusing the Gaussian is, and how dangerous it can be, consider a (dull) book called *Catastrophe* by Judge Richard Posner, a prolific writer. Posner bemoans civil servants' misunderstandings of randomness and recommends, among other things, that government policy makers learn statistics . . . from economists. Judge Posner appears to be trying to foment catastrophes. Yet, in spite of being one of those people who should spend more time reading and less time writing, he can be an insightful, deep, and original thinker; like many people, he just isn't aware of the distinction between Mediocristan and Extremistan, and he believes that statistics is a "science," never a fraud. If you run into him, please make him aware of these things.

QUÉTELET'S AVERAGE MONSTER

This monstrosity called the Gaussian bell curve is not Gauss's doing. Although he worked on it, he was a mathematician dealing with a theoretical point, not making claims about the structure of reality like statistical-minded scientists. G. H. Hardy wrote in "A Mathematician's Apology":

> The "real" mathematics of the "real" mathematicians, the mathematics of Fermat and Euler and Gauss and Abel and Riemann, is almost wholly "useless" (and this is as true of "applied" as of "pure" mathematics).

As I mentioned earlier, the bell curve was mainly the concoction of a gambler, Abraham de Moivre (1667–1754), a French Calvinist refugee who spent much of his life in London, though speaking heavily accented English. But it is Quételet, not Gauss, who counts as one of the most destructive fellows in the history of thought, as we will see next.

Adolphe Quételet (1796–1874) came up with the notion of a physically average human, *l'homme moyen.* There was nothing *moyen* about Quételet, "a man of great creative passions, a creative man full of energy." He wrote poetry and even coauthored an opera. The basic problem with Quételet was that he was a mathematician, not an empirical scientist, but he did not know it. He found harmony in the bell curve.

The problem exists at two levels. *Primo,* Quételet had a normative idea, to make the world fit his average, in the sense that the average, to him, was the "normal." It would be wonderful to be able to ignore the contribution of the unusual, the "nonnormal," the Black Swan, to the total. But let us leave that dream for utopia.

Secondo, there was a serious associated empirical problem. Quételet saw bell curves everywhere. He was blinded by bell curves and, I have learned, again, once you get a bell curve in your head it is hard to get it out. Later, Frank Ysidro Edgeworth would refer to Quételesmus as the grave mistake of seeing bell curves everywhere.

Golden Mediocrity

Quételet provided a much needed product for the ideological appetites of his day. As he lived between 1796 and 1874, so consider the roster of his contemporaries: Saint-Simon (1760–1825), Pierre-Joseph Proudhon (1809–1865), and Karl Marx (1818–1883), each the source of a different version of socialism. Everyone in this post-Enlightenment moment was longing for the aurea mediocritas, the golden mean: in wealth, height, weight, and so on. This longing contains some element of wishful thinking mixed with a great deal of harmony and . . . Platonicity.

I always remember my father's injunction that *in medio stat virtus,* "virtue lies in moderation." Well, for a long time that was the ideal; mediocrity, in that sense, was even deemed golden. All-embracing mediocrity.

But Quételet took the idea to a different level. Collecting statistics, he started creating standards of "means." Chest size, height, the weight of babies at birth, very little escaped his *standards*. Deviations from the norm, he found, became exponentially more rare as the magnitude of the deviation increased. Then, having conceived of this idea of the physical characteristics of *l'homme moyen,* Monsieur Quételet switched to social matters. *L'homme moyen* had his habits, his consumption, his methods.

Through his construct of *l'homme moyen physique* and *l'homme moyen moral*, the physically and morally average man, Quételet created a range of deviance from the average that positions all people either to the left or right of center and, truly, punishes those who find themselves occupying the extreme left or right of the statistical bell curve. They became *abnormal*. How this inspired Marx, who cites Quételet regarding this concept of an average or normal man, is obvious: "Societal deviations in terms of the distribution of wealth for example, must be minimized," he wrote in *Das Kapital*.

One has to give some credit to the scientific establishment of Quételet's day. They did not buy his arguments at once. The philosopher/mathematician/economist Augustin Cournot, for starters, did not believe that one could establish a standard human on purely quantitative grounds. Such a standard would be dependent on the attribute under consideration. A measurement in one province may differ from that in another province. Which one should be the standard? *L'homme moyen* would be a monster, said Cournot. I will explain his point as follows.

Assuming there is something desirable in being an average man, he must have an unspecified specialty in which he would be more gifted than other people—he cannot be average in everything. A pianist would be better on average at playing the piano, but worse than the norm at, say, horseback riding. A draftsman would have better drafting skills, and so on. *The notion of a man deemed average is different from that of a man who is average in everything he does.* In fact, an exactly average human would have to be half male and half female. Quételet completely missed that point.

God's Error

A much more worrisome aspect of the discussion is that in Quételet's day, the name of the Gaussian distribution was *la loi des erreurs*, the law of errors, since one of its earliest applications was the distribution of errors in astronomic measurements. Are you as worried as I am? Divergence from the mean (here the median as well) was treated precisely as an error! No wonder Marx fell for Quételet's ideas.

This concept took off very quickly. The *ought* was confused with the *is,* and this with the imprimatur of science. The notion of the average man is steeped in the culture attending the birth of the European middle class, the nascent post-Napoleonic shopkeeper's culture, chary of excessive wealth and intellectual brilliance. In fact, the dream of a society with compressed outcomes is assumed to correspond to the aspirations of a rational human being facing a genetic lottery. If you had to pick a society to be born into for your next life, but could not know which outcome awaited you, it is assumed you would probably take no gamble; you would like to belong to a society without divergent outcomes.

One entertaining effect of the glorification of mediocrity was the creation of a political party in France called Poujadism, composed initially of a grocery-store movement. It was the warm huddling together of the semi-favored hoping to see the rest of the universe compress itself into their rank—a case of non-proletarian revolution. It had a grocery-store-owner mentality, down to the employment of the mathematical tools. Did Gauss provide the mathematics for the shopkeepers?

Poincaré to the Rescue

Poincaré himself was quite suspicious of the Gaussian. I suspect that he felt queasy when it and similar approaches to modeling uncertainty were presented to him. Just consider that the Gaussian was initially meant to measure astronomic errors, and that Poincaré's ideas of modeling celestial mechanics were fraught with a sense of deeper uncertainty.

Poincaré wrote that one of his friends, an unnamed "eminent physicist," complained to him that physicists tended to use the Gaussian curve because they thought mathematicians believed it a mathematical necessity; mathematicians used it because they believed that physicists found it to be an empirical fact.

Eliminating Unfair Influence

Let me state here that, except for the grocery-store mentality, I truly believe in the value of middleness and mediocrity—what humanist does not want to minimize the discrepancy between humans? Nothing is more repugnant than the inconsiderate ideal of the Übermensch! My true problem is epistemological. Reality is not Mediocristan, so we should learn to live with it.

"The Greeks Would Have Deified It"

The list of people walking around with the bell curve stuck in their heads, thanks to its Platonic purity, is incredibly long.

Sir Francis Galton, Charles Darwin's first cousin and Erasmus Darwin's grandson, was perhaps, along with his cousin, one of the last independent gentlemen scientists—a category that also included Lord Cavendish, Lord Kelvin, Ludwig Wittgenstein (in his own way), and to some extent, our überphilosopher Bertrand Russell. Although John Maynard Keynes was not quite in that category, his thinking epitomizes it. Galton lived in the Victorian era when heirs and persons of leisure could, among other choices, such as horseback riding or hunting, become thinkers, scientists, or (for those less gifted) politicians. There is much to be wistful about in that era: the authenticity of someone doing science for science's sake, without direct career motivations.

Unfortunately, doing science for the love of knowledge does not necessarily mean you will head in the right direction. Upon encountering and absorbing the "normal" distribution, Galton fell in love with it. He was said to have exclaimed that if the Greeks had known about it, they would have deified it. His enthusiasm may have contributed to the prevalence of the use of the Gaussian.

Galton was blessed with no mathematical baggage, but he had a rare obsession with measurement. He did not know about the law of large numbers, but rediscovered it from the data itself. He built the quincunx, a pinball machine that shows the development of the bell curve—on which, more in a few paragraphs. True, Galton applied the bell curve to areas like genetics and heredity, in which its use was justified. But his enthusiasm helped thrust nascent statistical methods into social issues.

"Yes/No" Only Please

Let me discuss here the extent of the damage. If you're dealing with qualitative inference, such as in psychology or medicine, looking for yes/no answers to which magnitudes don't apply, then you can assume you're in Mediocristan without serious problems. The impact of the improbable cannot be too large. You have cancer or you don't, you are pregnant or you are not, et cetera. Degrees of deadness or pregnancy are not relevant (unless you are dealing with epidemics). But if you are dealing with aggregates, where magnitudes do matter, such as income, your wealth, return on a portfolio, or book sales, then you will have a problem and get the wrong distribution if you use the Gaussian, as it does not belong there. One single number can disrupt all your averages; one single loss can eradicate a century of profits. You can no longer say "this is an exception." The statement "Well, I can lose money" is not informational unless you can attach a quantity to that loss. You can lose all your net worth or you can lose a fraction of your daily income; there is a difference.

This explains why empirical psychology and its insights on human nature, which I presented in the earlier parts of this book, are robust to the mistake of using the bell curve; they are also lucky, since most of their variables allow for the application of conventional Gaussian statistics. When measuring how many people in a sample have a bias, or make a mistake, these studies generally elicit a yes/no type of result. No single observation, by itself, can disrupt their overall findings.

I will next proceed to a sui generis presentation of the bell-curve idea from the ground up.

A (LITERARY) THOUGHT EXPERIMENT ON WHERE THE BELL CURVE COMES FROM

Consider a pinball machine like the one shown in Figure 8. Launch 32 balls, assuming a well-balanced board so that the ball has equal odds of falling right or left at any juncture when hitting a pin. Your expected outcome is that many balls will land in the center columns and that the number of balls will decrease as you move to the columns away from the center.

Next, consider a gedanken, a thought experiment. A man flips a coin and after each toss he takes a step to the left or a step to the right, depending on whether the coin came up heads or tails. This is called the random walk, but it does not necessarily concern itself with walking. You could identically say that instead of taking a step to the left or to the right, you would win or lose $1 at every turn, and you will keep track of the cumulative amount that you have in your pocket.

Assume that I set you up in a (legal) wager where the odds are neither in your favor nor against you. Flip a coin. Heads, you make $1, tails, you lose $1.

At the first flip, you will either win or lose.

At the second flip, the number of possible outcomes doubles. Case one:

FIGURE 8: THE QUINCUNX (SIMPLIFIED)—A PINBALL MACHINE

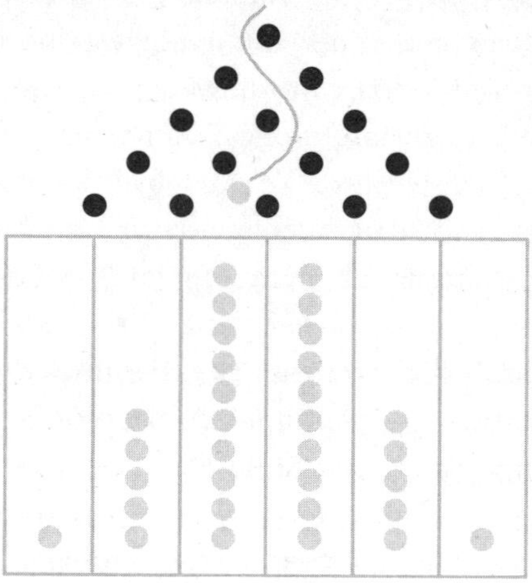

Drop balls that, at every pin, randomly fall right or left. Above is the most probable scenario, which greatly resembles the bell curve (a.k.a. Gaussian disribution). *Courtesy of Alexander Taleb.*

win, win. Case two: win, lose. Case three: lose, win. Case four: lose, lose. Each of these cases has equivalent odds, the combination of a single win and a single loss has an incidence twice as high because cases two and three, win-lose and lose-win, amount to the same outcome. And that is the key for the Gaussian. So much in the middle washes out—and we will see that there is a lot in the middle. So, if you are playing for $1 a round, after two rounds you have a 25 percent chance of making or losing $2, but a 50 percent chance of breaking even.

Let us do another round. The third flip again doubles the number of cases, so we face eight possible outcomes. Case 1 (it was win, win in the second flip) branches out into win, win, win and win, win, lose. We add a win or lose to the end of each of the previous results. Case 2 branches out into win, lose, win and win, lose, lose. Case 3 branches out into lose, win, win and lose, win, lose. Case 4 branches out into lose, lose, win and lose, lose, lose.

We now have eight cases, all equally likely. Note that again you can group the middling outcomes where a win cancels out a loss. (In Galton's quincunx, situations where the ball falls left and then falls right, or vice versa, dominate so you end up with plenty in the middle.) The net, or cumulative, is the following: 1) *three wins;* 2) two wins, one loss, net *one win;* 3) two wins, one loss, net *one win;* 4) one win, two losses, net *one loss;* 5) two wins, one loss, net *one win;* 6) two losses, one win, net *one loss;* 7) two losses, one win, net *one loss;* and, finally, 8) *three losses.*

Out of the eight cases, the case of three wins occurs once. The case of three losses occurs once. The case of one net loss (one win, two losses) occurs three times. The case of one net win (one loss, two wins) occurs three times.

Play one more round, the fourth. There will be sixteen equally likely outcomes. You will have one case of four wins, one case of four losses, four cases of two wins, four cases of two losses, and six break-even cases.

The quincunx (its name is derived from the Latin for five) in the pinball example shows the fifth round, with sixty-four possibilities, easy to track. Such was the concept behind the quincunx used by Francis Galton. Galton was both insufficiently lazy and a bit too innocent of mathematics; instead of building the contraption, he could have worked with simpler algebra, or perhaps undertaken a thought experiment like this one.

Let's keep playing. Continue until you have forty flips. You can perform them in minutes, but we will need a calculator to work out the number of outcomes, which are taxing to our simple thought method. You will have about 1,099,511,627,776 possible combinations—more than one thousand billion. Don't bother doing the calculation manually, it is two multiplied by itself forty times, since each branch doubles at every juncture. (Recall that we added a win and a lose at the end of the alternatives of the third round to go to the fourth round, thus doubling the number of alternatives.) Of these combinations, only one will be up forty, and only one will be down forty. The rest will hover around the middle, here zero.

We can already see that in this type of randomness extremes are exceedingly rare. One in 1,099,511,627,776 is up forty out of forty tosses. If you perform the exercise of forty flips once per hour, the odds of getting 40 ups in a row are so small that it would take quite a bit of forty-flip trials to see it. Assuming you take a few breaks to eat, argue with your friends and roommates, have a beer, and sleep, you can expect to wait close to four million lifetimes to get a 40-up outcome (or a 40-down outcome) just once. And consider the following. Assume you play one additional round, for a total of 41; to get 41 straight heads would take eight million lifetimes! Going from 40 to 41 halves the odds. This is a key at-

FIGURE 9: NUMBERS OF WINS TOSSED

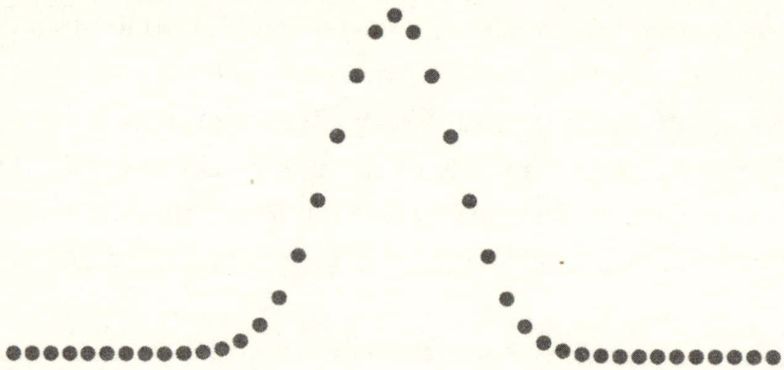

Result of forty tosses. We see the proto-bell curve emerging.

tribute of the nonscalable framework to analyzing randomness: extreme deviations decrease at an increasing rate. You can expect to toss 50 heads in a row once in four billion lifetimes!

We are not yet fully in a Gaussian bell curve, but we are getting dangerously close. This is still proto-Gaussian, but you can see the gist. (Actually, you will never encounter a Gaussian in its purity since it is a Platonic form—you just get closer but cannot attain it.) However, as you can see in Figure 9, the familiar bell shape is starting to emerge.

How do we get even closer to the perfect Gaussian bell curve? By refining the flipping process. We can either flip 40 times for $1 a flip or 4,000 times for ten cents a flip, and add up the results. Your expected risk is about the same in both situations—and that is a trick. The equivalence in the two sets of flips has a little nonintuitive hitch. We multiplied the number of bets by 100, but divided the bet size by 10—don't look for a reason now, just assume that they are "equivalent." The overall risk is equivalent, but now we have opened up the possibility of winning or losing 400 times in a row. The odds are about one in 1 with 120 zeroes after it, that is, one in 1,000,000,000,000,000,000,000,000,000,000,000,000, 000,000,000,000,000,000,000,000,000,000,000,000,000,000,000, 000,000,000,000,000,000,000,000,000,000,000,000 times.

Continue the process for a while. We go from 40 tosses for $1 each to 4,000 tosses for 10 cents, to 400,000 tosses for 1 cent, getting close and closer to a Gaussian. Figure 10 shows results spread between -40 and 40, namely eighty plot points. The next one would bring that up to 8,000 points.

FIGURE 10: A MORE ABSTRACT VERSION: PLATO'S CURVE

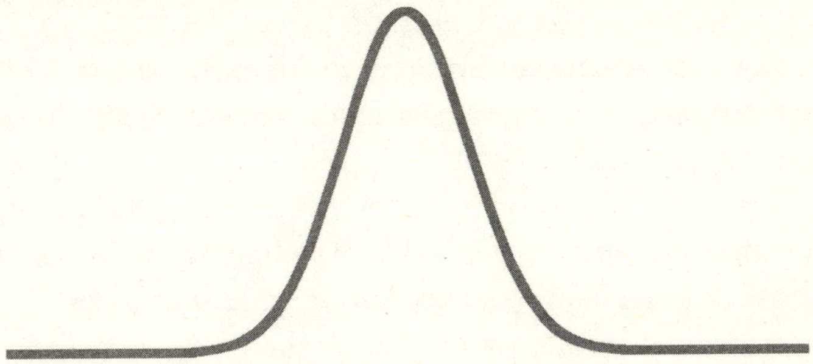

An infinite number of tosses.

Let's keep going. We can flip 4,000 times staking a tenth of a penny. How about 400,000 times at 1/1000 of a penny? As a Platonic form, the pure Gaussian curve is principally what happens when he have an infinity of tosses per round, with each bet infinitesimally small. Do not bother trying to visualize the results, or even make sense out of them. We can no longer talk about an "infinitesimal" bet size (since we have an infinity of these, and we are in what mathematicians call a continuous framework). The good news is that there is a substitute.

We have moved from a simple bet to something completely abstract. We have moved from observations into the realm of mathematics. In mathematics things have a purity to them.

Now, something completely abstract is not supposed to exist, so *please do not even make an attempt to understand Figure 10.* Just be aware of its use. Think of it as a thermometer: you are not supposed to understand what the temperature *means* in order to talk about it. You just need to know the correspondence between temperature and comfort (or some other empirical consideration). Sixty degrees corresponds to pleasant weather; ten below is not something to look forward to. You don't necessarily care about the actual speed of the collisions among particles that more technically explains temperature. Degrees are, in a way, a means for your mind to translate some external phenomena into a number. Likewise, the Gaussian bell curve is set so that 68.2 percent of the observations fall between minus one and plus one standard deviations away from the average. I repeat: do not even try to understand whether *standard deviation* is *average deviation*—it is not, and a large (too large) number of people using the word *standard deviation* do not understand this point. Standard deviation is just a number that you scale things to, a matter of mere correspondence *if phenomena were Gaussian*.

These standard deviations are often nicknamed "sigma." People also talk about "variance" (same thing: variance is the square of the sigma, i.e., of the standard deviation).

Note the symmetry in the curve. You get the same results whether the sigma is positive or negative. The odds of falling below -4 sigmas are the same as those of exceeding 4 sigmas, here 1 in 32,000 times.

As the reader can see, the main point of the Gaussian bell curve is, as I have been saying, that most observations hover around the mediocre, the mean, while the odds of a deviation decline faster and faster (exponentially) as you move away from the mean. If you need to retain one single piece of information, just remember this dramatic speed of decrease in the odds as you move away from the average. Outliers are increasingly unlikely. You can safely ignore them.

This property also generates the supreme law of Mediocristan: given the paucity of large deviations, their contribution to the total will bc vanishingly small.

In the height example earlier in this chapter, I used units of deviations of ten centimeters, showing how the incidence declined as the height increased. These were one sigma deviations; the height table also provides an example of the operation of "scaling to a sigma" by using the sigma as a unit of measurement.

Those Comforting Assumptions

Note the central assumptions we made in the coin-flip game that led to the proto-Gaussian, or mild randomness.

First central assumption: the flips are independent of one another. The coin has no memory. The fact that you got heads or tails on the previous flip does not change the odds of your getting heads or tails on the next one. You do not become a "better" coin flipper over time. If you introduce memory, or skills in flipping, the entire Gaussian business becomes shaky.

Recall our discussions in Chapter 14 on preferential attachment and cumulative advantage. Both theories assert that winning today makes you more likely to win in the future. Therefore, probabilities are dependent on history, and the first central assumption leading to the Gaussian bell curve fails in reality. In games, of course, past winnings are not supposed to translate into an increased probability of future gains—but not so in real life, which is why I worry about teaching probability from games. But when winning leads to more winning, you are far more likely to see forty wins in a row than with a proto-Gaussian.

Second central assumption: no "wild" jump. The step size in the building block of the basic random walk is always known, namely one step. There is no uncertainty as to the size of the step. We did not encounter situations in which the move varied wildly.

Remember that if either of these two central assumptions is not met, your moves (or coin tosses) will not cumulatively lead to the bell curve. Depending on what happens, they can lead to the wild Mandelbrotian-style scale-invariant randomness.

"The Ubiquity of the Gaussian"

One of the problems I face in life is that whenever I tell people that the Gaussian bell curve is not ubiquitous in real life, only in the minds of statisticians, they require me to "prove it"—which is easy to do, as we will see in the next two chapters, yet nobody has managed to prove the opposite. Whenever I suggest a process that is not Gaussian, I am asked to justify my suggestion and to, beyond the phenomena, "give them the theory behind it." We saw in Chapter 14 the rich-get-richer models that were proposed in order to justify not using a Gaussian. Modelers were forced to spend their time writing theories on possible models that generate the scalable—as if they needed to be apologetic about it. Theory shmeory! I have an epistemological problem with that, with the need to justify the world's failure to resemble an idealized model that someone blind to reality has managed to promote.

My technique, instead of studying the possible models generating non–bell curve randomness, hence making the same errors of blind theorizing, is to do the opposite: to know the bell curve as intimately as I can and identify where it can and cannot hold. I know where Mediocristan is. To me it is frequently (nay, almost always) the users of the bell curve who do not understand it well, and have to justify it, and not the opposite.

This ubiquity of the Gaussian is not a property of the world, but a problem in our minds, stemming from the way we look at it.

. . .

The next chapter will address the scale invariance of nature and address the properties of the fractal. The chapter after that will probe the misuse of the Gaussian in socioeconomic life and "the need to produce theories."

I sometimes get a little emotional because I've spent a large part of my life thinking about this problem. Since I started thinking about it, and conducting a variety of thought experiments as I have above, I have not for the life of me been able to find anyone around me in the business and statistical world who was intellectually consistent in that he both accepted the Black Swan and rejected the Gaussian and Gaussian tools. Many people accepted my Black Swan idea but could not take it to its logical conclusion, which is that you cannot use one single measure for randomness called standard deviation (and call it "risk"); you cannot expect a *simple* answer to characterize uncertainty. To go the extra step requires courage, commitment, an ability to connect the dots, a desire to understand randomness fully. It also means not accepting other people's wisdom as gospel. Then I started finding physicists who had rejected the Gaussian tools but fell for another sin: gullibility about precise predictive models, mostly elaborations around the preferential attachment of Chapter 14—another form of Platonicity. I could not find anyone with depth and scientific technique who looked at the world of randomness and understood its nature, who looked at calculations as an aid, not a principal aim. It took me close to a decade and a half to find that thinker, the man who made many swans gray: Mandelbrot—the great Benoît Mandelbrot.

Chapter Sixteen

THE AESTHETICS OF RANDOMNESS

Mandelbrot's library—Was Galileo blind?—Pearls to swine—Self-affinity—How the world can be complicated in a simple way, or, perhaps, simple in a very complicated way

THE POET OF RANDOMNESS

It was a melancholic afternoon when I smelled the old books in Benoît Mandelbrot's library. This was on a hot day in August 2005, and the heat exacerbated the musty odor of the glue of old French books bringing on powerful olfactory nostalgia. I usually succeed in repressing such nostalgic excursions, but not when they sneak up on me as music or smell. The odor of Mandelbrot's books was that of French literature, of my parents' library, of the hours spent in bookstores and libraries when I was a teenager when many books around me were (alas) in French, when I thought that Literature was above anything and everything. (I haven't been in contact with many French books since my teenage days.) However abstract I wanted it to be, Literature had a physical embodiment, it had a smell, and this was it.

The afternoon was also gloomy because Mandelbrot was moving away, exactly when I had become entitled to call him at crazy hours just because I had a question, such as why people didn't realize that the 80/20 could be 50/01. Mandelbrot had decided to move to the Boston area, not to retire, but to work for a research center sponsored by a national laboratory. Since he was moving to an apartment in Cambridge, and leaving his oversize house in the Westchester suburbs of New York, he had invited me to come take my pick of his books.

Even the titles of the books had a nostalgic ring. I filled up a box with French titles, such as a 1949 copy of Henri Bergson's *Matière et mémoire,* which it seemed Mandelbrot bought when he was a student (the smell!).

After having mentioned his name left and right throughout this book, I will finally introduce Mandelbrot, principally as the first person with an academic title with whom I ever spoke about randomness without feeling defrauded. Other mathematicians of probability would throw at me theorems with Russian names such as "Sobolev," "Kolmogorov," Wiener measure, without which they were lost; they had a hard time getting to the heart of the subject or exiting their little box long enough to consider its empirical flaws. With Mandelbrot, it was different: it was as if we both originated from the same country, meeting after years of frustrating exile, and were finally able to speak in our mother tongue without straining. He is the only flesh-and-bones teacher I ever had—my teachers are usually books in my library. I had way too little respect for mathematicians dealing with uncertainty and statistics to consider any of them my teachers—in my mind mathematicians, trained for certainties, had no business dealing with randomness. Mandelbrot proved me wrong.

He speaks an unusually precise and formal French, much like that spoken by Levantines of my parents' generation or Old World aristocrats. This made it odd to hear, on occasion, his accented, but very standard, colloquial American English. He is tall, overweight, which makes him look baby-faced (although I've never seen him eat a large meal), and has a strong physical presence.

From the outside one would think that what Mandelbrot and I have in common is wild uncertainty, Black Swans, and dull (and sometimes less dull) statistical notions. But, although we are collaborators, this is not what our major conversations revolve around. It is mostly matters literary and aesthetic, or historical gossip about people of extraordinary intellectual refinement. I mean refinement, not achievement. Mandelbrot could tell stories about the phenomenal array of hotshots he has worked with over the past century, but somehow I am programmed to consider scientists' personae far less interesting than those of colorful erudites. Like me, Mandelbrot takes an interest in urbane individuals who combine traits generally thought not to coexist together. One person he often mentions is Baron Pierre Jean de Menasce, whom he met at Princeton in the 1950s, where de Menasce was the roommate of the physicist Oppenheimer. De Menasce was exactly the kind of person I am interested in, the embodiment of a Black Swan. He came from an opulent Alexandrian Jewish merchant family, French and Italian–speaking like all sophisticated Levantines. His forebears had taken a Venetian spelling for their Arabic name, added a Hungarian noble title along the way, and socialized with royalty. De Menasce not only converted to Christianity, but became a Dominican priest and a great scholar of Semitic and Persian languages. Mandelbrot kept questioning me about Alexandria, since he was always looking for such characters.

True, intellectually sophisticated characters were exactly what I looked for in life. My erudite and polymathic father—who, were he still alive, would have only been two weeks older than Benoît M.—liked the company of extremely cultured Jesuit priests. I remember these Jesuit visitors occupying my chair at the dining table. I recall that one had a medical degree and a PhD in physics, yet taught Aramaic to locals in Beirut's Institute of Eastern Languages. His previous assignment could have been teaching high school physics, and the one before that was perhaps in the medical school. This kind of erudition impressed my father far more than scientific assembly-line work. I may have something in my genes driving me away from *bildungsphilisters*.

Although Mandelbrot often expressed amazement at the temperament of high-flying erudites and remarkable but not-so-famous scientists, such as his old friend Carleton Gajdusek, a man who impressed him with his ability to uncover the causes of tropical diseases, he did not seem eager to trumpet his association with those we consider great scientists. It took me a while to discover that he had worked with an impressive list of scientists in seemingly every field, something a name-dropper would have brought up continuously. Although I have been working with him for a few years now, only the other day, as I was chatting with his wife, did I discover that he spent two years as the mathematical collaborator of the psychologist Jean Piaget. Another shock came when I discovered that he had also worked with the great historian Fernand Braudel, but Mandelbrot did not seem to be interested in Braudel. He did not care to discuss John von Neuman with whom he had worked as a postdoctoral fellow. His scale was inverted. I asked him once about Charles Tresser, an unknown physicist I met at a party who wrote papers on chaos theory and supplemented his researcher's income by making pastry for a shop he ran near New York City. He was emphatic: *"un homme extraordinaire,"* he called Tresser, and could not stop praising him. But when I asked him about a particular famous hotshot, he replied, "He is the prototypical *bon élève,* a student with good grades, no depth, and no vision." That hotshot was a Nobel laureate.

THE PLATONICITY OF TRIANGLES

Now, why am I calling this business Mandelbrotian, or fractal, randomness? Every single bit and piece of the puzzle has been previously mentioned by someone else, such as Pareto, Yule, and Zipf, but it was Mandelbrot who a) connected the dots, b) linked randomness to geometry (and a special brand at that), and c) took the subject to its natural conclusion. Indeed many mathematicians are famous today partly because he dug out their works to back up his claims—the strategy I am following here in this book. "I had to invent my predecessors, so people take me seriously," he once told me, and he used the credibility of big guns as a rhetorical device. One can almost always ferret out predecessors for any thought. You can always find someone who worked on a part of your argument and use his contribution as your backup. The scientific association with a big idea, the "brand name," goes to the one who connects the dots, not the one who makes a casual observation—even Charles Darwin, who uncultured scientists claim "invented" the survival of the fittest, was not the first to mention it. He wrote in the introduction of *The Origin of Species* that the facts he presented were not necessarily original; it was the consequences that he thought were "interesting" (as he put it with characteristic Victorian modesty). In the end it is those who derive consequences and seize the importance of the ideas, seeing their real value, who win the day. They are the ones who can talk about the subject.

So let me describe Mandelbrotian geometry.

The Geometry of Nature

Triangles, squares, circles, and the other geometric concepts that made many of us yawn in the classroom may be beautiful and pure notions, but they seem more present in the minds of architects, design artists, modern art buildings, and schoolteachers than in nature itself. That's fine, except that most of us aren't aware of this. Mountains are not triangles or pyramids; trees are not circles; straight lines are almost never seen anywhere. Mother Nature did not attend high school geometry courses or read the books of Euclid of Alexandria. Her geometry is jagged, but with a logic of its own and one that is easy to understand.

I have said that we seem naturally inclined to Platonify, and to think exclusively in terms of studied material: nobody, whether a bricklayer or a natural philosopher, can easily escape the enslavement of such conditioning. Consider that the great Galileo, otherwise a debunker of falsehoods, wrote the following:

> The great book of Nature lies ever open before our eyes and the true philosophy is written in it. . . . But we cannot read it unless we have first learned the language and the characters in which it is written. . . . It is written in mathematical language and the characters are triangles, circles and other geometric figures.

Was Galileo legally blind? Even the great Galileo, with all his alleged independence of mind, was not capable of taking a clean look at Mother Nature. I am confident that he had windows in his house and that he ventured outside from time to time: he should have known that triangles are not easily found in nature. We are so easily brainwashed.

We are either blind, or illiterate, or both. That nature's geometry is not Euclid's was so obvious, and nobody, almost nobody, saw it.

This (physical) blindness is identical to the ludic fallacy that makes us think casinos represent randomness.

Fractality

But first, a description of fractals. Then we will show how they link to what we call power laws, or scalable laws.

Fractal is a word Mandelbrot coined to describe the geometry of the rough and *broken*—from the Latin *fractus,* the origin of *fractured. Fractality* is the repetition of geometric patterns at different scales, revealing smaller and smaller versions of themselves. Small parts resemble, to some degree, the whole. I will try to show in this chapter how the fractal applies to the brand of uncertainty that should bear Mandelbrot's name: Mandelbrotian randomness.

The veins in leaves look like branches; branches look like trees; rocks look like small mountains. There is no qualitative change when an object changes size. If you look at the coast of Britain from an airplane, it resembles what you see when you look at it with a magnifying glass. This character of self-affinity implies that one deceptively short and simple rule of iteration can be used, either by a computer or, more randomly, by Mother Nature, to build shapes of seemingly great complexity. This can come in handy for computer graphics, but, more important, it is how nature works. Mandelbrot designed the mathematical object now known as the Mandelbrot set, the most famous object in the history of mathematics. It became popular with followers of chaos theory because it generates pictures of ever increasing complexity by using a deceptively minuscule recursive rule; *recursive* means that something can be reapplied to itself infinitely. You can look at the set at smaller and smaller resolutions without *ever* reaching the limit; you will continue to see recognizable shapes. The shapes are never the same, yet they bear an affinity to one another, a strong family resemblance.

These objects play a role in aesthetics. Consider the following applications:

Visual arts: Most computer-generated objects are now based on some version of the Mandelbrotian fractal. We can also see fractals in architecture, paintings, and many works of visual art—of course, not consciously incorporated by the work's creator.

Music: Slowly hum the four-note opening of Beethoven's Fifth Symphony: *ta-ta-ta-ta*. Then replace each individual note with the same four-note opening, so that you end up with a measure of sixteen notes. You will see (or, rather, hear) that each smaller wave resembles the original larger one. Bach and Mahler, for instance, wrote submovements that resemble the larger movements of which they are a part.

Poetry: Emily Dickinson's poetry, for instance, is fractal: the large resembles the small. It has, according to a commentator, "a consciously made assemblage of dictions, metres, rhetorics, gestures, and tones."

Fractals initially made Benoît M. a pariah in the mathematical establishment. French mathematicians were horrified. What? Images? *Mon dieu!* It was like showing a porno movie to an assembly of devout Eastern Orthodox grandmothers in my ancestral village of Amioun. So Mandelbrot spent time as an intellectual refugee at an IBM research center in upstate New York. It was a *f*** you money* situation, as IBM let him do whatever he felt like doing.

But the general public (mostly computer geeks) got the point. Mandelbrot's book *The Fractal Geometry of Nature* made a splash when it came out a quarter century ago. It spread through artistic circles and led to studies in aesthetics, architectural design, even large industrial applications. Benoît M. was even offered a position as a professor of medicine! Supposedly the lungs are self-similar. His talks were invaded by all sorts of artists, earning him the nickname the Rock Star of Mathematics. The computer age helped him become one of the most influential mathematicians in history, in terms of the applications of his work, way before his acceptance by the ivory tower. We will see that, in addition to its universality, his work offers an unusual attribute: it is remarkably easy to understand.

A few words on his biography. Mandelbrot came to France from Warsaw in 1936, at the age of twelve. Owing to the vicissitudes of a clandestine life during Nazi-occupied France, he was spared some of the conventional Gallic education with its uninspiring algebraic drills, becoming largely self-taught. He was later deeply influenced by his uncle Szolem, a prominent member of the French mathematical establishment and holder of a chair at the Collège de France. Benoît M. later settled in the United States, working most of his life as an industrial scientist, with a few transitory and varied academic appointments.

The computer played two roles in the new science Mandelbrot helped conceive. First, fractal objects, as we have seen, can be generated with a simple rule applied to itself, which makes them ideal for the automatic activity of a computer (or Mother Nature). Second, in the generation of visual intuitions lies a dialectic between the mathematician and the objects generated.

Now let us see how this takes us to randomness. In fact, it is with probability that Mandelbrot started his career.

A Visual Approach to Extremistan/Mediocristan

I am looking at the rug in my study. If I examine it with a microscope, I will see a very rugged terrain. If I look at it with a magnifying glass, the terrain will be smoother but still highly uneven. But when I look at it from a standing position, it appears uniform—it is almost as smooth as a sheet of paper. The rug at eye level corresponds to Mediocristan and the law of large numbers: I am seeing the sum of undulations, and *these iron out.* This is like Gaussian randomness: the reason my cup of coffee does not jump is that the sum of all of its moving particles becomes smooth. Likewise, you reach certainties by adding up small Gaussian uncertainties: this is the law of large numbers.

The Gaussian is not self-similar, and that is why my coffee cup does not jump on my desk.

Now, consider a trip up a mountain. No matter how high you go on the surface of the earth, it will remain jagged. This is even true at a height of 30,000 feet. When you are flying above the Alps, you will still see jagged mountains in place of small stones. So some surfaces are not from Mediocristan, and changing the resolution does not make them much smoother. (Note that this effect only disappears when you go up to more extreme heights. Our planet looks smooth to an observer from space, but this is because it is too small. If it were a bigger planet, then it would have mountains that would dwarf the Himalayas, and it would require observation from a greater distance for it to look smooth. Likewise, if the planet had a larger population, even maintaining the same average wealth, we would be likely to find someone whose net worth would vastly surpass that of Bill Gates.)

Figures 11 and 12 illustrate the above point: an observer looking at the first picture might think that a lens cap has fallen on the ground.

Recall our brief discussion of the coast of Britain. If you look at it from an airplane, its contours are not so different from the contours you see on the shore. The change in scaling does not alter the shapes or their degree of smoothness.

Pearls to Swine

What does fractal geometry have to do with the distribution of wealth, the size of cities, returns in the financial markets, the number of casualties in war, or the size of planets? Let us connect the dots.

The key here is that *the fractal has numerical or statistical measures that are (somewhat) preserved across scales*—the ratio is the same, unlike the Gaussian. Another view of such self-similarity is presented in Figure 13. As we saw in Chapter 15, the superrich are similar to the rich, only richer—wealth is scale independent, or, more precisely, of unknown scale dependence.

In the 1960s Mandelbrot presented his ideas on the prices of commodities and financial securities to the economics establishment, and the financial economists got all excited. In 1963 the then dean of the Univer-

FIGURE 11: Apparently, a lens cap has been dropped on the ground. Now turn the page.

sity of Chicago Graduate School of Business, George Shultz, offered him a professorship. This is the same George Shultz who later became Ronald Reagan's secretary of state.

Shultz called him one evening to rescind the offer.

At the time of writing, forty-four years later, nothing has happened in economics and social science statistics—except for some cosmetic fiddling that treats the world as if we were subject only to mild randomness—and yet Nobel medals were being distributed. Some papers were written offering "evidence" that Mandelbrot was wrong by people who do not get the central argument of this book—you can always produce data "corroborating" that the underlying process is Gaussian by finding periods that do not have rare events, just like you can find an afternoon during which no one killed anyone and use it as "evidence" of honest behavior. I will repeat that, because of the asymmetry with induction, just as it is easier to reject innocence than accept it, it is easier to reject a bell curve than accept it; conversely, it is more difficult to reject a fractal than to accept it. Why? Because a single event can destroy the argument that we face a Gaussian bell curve.

In sum, four decades ago, Mandelbrot gave pearls to economists and résumé-building philistines, which they rejected because the ideas were

FIGURE 12: The object is not in fact a lens cap. These two photos illustrate scale invariance: the terrain is fractal. Compare it to man-made objects such as a car or a house. *Source: Professor Stephen W. Wheatcraft, University of Nevada, Reno.*

too good for them. It was, as the saying goes, *margaritas ante porcos*, pearls before swine.

In the rest of this chapter I will explain how I can endorse Mandelbrotian fractals as a representation of much of randomness without necessarily accepting their precise use. Fractals should be the default, the approximation, the framework. They do not solve the Black Swan problem and do not turn all Black Swans into predictable events, but they significantly mitigate the Black Swan problem by making such large events conceivable. (It makes them gray. Why gray? Because only the Gaussian give you certainties. More on that, later.)

THE LOGIC OF FRACTAL RANDOMNESS (WITH A WARNING)*

I have shown in the wealth lists in Chapter 15 the logic of a fractal distribution: if wealth doubles from 1 million to 2 million, the incidence of peo-

* The nontechnical reader can skip from here until the end of the chapter.

FIGURE 13: THE PURE FRACTAL STATISTICAL MOUNTAIN

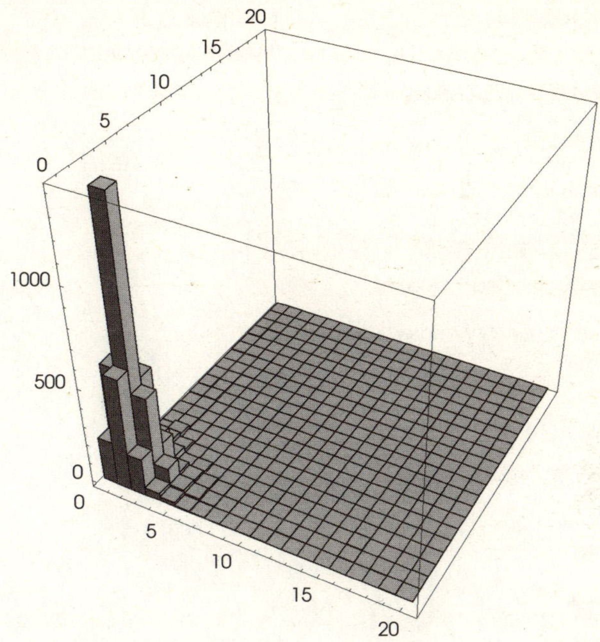

The degree of inequality will be the same in all sixteen subsections of the graph. In the Gaussian world, disparities in wealth (or any other quantity) decrease when you look at the upper end—so billionaires should be more equal in relation to one another than millionaires are, and millionaires more equal in relation to one another than the middle class. This lack of equality at all wealth levels, in a nutshell, is statistical self-similarity.

ple with at least that much money is cut in four, which is an exponent of two. If the exponent were one, then the incidence of that wealth or more would be cut in two. The exponent is called the "power" (which is why some people use the term *power law*). Let us call the number of occurrences higher than a certain level an "exceedance"—an exceedance of two million is the number of persons with wealth more than two million. One main property of these fractals (or another way to express their main property, scalability) is that the ratio of two exceedances* is going to be the ratio of the two numbers to the negative power of the power exponent.

* By using symmetry we could also examine the incidences below the number.

TABLE 2: ASSUMED EXPONENTS FOR VARIOUS PHENOMENA*

Phenomenon	*Assumed Exponent (vague approximation)*
Frequency of use of words	1.2
Number of hits on websites	1.4
Number of books sold in the U.S.	1.5
Telephone calls received	1.22
Magnitude of earthquakes	2.8
Diameter of moon craters	2.14
Intensity of solar flares	0.8
Intensity of wars	0.8
Net worth of Americans	1.1
Number of persons per family name	1
Population of U.S. cities	1.3
Market moves	3 (or lower)
Company size	1.5
People killed in terrorist attacks	2 (but possibly a much lower exponent)

* *Source: M.E.J. Newman (2005) and the author's own calculations.*

Let us illustrate this. Say that you "think" that only 96 books a year will sell more than 250,000 copies (which is what happened last year), and that you "think" that the exponent is around 1.5. You can extrapolate to estimate that around 34 books will sell more than 500,000 copies—simply 96 times $(500{,}000/250{,}000)^{-1.5}$. We can continue, and note that around 8 books should sell more than a million copies, here 96 times $(1{,}000{,}000/250{,}000)^{-1.5}$.

Let me show the different measured exponents for a variety of phenomena.

Let me tell you upfront that these exponents mean very little in terms of numerical precision. We will see why in a minute, but just note for now that we do not *observe* these parameters; we simply guess them, or infer them for statistical information, which makes it hard at times to know the

TABLE 3: THE MEANING OF THE EXPONENT

Exponent	*Share of the top 1%*	*Share of the top 20%*
1	99.99%*	99.99%
1.1	66%	86%
1.2	47%	76%
1.3	34%	69%
1.4	27%	63%
1.5	22%	58%
2	10%	45%
2.5	6%	38%
3	4.6%	34%

* *Clearly, you do not observe 100 percent in a finite sample.*

true parameters—if it in fact exists. Let us first examine the practical consequences of an exponent.

Table 2 illustrates the impact of the highly improbable. It shows the contributions of the top 1 percent and 20 percent to the total. The lower the exponent, the higher those contributions. But look how sensitive the process is: between 1.1 and 1.3 you go from 66 percent of the total to 34 percent. Just a 0.2 difference in the exponent changes the result dramatically—and such a difference can come from a simple measurement error. This difference is not trivial: just consider that we have no precise idea what the exponent is because we cannot measure it directly. All we do is estimate from past data or rely on theories that allow for the building of some model that would give us some idea—but these models may have hidden weaknesses that prevent us from blindly applying them to reality.

So keep in mind that the 1.5 exponent is an approximation, that it is hard to compute, that you do not get it from the gods, at least not easily, and that you will have a monstrous sampling error. You will observe that the number of books selling above a million copies is not always going to be 8—It could be as high as 20, or as low as 2.

More significantly, this exponent begins to apply at some number called "crossover," and addresses numbers larger than this crossover. It may start at 200,000 books, or perhaps only 400,000 books. Likewise, wealth has different properties before, say, $600 million, when inequality grows, than it does below such a number. How do you know where the crossover point is? This is a problem. My colleagues and I worked with around 20 million pieces of financial data. We all had the same data set, yet we never agreed on exactly what the exponent was in our sets. We knew the data revealed a fractal power law, but we learned that one could not produce a precise number. But what we did know—*that the distribution is scalable and fractal*—was sufficient for us to operate and make decisions.

The Problem of the Upper Bound

Some people have researched and accepted the fractal "up to a point." They argue that wealth, book sales, and market returns all have a certain level when things stop being fractal. "Truncation" is what they propose. I agree that there is a level where fractality *might* stop, but where? Saying that there is an upper limit *but I don't know how high it is,* and saying *there is no limit* carry the same consequences in practice. Proposing an upper limit is highly unsafe. You may say, Let us cap wealth at $150 billion in our analyses. Then someone else might say, Why not $151 billion? Or why not $152 billion? We might as well consider that the variable is unlimited.

Beware the Precision

I have learned a few tricks from experience: whichever exponent I try to measure will be likely to be overestimated (recall that a higher exponent implies a smaller role for large deviations)—what you see is likely to be less Black Swannish than what you do not see. I call this the masquerade problem.

Let's say I generate a process that has an exponent of 1.7. You do not see what is inside the engine, only the data coming out. If I ask you what the exponent is, odds are that you will compute something like 2.4. You would do so even if you had a million data points. The reason is that it takes a long time for some fractal processes to reveal their properties, and you underestimate the severity of the shock.

Sometimes a fractal can make you believe that it is Gaussian, particularly when the cutpoint starts at a high number. With fractal distributions, extreme deviations of that kind are rare enough to smoke you: you don't recognize the distribution as fractal.

The Water Puddle Revisited

As you have seen, we have trouble knowing the parameters of whichever model we assume runs the world. So with Extremistan, the problem of induction pops up again, this time even more significantly than at any previous time in this book. Simply, if a mechanism is fractal it can deliver large values; therefore the incidence of large deviations is possible, but how possible, how often they should occur, will be hard to know with any precision. This is similar to the water puddle problem: plenty of ice cubes could have generated it. As someone who goes from reality to possible explanatory models, I face a completely different spate of problems from those who do the opposite.

I have just read three "popular science" books that summarize the research in complex systems: Mark Buchanan's *Ubiquity,* Philip Ball's *Critical Mass,* and Paul Ormerod's *Why Most Things Fail.* These three authors present the world of social science as full of power laws, a view with which I most certainly agree. They also claim that there is *universality* of many of these phenomena, that there is a wonderful similarity between various processes in nature and the behavior of social groups, which I agree with. They back their studies with the various theories on networks and show the wonderful correspondence between the so-called critical phenomena in natural science and the self-organization of social groups. They bring together processes that generate avalanches, social contagions, and what they call informational cascades, which I agree with.

Universality is one of the reasons physicists find power laws associated with critical points particularly interesting. There are many situations, both in dynamical systems theory and statistical mechanics, where many of the properties of the dynamics around critical points are independent of the details of the underlying dynamical system. The exponent at the critical point may be the same for many systems in the same group, even though many other aspects of the system are different. I almost agree with this notion of universality. Finally, all three authors encourage us to apply techniques from statistical physics, avoiding econometrics and Gaussian-style nonscalable distributions like the plague, and I couldn't agree more.

But all three authors, by producing, or promoting precision, fall into the trap of not differentiating between the forward and the backward processes (between the problem and the inverse problem)—to me, the greatest scientific and epistemological sin. They are not alone; nearly everyone who works with data but doesn't make decisions on the basis of these data tends to be guilty of the same sin, a variation of the narrative fallacy. In the absence of a feedback process you look at models and think that they confirm reality. I believe in the ideas of these three books, but not in the way they are being used—and certainly not with the precision the authors ascribe to them. As a matter of fact, complexity theory should make us *more* suspicious of scientific claims of precise models of reality. It does not make all the swans white; that is predictable: it makes them gray, and only gray.

As I have said earlier, the world, epistemologically, is literally a different place to a bottom-up empiricist. We don't have the luxury of sitting down to read the equation that governs the universe; we just observe data and make an assumption about what the real process might be, and "calibrate" by adjusting our equation in accordance with additional information. As events present themselves to us, we compare what we see to what we expected to see. It is usually a humbling proccss, particularly for someone aware of the narrative fallacy, to discover that history runs forward, not backward. As much as one thinks that businessmen have big egos, these people are often humbled by reminders of the differences between decision and results, between precise models and reality.

What I am talking about is opacity, incompleteness of information, the invisibility of the generator of the world. History does not reveal its mind to us—we need to guess what's inside of it.

From Representation to Reality

The above idea links all the parts of this book. While many study psychology, mathematics, or evolutionary theory and look for ways to take it to the bank by applying their ideas to business, I suggest the exact opposite: study the intense, uncharted, humbling uncertainty in the markets as a means to get insights about the nature of randomness that is applicable to psychology, probability, mathematics, decision theory, and even statistical physics. You will see the sneaky manifestations of the narrative fallacy, the ludic fallacy, and the great errors of Platonicity, of going from representation to reality.

When I first met Mandelbrot I asked him why an established scientist like him who should have more valuable things to do with his life would take an interest in such a vulgar topic as finance. I thought that finance and economics were just a place where one learned from various empirical phenomena and filled up one's bank account with *f*** you* cash before leaving for bigger and better things. Mandelbrot's answer was, "*Data,* a gold mine of data." Indeed, everyone forgets that he started in economics before moving on to physics and the geometry of nature. Working with such abundant data humbles us; it provides the intuition of the following error: traveling the road between representation and reality in the wrong direction.

The problem of the *circularity of statistics* (which we can also call the statistical regress argument) is as follows. Say you need past data to discover whether a probability distribution is Gaussian, fractal, or something else. You will need to establish whether you have enough data to back up your claim. How do we know if we have enough data? From the probability distribution—a distribution does tell you whether you have enough data to "build confidence" about what you are inferring. If it is a Gaussian bell curve, then a few points will suffice (the law of large numbers once again). And how do you know if the distribution is Gaussian? Well, from the data. So we need the data to tell us what the probability distribution is, and a probability distribution to tell us how much data we need. This causes a severe regress argument.

This regress does not occur if you *assume beforehand* that the distribution is Gaussian. It happens that, for some reason, the Gaussian yields its properties rather easily. Extremistan distributions do not do so. So selecting the Gaussian while invoking some general law appears to be convenient. The Gaussian is used as a default distribution for that very reason. As I keep repeating, assuming its application beforehand may work with a small number of fields such as crime statistics, mortality rates, matters from Mediocristan. But not for historical data of unknown attributes and not for matters from Extremistan.

Now, why aren't statisticians who work with historical data aware of this problem? First, they do not like to hear that their entire business has been canceled by the problem of induction. Second, they are not confronted with the results of their predictions in rigorous ways. As we saw with the Makridakis competition, they are grounded in the narrative fallacy, and they do not want to hear it.

ONCE AGAIN, BEWARE THE FORECASTERS

Let me take the problem one step higher up. As I mentioned earlier, plenty of fashionable models attempt to explain the genesis of Extremistan. In fact, they are grouped into two broad classes, but there are occasionally more approaches. The first class includes the simple rich-get-richer (or big-get-bigger) style model that is used to explain the lumping of people around cities, the market domination of Microsoft and VHS (instead of Apple and Betamax), the dynamics of academic reputations, etc. The second class concerns what are generally called "percolation models," which address not the behavior of the individual, but rather the terrain in which he operates. When you pour water on a porous surface, the structure of that surface matters more than does the liquid. When a grain of sand hits a pile of other grains of sand, how the terrain is organized is what determines whether there will be an avalanche.

Most models, of course, attempt to be precisely predictive, not just descriptive; I find this infuriating. They are nice tools for illustrating the genesis of Extremistan, but I insist that the "generator" of reality does not appear to obey them closely enough to make them helpful in precise forecasting. At least to judge by anything you find in the current literature on the subject of Extremistan. Once again we face grave calibration problems, so it would be a great idea to avoid the common mistakes made while calibrating a nonlinear process. Recall that nonlinear processes have greater degrees of freedom than linear ones (as we saw in Chapter 11), with the implication that you run a great risk of using the wrong model. Yet once in a while you run into a book or articles advocating the application of models from statistical physics to reality. Beautiful books like Philip Ball's illustrate and inform, but they should not lead to precise quantitative models. Do not take them at face value.

But let us see what we *can* take home from these models.

Once Again, a Happy Solution

First, in assuming a scalable, I accept that an arbitrarily large number is possible. In other words, inequalities should not stop above some *known* maximum bound.

Say that the book *The Da Vinci Code* sold around 60 million copies. (The Bible sold about a billion copies but let's ignore it and limit our analysis to lay books written by individual authors.) Although we have never known a lay book to sell 200 million copies, we can consider that the possibility is not zero. It's small, but it's not zero. For every three *Da Vinci Code*–style bestsellers, there might be one superbestseller, and though one has not happened so far, we cannot rule it out. And for every fifteen *Da Vinci Code*s there will be one superbestseller selling, say, 500 million copies.

Apply the same logic to wealth. Say the richest person on earth is worth $50 billion. There is a nonnegligible probability that next year someone with $100 billion or more will pop out of nowhere. For every three people with more than $50 billion, there could be one with $100 billion or more. There is a much smaller probability of there being someone with more than $200 billion—one third of the previous probability, but nevertheless not zero. There is even a minute, but not zero probability of there being someone worth more than $500 billion.

This tells me the following: I can make inferences about things that I do not see in my data, but these things should still belong to the realm of possibilities. There is an invisible bestseller out there, one that is absent from the past data but that you need to account for. Recall my point in Chapter 13: it makes investment in a book or a drug better than statistics on past data might suggest. But it can make stock market losses worse than what the past shows.

Wars are fractal in nature. A war that kills more people than the devastating Second World War is possible—not likely, but not a zero probability, although such a war has never happened in the past.

Second, I will introduce an illustration from nature that will help to make the point about precision. A mountain is somewhat similar to a stone: it has an affinity with a stone, a family resemblance, but it is not identical. The word to describe such resemblances is *self-affine,* not the precise *self-similar,* but Mandelbrot had trouble communicating the notion of affinity, and the term *self-similar* spread with its connotation of precise resemblance rather than family resemblance. As with the mountain and the stone, the distribution of wealth above $1 billion is not exactly the same as that below $1 billion, but the two distributions have "affinity."

Third, I said earlier that there have been plenty of papers in the world of econophysics (the application of statistical physics to social and economic phenomena) aiming at such calibration, at pulling numbers from the world of phenomena. Many try to be predictive. Alas, we are not able to predict "transitions" into crises or contagions. My friend Didier Sornette attempts to build predictive models, which I love, except that I cannot use them to make predictions—but please don't tell him; he might stop building them. That I can't use them as he intends does not invalidate his work, it just makes the interpretations require broad-minded thinking, unlike models in conventional economics that are fundamentally flawed. We may be able to do well with some of Sornette's phenomena, but not all.

WHERE IS THE GRAY SWAN?

I have written this entire book about the Black Swan. This is not because I am in love with the Black Swan; as a humanist, I hate it. I hate most of the unfairness and damage it causes. Thus I would like to eliminate many Black Swans, or at least to mitigate their effects and be protected from them. Fractal randomness is a way to reduce these surprises, to make some of the swans appear possible, so to speak, to make us aware of their consequences, to make them gray. *But fractal randomness does not yield precise answers.* The benefits are as follows. If you know that the stock market *can* crash, as it did in 1987, then such an event is not a Black Swan. The crash of 1987 is not an outlier if you use a fractal with an exponent of three. If you know that biotech companies can deliver a megablockbuster drug, bigger than all we've had so far, then it won't be a Black Swan, and you will not be surprised, should that drug appear.

Thus Mandelbrot's fractals allow us to account for a few Black Swans, but not all. I said earlier that some Black Swans arise because we ignore sources of randomness. Others arise when we overestimate the fractal exponent. A gray swan concerns modelable extreme events, a black swan is about unknown unknowns.

I sat down and discussed this with the great man, and it became, as usual, a linguistic game. In Chapter 9 I presented the distinction economists make between Knightian uncertainty (incomputable) and Knightian risk (computable); this distinction cannot be so original an idea to be absent in our vocabulary, and so we looked for it in French. Mandelbrot mentioned one of his friends and prototypical heroes, the aristocratic mathematician Marcel-Paul Schützenberger, a fine erudite who (like this author) was easily bored and could not work on problems beyond their point of diminishing returns. Schützenberger insisted on the clear-cut distinction in the French language between *hasard* and *fortuit*. *Hasard*, from the Arabic *az-zahr*, implies, like *alea*, dice—tractable randomness; *fortuit* is my Black Swan—the purely accidental and unforeseen. We went to the *Petit Robert* dictionary; the distinction effectively exists there. *Fortuit* seems to correspond to my epistemic opacity, *l'imprévu et non quantifiable; hasard* to the more ludic type of uncertainty that was proposed by the Chevalier de Méré in the early gambling literature. Remarkably, the Arabs may have introduced another word to the business of uncertainty: *rizk,* meaning property.

I repeat: Mandelbrot deals with gray swans; I deal with the Black Swan. So Mandelbrot domesticated many of my Black Swans, but not all of them, not completely. But he shows us a glimmer of hope with his method, a way to start thinking about the problems of uncertainty. You are indeed much safer if you know where the wild animals are.

Chapter Seventeen

LOCKE'S MADMEN, OR BELL CURVES IN THE WRONG PLACES*

What?—Anyone can become president—Alfred Nobel's legacy—Those medieval days

I have in my house two studies: one real, with interesting books and literary material; the other nonliterary, where I do not enjoy working, where I relegate matters prosaic and narrowly focused. In the nonliterary study is a wall full of books on statistics and the history of statistics, books I never had the fortitude to burn or throw away; though I find them largely useless outside of their academic applications (Carneades, Cicero, and Foucher know a lot more about probability than all these pseudosophisticated volumes). I cannot use them in class because I promised myself never to teach trash, even if dying of starvation. Why can't I use them? Not one of these books deals with Extremistan. Not one. The few books that do are not by statisticians but by statistical physicists. We are teaching people methods from Mediocristan and turning them loose in Extremistan. It is like developing a medicine for plants and applying it to humans. It is no wonder that we run the biggest risk of all: we handle *matters that belong*

* This is a simple illustration of the general point of this book in finance and economics. If you do not believe in applying the bell curve to social variables, and if, like many professionals, you are already convinced that "modern" financial theory is dangerous junk science, you can safely skip this chapter.

to Extremistan, but treated as if they belonged to Mediocristan, as an "approximation."

Several hundred thousand students in business schools and social science departments from Singapore to Urbana-Champaign, as well as people in the business world, continue to study "scientific" methods, all grounded in the Gaussian, all embedded in the ludic fallacy.

This chapter examines disasters stemming from the application of phony mathematics to social science. The real topic might be the dangers to our society brought about by the Swedish academy that awards the Nobel Prize.

Only Fifty Years

Let us return to the story of my business life. Look at the graph in Figure 14. In the last fifty years, the ten most extreme days in the financial markets represent half the returns. Ten days in fifty years. Meanwhile, we are mired in chitchat.

Clearly, anyone who wants more than the high number of six sigma as proof that markets are from Extremistan needs to have his head examined. Dozens of papers show the inadequacy of the Gaussian family of distributions and the scalable nature of markets. Recall that, over the years, I myself have run statistics backward and forward on 20 million pieces of data that made me despise anyone talking about markets in Gaussian terms. But people have a hard time making the leap to the consequences of this knowledge.

The strangest thing is that people in business usually agree with me when they listen to me talk or hear me make my case. But when they go to the office the next day they revert to the Gaussian tools so entrenched in their habits. Their minds are domain-dependent, so they can exercise critical thinking at a conference while not doing so in the office. Furthermore, the Gaussian tools give them numbers, which seem to be "better than nothing." The resulting measure of future uncertainty satisfies our ingrained desire to simplify even if that means squeezing into one single number matters that are too rich to be described that way.

The Clerks' Betrayal

I ended Chapter 1 with the stock market crash of 1987, which allowed me to aggressively pursue my Black Swan idea. Right after the crash, when I

FIGURE 14

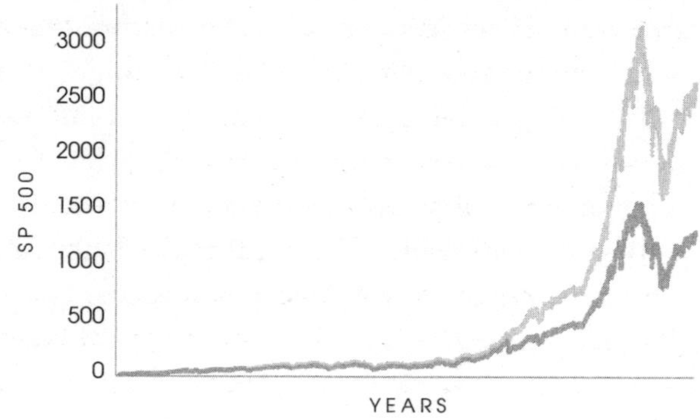

By removing the ten biggest one-day moves from the U.S. stock market over the past fifty years, we see a huge difference in returns—and yet conventional finance sees these one-day jumps as mere anomalies. (This is only one of many such tests. While it is quite convincing on a casual read, there are many more-convincing ones from a mathematical standpoint, such as the incidence of 10 sigma events.)

stated that those using sigmas (i.e., standard deviations) as a measure of the degree of risk and randomness were charlatans, everyone agreed with me. If the world of finance were Gaussian, an episode such as the crash (more than twenty standard deviations) would take place every several billion lifetimes of the universe (look at the height example in Chapter 15). According to the circumstances of 1987, people accepted that rare events take place and are the main source of uncertainty. They were just unwilling to give up on the Gaussian as a central measurement tool—"Hey, we have nothing else." People want a number to anchor on. Yet the two methods are logically incompatible.

Unbeknownst to me, 1987 was not the first time the idea of the Gaussian was shown to be lunacy. Mandelbrot proposed the scalable to the economics establishment around 1960, and showed them how the Gaussian curve did not fit prices *then*. But after they got over their excitement, they realized that they would have to relearn their trade. One of the influential economists of the day, the late Paul Cootner, wrote, "Mandelbrot, like Prime Minister Churchill before him, promised us not utopia, but blood, sweat, toil, and tears. If he is right, almost all our statistical tools are obsolete [or] meaningless." I propose two corrections to Cootner's statement. First, I would replace *almost all* with *all*. Second, I disagree with the blood and sweat business. I find Mandelbrot's randomness considerably easier to understand than the conventional statistics. If you come fresh to the business, do not rely on the old theoretical tools, and do not have a high expectation of certainty.

Anyone Can Become President

And now a brief history of the "Nobel" Prize in economics, which was established by the Bank of Sweden in honor of Alfred Nobel, who may be, according to his family who wants the prize abolished, now rolling in his grave with disgust. An activist family member calls the prize a public relations coup by economists aiming to put their field on a higher footing than it deserves. True, the prize has gone to some valuable thinkers, such as the empirical psychologist Daniel Kahneman and the thinking economist Friedrich Hayek. But the committee has gotten into the habit of handing out Nobel Prizes to those who "bring rigor" to the process with pseudoscience and phony mathematics. *After* the stock market crash, they rewarded two theoreticians, Harry Markowitz and William Sharpe, who built beautifully Platonic models on a Gaussian base, contributing to what is called Modern Portfolio Theory. Simply, if you remove their Gaussian assumptions and treat prices as scalable, you are left with hot air. The Nobel Committee could have tested the Sharpe and Markowitz models—they work like quack remedies sold on the Internet—but nobody in Stockholm seems to have thought of it. Nor did the committee come to us practitioners to ask us our opinions; instead it relied on an academic vetting process that, in some disciplines, can be corrupt all the way to the marrow. After that award I made a prediction: "In a world in which these two get the Nobel, anything can happen. Anyone can become president."

So the Bank of Sweden and the Nobel Academy are largely responsible for giving credence to the use of the Gaussian Modern Portfolio Theory as institutions have found it a great cover-your-behind approach. Software vendors have sold "Nobel crowned" methods for millions of dollars. How could you go wrong using it? Oddly enough, everyone in the business world initially knew that the idea was a fraud, but people get used to such methods. Alan Greenspan, the chairman of the Federal Reserve bank, supposedly blurted out, "I'd rather have the opinion of a trader than a mathematician." Meanwhile, the Modern Portfolio Theory started spreading. I will repeat the following until I am hoarse: it is contagion that determines the fate of a theory in social science, not its validity.

I only realized later that Gaussian-trained finance professors were taking over business schools, and therefore MBA programs, and producing close to a hundred thousand students a year in the United States alone, all brainwashed by a phony portfolio theory. No empirical observation could halt the epidemic. It seemed better to teach students a theory based on the Gaussian than to teach them no theory at all. It looked more "scientific" than giving them what Robert C. Merton (the son of the sociologist Robert K. Merton we discussed earlier) called the "anecdote." Merton wrote that before portfolio theory, finance was "a collection of anecdotes, rules of thumb, and manipulation of accounting data." Portfolio theory allowed "the subsequent evolution from this conceptual potpourri to a rigorous economic theory." For a sense of the degree of intellectual seriousness involved, and to compare neoclassical economics to a more honest science, consider this statement from the nineteenth-century father of modern medicine, Claude Bernard: "Facts for now, but with scientific aspirations for later." You should send economists to medical school.

So the Gaussian* pervaded our business and scientific cultures, and terms such as *sigma, variance, standard deviation, correlation, R square,* and the eponymous *Sharpe ratio,* all directly linked to it, pervaded the lingo. If you read a mutual fund prospectus, or a description of a hedge fund's exposure, odds are that it will supply you, among other information, with some quantitative summary claiming to measure "risk." That measure will be based on one of the above buzzwords derived from the bell curve and its kin. Today, for instance, pension funds' investment policy and choice of funds are vetted by "consultants" who rely on portfolio theory. If there is a problem, they can claim that they relied on standard scientific method.

More Horror

Things got a lot worse in 1997. The Swedish academy gave another round of Gaussian-based Nobel Prizes to Myron Scholes and Robert C. Merton, who had improved on an old mathematical formula and made it compatible with the existing grand Gaussian general financial equilibrium

* Granted, the Gaussian has been tinkered with, using such methods as complementary "jumps," stress testing, regime switching, or the elaborate methods known as GARCH, but while these methods represent a good effort, they fail to address the bell curve's fundamental flaws. Such methods are not scale-invariant. This, in my opinion, can explain the failures of sophisticated methods in real life as shown by the Makridakis competition.

theories—hence acceptable to the economics establishment. The formula was now "useable." It had a list of long forgotten "precursors," among whom was the mathematician and gambler Ed Thorp, who had authored the bestselling *Beat the Dealer,* about how to get ahead in blackjack, but somehow people believe that Scholes and Merton invented it, when in fact they just made it acceptable. The formula was my bread and butter. Traders, bottom-up people, know its wrinkles better than academics by dint of spending their nights worrying about their risks, except that few of them could express their ideas in technical terms, so I felt I was representing them. Scholes and Merton made the formula dependent on the Gaussian, but their "precursors" subjected it to no such restriction.*

The postcrash years were entertaining for me, intellectually. I attended conferences in finance and mathematics of uncertainty; not once did I find a speaker, Nobel or no Nobel, who understood what he was talking about when it came to probability, so I could freak them out with my questions. They did "deep work in mathematics," but when you asked them where they got their probabilities, their explanations made it clear that they had fallen for the ludic fallacy—there was a strange cohabitation of technical skills and absence of understanding that you find in idiot savants. Not once did I get an intelligent answer or one that was not ad hominem. Since I was questioning their entire business, it was understandable that I drew all manner of insults: "obsessive," "commercial," "philosophical," "essayist," "idle man of leisure," "repetitive," "practitioner" (this is an insult in academia), "academic" (this is an insult in business). Being on the receiving end of angry insults is not that bad; you can get quickly used to it and focus on what is *not* said. Pit traders are trained to handle angry rants. If you work in the chaotic pits, someone in a particularly bad mood from losing money might start cursing at you until he injures his vocal cords, then forget about it and, an hour later, invite you to his Christmas party. So you become numb to insults, particularly if you teach yourself to imagine that the person uttering them is a variant of a noisy ape with little personal control. Just keep your composure, smile, focus on analyzing the speaker not the message, and you'll win the argument. An ad hominem

* More technically, remember my career as an option professional. Not ony does an option on a very long shot benefit from Black Swans, but it benefits disproportionately from them—something Scholes and Merton's "formula" misses. The option payoff is so powerful that you do not have to be right on the odds: you can be wrong on the probability, but get a monstrously large payoff. I've called this the "double bubble": the mispricing of the probability and that of the payoff.

attack against an intellectual, not against an idea, is highly flattering. It indicates that the person does not have anything intelligent to say about your message.

The psychologist Philip Tetlock (the expert buster in Chapter 10), after listening to one of my talks, reported that he was struck by the presence of an acute state of cognitive dissonance in the audience. But how people resolve this cognitive tension, as it strikes at the core of everything they have been taught and at the methods they practice, and realize that they will continue to practice, can vary a lot. It was symptomatic that almost all people who attacked my thinking attacked a deformed version of it, like "it is all random and unpredictable" rather than "it is largely random," or got mixed up by showing me how the bell curve works in some physical domains. Some even had to change my biography. At a panel in Lugano, Myron Scholes once got in to a state of rage, and went after a transformed version of my ideas. I could see pain in his face. Once, in Paris, a prominent member of the mathematical establishment, who invested part of his life on some minute sub-sub-property of the Gaussian, blew a fuse—right when I showed empirical evidence of the role of Black Swans in markets. He turned red with anger, had difficulty breathing, and started hurling insults at me for having desecrated the institution, lacking *pudeur* (modesty); he shouted "I am a member of the Academy of Science!" to give more strength to his insults. (The French translation of my book was out of stock the next day.) My best episode was when Steve Ross, an economist perceived to be an intellectual far superior to Scholes and Merton, and deemed a formidable debater, gave a rebuttal to my ideas by signaling small errors or approximations in my presentation, such as "Markowitz was not the first to . . ." thus certifying that he had no answer to my main point. Others who had invested much of their lives in these ideas resorted to vandalism on the Web. Economists often invoke a strange argument by Milton Friedman that states that models do not have to have realistic assumptions to be acceptable—giving them license to produce severely defective mathematical representations of reality. The problem of course is that these Gaussianizations do not have realistic assumptions and do not produce reliable results. They are neither realistic nor predictive. Also note a mental bias I encounter on the occasion: people mistake an event with a small probability, say, one in twenty years for a periodically occurring one. They think that they are safe if they are only exposed to it for ten years.

I had trouble getting the message about the difference between Mediocristan and Extremistan through—many arguments presented to me were about how society has done well with the bell curve—just look at credit bureaus, etc.

The only comment I found unacceptable was, "You are right; we need you to remind us of the weakness of these methods, but you cannot throw the baby out with the bath water," meaning that I needed to accept their reductive Gaussian distribution while also accepting that large deviations could occur—they didn't realize the incompatibility of the two approaches. It was as if one could be half dead. Not one of these users of portfolio theory in twenty years of debates, explained *how* they could accept the Gaussian framework as well as large deviations. Not one.

Confirmation

Along the way I saw enough of the confirmation error to make Karl Popper stand up with rage. People would find data in which there were no jumps or extreme events, and show me a "proof" that one could use the Gaussian. This was exactly like my example of the "proof" that O. J. Simpson is not a killer in Chapter 5. The entire statistical business confused absence of proof with proof of absence. Furthermore, people did not understand the elementary asymmetry involved: you need one single observation to reject the Gaussian, but millions of observations will not fully confirm the validity of its application. Why? Because the Gaussian bell curve disallows large deviations, but tools of Extremistan, the alternative, do not disallow long quiet stretches.

I did not know that Mandelbrot's work mattered outside aesthetics and geometry. Unlike him, I was not ostracized: I got a lot of approval from practitioners and decision makers, though not from their research staffs.

But suddenly I got the most unexpected vindication.

IT WAS JUST A BLACK SWAN

Robert Merton, Jr., and Myron Scholes were founding partners in the large speculative trading firm called Long-Term Capital Management, or LTCM, which I mentioned in Chapter 4. It was a collection of people with top-notch résumés, from the highest ranks of academia. They were considered geniuses. The ideas of portfolio theory inspired their risk management of possible outcomes—thanks to their sophisticated "calculations." They managed to enlarge the ludic fallacy to industrial proportions.

Then, during the summer of 1998, a combination of large events, triggered by a Russian financial crisis, took place that lay outside their models. It was a Black Swan. LTCM went bust and almost took down the entire financial system with it, as the exposures were massive. Since their models ruled out the possibility of large deviations, they allowed themselves to take a monstrous amount of risk. The ideas of Merton and Scholes, as well as those of Modern Portfolio Theory, were starting to go bust. The magnitude of the losses was spectacular, too spectacular to allow us to ignore the intellectual comedy. Many friends and I thought that the portfolio theorists would suffer the fate of tobacco companies: they were endangering people's savings and would soon be brought to account for the consequences of their Gaussian-inspired methods.

None of that happened.

Instead, MBAs in business schools went on learning portfolio theory. And the option formula went on bearing the name Black-Scholes-Merton, instead of reverting to its true owners, Louis Bachelier, Ed Thorp, and others.

How to "Prove" Things

Merton the younger is a representative of the school of neoclassical economics, which, as we have seen with LTCM, represents most powerfully the dangers of Platonified knowledge.* Looking at his methodology, I see the following pattern. He starts with rigidly Platonic assumptions, completely unrealistic—such as the Gaussian probabilities, along with many more equally disturbing ones. Then he generates "theorems" and "proofs" from these. The math is tight and elegant. The theorems are compatible with other theorems from Modern Portfolio Theory, themselves compatible with still other theorems, building a grand theory of how people consume, save, face uncertainty, spend, and project the future. He assumes that we know the likelihood of events. The beastly word *equilibrium* is always present. But the whole edifice is like a game that is entirely closed, like Monopoly with all of its rules.

* I am selecting Merton because I found him very illustrative of academically stamped obscurantism. I discovered Merton's shortcomings from an angry and threatening seven-page letter he sent me that gave me the impression that he was not too familiar with how we trade options, his very subject matter. He seemed to be under the impression that traders rely on "rigorous" economic theory—as if birds had to study (bad) engineering in order to fly.

A scholar who applies such methodology resembles Locke's definition of a madman: someone "reasoning correctly from erroneous premises."

Now, elegant mathematics has this property: it is perfectly right, not 99 percent so. This property appeals to mechanistic minds who do not want to deal with ambiguities. Unfortunately you have to cheat somewhere to make the world fit perfect mathematics; and you have to fudge your assumptions somewhere. We have seen with the Hardy quote that professional "pure" mathematicians, however, are as honest as they come.

So where matters get confusing is when someone like Merton tries to be mathematical and airtight rather than focus on fitness to reality.

This is where you learn from the minds of military people and those who have responsibilities in security. They do not care about "perfect" ludic reasoning; they want realistic ecological assumptions. In the end, they care about lives.

I mentioned in Chapter 11 how those who started the game of "formal thinking," by manufacturing phony premises in order to generate "rigorous" theories, were Paul Samuelson, Merton's tutor, and, in the United Kingdom, John Hicks. These two wrecked the ideas of John Maynard Keynes, which they tried to formalize (Keynes was interested in uncertainty, and complained about the mind-closing certainties induced by models). Other participants in the formal thinking venture were Kenneth Arrow and Gerard Debreu. All four were Nobeled. All four were in a delusional state under the effect of mathematics—what Dieudonné called "the music of reason," and what I call Locke's madness. All of them can be safely accused of having invented an imaginary world, one that lent itself to their mathematics. The insightful scholar Martin Shubik, who held that the degree of excessive abstraction of these models, a few steps beyond necessity, makes them totally unusable, found himself ostracized, a common fate for dissenters.*

If you question what they do, as I did with Merton Jr., they will ask for "tight proof." So they set the rules of the game, and you need to play by them. Coming from a practitioner background in which the principal asset is being able to work with messy, but empirically acceptable, mathematics,

* Medieval medicine was also based on equilibrium ideas when it was top-down and similar to theology. Luckily its practitioners went out of business, as they could not compete with the bottom-up surgeons, ecologically driven former barbers who gained clinical experience, and after whom a-Platonic clinical science was born. If I am alive, today, it is because scholastic top-down medicine went out of business a few centuries ago.

TABLE 4: TWO WAYS TO APPROACH RANDOMNESS

Skeptical Empiricism and the a-Platonic School	***The Platonic Approach***
Interested in what lies outside the Platonic fold	Focuses on the inside of the Platonic fold
Respect for those who have the guts to say "I don't know"	"You keep criticizing these models. These models are *all* we have."
Fat Tony	Dr. John
Thinks of Black Swans as a dominant source of randomness	Thinks of ordinary fluctuations as a dominant source of randomness, with jumps as an afterthought
Bottom-up	Top-down
Would ordinarily not wear suits (except to funerals)	Wears dark suits, white shirts; speaks in a boring tone
Prefers to be broadly right	Precisely wrong
Minimal theory, consides theorizing as a disease to resist	Everything needs to fit some grand, general socioeconomic model and "the rigor of economic theory"; frowns on the "descriptive"
Does not believe that we can easily compute probabilities	Built their entire apparatus on the assumptions that we can compute probabilities
Model: Sextus Empiricus and the school of evidence-based, minimum-theory empirical medicine	Model: Laplacian mechanics, the world and the economy like a clock
Develops intuitions from practice, goes from observations to books	Relies on scientific papers, goes from books to practice
Not inspired by any science, uses messy mathematics and computational methods	Inspired by physics, relies on abstract mathematics
Ideas based on skepticism, on the unread books in the library	Ideas based on beliefs, on what they think they know
Assumes Extremistan as a starting point	Assumes Mediocristan as a starting point
Sophisticated craft	Poor science
Seeks to be approximately right across a broad set of eventualities	Seeks to be perfectly right in a narrow model, under precise assumptions

I cannot accept a pretense of science. I much prefer a sophisticated craft, focused on tricks, to a failed science looking for certainties. Or could these neoclassical model builders be doing something worse? Could it be that they are involved in what Bishop Huet calls the manufacturing of certainties?

Let us see.

Skeptical empiricism advocates the opposite method. I care about the premises more than the theories, and I want to minimize reliance on theories, stay light on my feet, and reduce my surprises. I want to be broadly right rather than precisely wrong. Elegance in the theories is often indicative of Platonicity and weakness—it invites you to seek elegance for elegance's sake. A theory is like medicine (or government): often useless, sometimes necessary, always self-serving, and on occasion lethal. So it needs to be used with care, moderation, and close adult supervision.

The distinction in the above table between my model modern, skeptical empiricist and what Samuelson's puppies represent can be generalized across disciplines.

I've presented my ideas in finance because that's where I refined them. Let us now examine a category of people expected to be more thoughtful: the philosophers.

Chapter Eighteen

THE UNCERTAINTY OF THE PHONY

Philosophers in the wrong places—Uncertainty about (mostly) lunch—What I don't care about—Education and intelligence

This final chapter of Part Three focuses on a major ramification of the ludic fallacy: how those whose job it is to make us aware of uncertainty fail us and divert us into bogus certainties through the back door.

LUDIC FALLACY REDUX

I have explained the ludic fallacy with the casino story, and have insisted that the sterilized randomness of games does not resemble randomness in real life. Look again at Figure 7 in Chapter 15. The dice average out so quickly that I can say with certainty that the casino will beat me in the very near long run at, say, roulette, as the noise will cancel out, though not the skills (here, the casino's advantage). The more you extend the period (or reduce the size of the bets) the more randomness, by virtue of averaging, drops out of these gambling constructs.

The ludic fallacy is present in the following chance setups: random walk, dice throwing, coin flipping, the infamous digital "heads or tails" expressed as 0 or 1, Brownian motion (which corresponds to the movement of pollen particles in water), and similar examples. These setups generate a quality of randomness that does not even qualify as randomness—*protorandomness* would be a more appropriate designation. At their core, all theories built around the ludic fallacy ignore a layer of uncertainty. Worse, their proponents do not know it!

One severe application of such focus on small, as opposed to large, uncertainty concerns the hackneyed *greater uncertainty principle.*

Find the Phony

The greater uncertainty principle states that in quantum physics, one cannot measure certain pairs of values (with arbitrary precision), such as the position and momentum of particles. You will hit a lower bound of measurement: what you gain in the precision of one, you lose in the other. So there is an incompressible uncertainty that, in theory, will defy science and forever remain an uncertainty. This minimum uncertainty was discovered by Werner Heisenberg in 1927. I find it ludicrous to present the uncertainty principle as having anything to do with uncertainty. Why? First, this uncertainty is Gaussian. On average, it will disappear—recall that no one person's weight will significantly change the total weight of a thousand people. We may always remain uncertain about the future positions of small particles, but these uncertainties are very small and very numerous, and they average out—for Pluto's sake, they average out! They obey the law of large numbers we discussed in Chapter 15. Most other types of randomness do not average out! If there is one thing on this planet that is not so uncertain, it is the behavior of a collection of subatomic particles! Why? Because, as I have said earlier, when you look at an object, composed of a collection of particles, the fluctuations of the particles tend to balance out.

But political, social, and weather events do not have this handy property, and we patently cannot predict them, so when you hear "experts" presenting the problems of uncertainty in terms of subatomic particles, odds are that the expert is a phony. As a matter of fact, this may be the best way to spot a phony.

I often hear people say, "Of course there are limits to our knowledge," then invoke the greater uncertainty principle as they try to explain that "we cannot model everything"—I have heard such types as the economist Myron Scholes say this at conferences. But I am sitting here in New York, in August 2006, trying to go to my ancestral village of Amioun, Lebanon. Beirut's airport is closed owing to the conflict between Israel and the Shiite militia Hezbollah. There is no published airline schedule that will inform me when the war will end, if it ends. I can't figure out if my house will be standing, if Amioun will still be on the map—recall that the family house was destroyed once before. I can't figure out whether the war is going to degenerate into something even more severe. Looking into the outcome of the war, with all my relatives, friends, and property exposed to it, I face *true* limits of knowledge. Can someone explain to me why I should care about subatomic particles that, anyway, converge to a Gaussian? People can't predict how long they will be happy with recently acquired objects, how long their marriages will last, how their new jobs will turn out, yet it's subatomic particles that they cite as "limits of prediction." They're ignoring a mammoth standing in front of them in favor of matter even a microscope would not allow them to see.

Can Philosophers Be Dangerous to Society?

I will go further: people who worry about pennies instead of dollars can be dangerous to society. They mean well, but, invoking my Bastiat argument of Chapter 8, they are a threat to us. They are wasting our studies of uncertainty by focusing on the insignificant. Our resources (both cognitive and scientific) are limited, perhaps too limited. Those who distract us increase the risk of Black Swans.

This commoditization of the notion of uncertainty as symptomatic of Black Swan blindness is worth discussing further here.

Given that people in finance and economics are seeped in the Gaussian to the point of choking on it, I looked for financial economists with philosophical bents to see how their critical thinking allows them to handle this problem. I found a few. One such person got a PhD in philosophy, then, four years later, another in finance; he published papers in both fields, as well as numerous textbooks in finance. But I was disheartened by him: he seemed to have compartmentalized his ideas on uncertainty so that he had two distinct professions: philosophy and quantitative finance. The problem of induction, Mediocristan, epistemic opacity, or the offensive assumption of the Gaussian—these did not hit him as true problems. His numerous textbooks drilled Gaussian methods into students' heads, as though their author had forgotten that he was a philosopher. Then he promptly remembered that he was when writing philosophy texts on seemingly scholarly matters.

The same context specificity leads people to take the escalator to the StairMasters, but the philosopher's case is far, far more dangerous since he uses up our storage for critical thinking in a sterile occupation. Philosophers like to practice philosophical thinking on me-too subjects that other philosophers call philosophy, and they leave their minds at the door when they are outside of these subjects.

The Problem of Practice

As much as I rail against the bell curve, Platonicity, and the ludic fallacy, my principal problem is not so much with statisticians—after all, these are computing people, not thinkers. We should be far less tolerant of philosophers, with their bureaucratic apparatchiks closing our minds. Philosophers, the watchdogs of critical thinking, have duties beyond those of other professions.

HOW MANY WITTGENSTEINS CAN DANCE ON THE HEAD OF A PIN?

A number of semishabbily dressed (but thoughtful-looking) people gather in a room, silently looking at a guest speaker. They are all professional philosophers attending the prestigious weekly colloquium at a New York–area university. The speaker sits with his nose drowned in a set of typewritten pages, from which he reads in a monotone voice. He is hard to follow, so I daydream a bit and lose his thread. I can vaguely tell that the discussion revolves around some "philosophical" debate about Martians invading your head and controlling your will, all the while preventing you from knowing it. There seem to be several theories concerning this idea, but the speaker's opinion differs from those of other writers on the subject. He spends some time showing where his research on these head-hijacking Martians is unique. After his monologue (fifty-five minutes of relentless reading of the typewritten material) there is a short break, then another fifty-five minutes of discussion about Martians planting chips and other outlandish conjectures. Wittgenstein is occasionally mentioned (you can always mention Wittgenstein since he is vague enough to always seem relevant).

Every Friday, at four P.M., the paychecks of these philosophers will hit their respective bank accounts. A fixed proportion of their earnings, about 16 percent on average, will go into the stock market in the form of an automatic investment into the university's pension plan. These people are professionally employed in the business of questioning what we take for granted; they are trained to argue about the existence of god(s), the definition of truth, the redness of red, the meaning of meaning, the difference between the semantic theories of truth, conceptual and nonconceptual representations . . . Yet they believe blindly in the stock market, and in the abilities of their pension plan manager. Why do they do so? Because they accept that this is what people should do with their savings, because "experts" tell them so. They doubt their own senses, but not for a second do they doubt their automatic purchases in the stock market. This domain dependence of skepticism is no different from that of medical doctors (as we saw in Chapter 8).

Beyond this, they may believe without question that we can predict societal events, that the Gulag will toughen you a bit, that politicians know more about what is going on than their drivers, that the chairman of the Federal Reserve saved the economy, and so many such things. They may also believe that nationality matters (they always stick "French," "German," or "American" in front of a philosopher's name, as if this has something to do with anything he has to say). Spending time with these people, whose curiosity is focused on regimented on-the-shelf topics, feels stifling.

Where Is Popper When You Need Him?

I hope I've sufficiently drilled home the notion that, as a practitioner, my thinking is rooted in the belief that you cannot go from books to problems, but the reverse, from problems to books. This approach incapacitates much of that career-building verbiage. A scholar should not be a library's tool for making another library, as in the joke by Daniel Dennett.

Of course, what I am saying here has been said by philosophers before, at least by the real ones. The following remark is one reason I have inordinate respect for Karl Popper; it is one of the few quotations in this book that I am not attacking.

> The degeneration of philosophical schools in its turn is the consequence of the mistaken belief that one can philosophize without having been compelled to philosophize by problems outside philosophy. . . . *Genuine philosophical problems are always rooted outside philosophy and they die if these roots decay.* . . . [emphasis mine] These roots are

> easily forgotten by philosophers who "study" philosophy instead of being forced into philosophy by the pressure of nonphilosophical problems.

Such thinking may explain Popper's success outside philosophy, particularly with scientists, traders, and decision makers, as well as his relative failure inside of it. (He is rarely studied by his fellow philosophers; they prefer to write essays on Wittgenstein.)

Also note that I do not want to be drawn into philosophical debates with my Black Swan idea. What I mean by Platonicity is not so metaphysical. Plenty of people have argued with me about whether I am against "essentialism" (i.e., things that I hold don't have a Platonic essence), if I believe that mathematics would work in an alternative universe, or some such thing. Let me set the record straight. I am a no-nonsense practitioner; I am not saying that mathematics does not correspond to an objective structure of reality; my entire point is that we are, epistemologically speaking, putting the cart before the horse and, of the space of possible mathematics, risk using the wrong one and being blinded by it. I truly believe that there are some mathematics that work, but that these are not as easily within our reach as it seems to the "confirmators."

The Bishop and the Analyst

I am most often irritated by those who attack the bishop but somehow fall for the securities analyst—those who exercise their skepticism against religion but not against economists, social scientists, and phony statisticians. Using the confirmation bias, these people will tell you that religion was horrible for mankind by counting deaths from the Inquisition and various religious wars. But they will not show you how many people were killed by nationalism, social science, and political theory under Stalinism or during the Vietnam War. Even priests don't go to bishops when they feel ill: their first stop is the doctor's. But we stop by the offices of many pseudoscientists and "experts" without alternative. We no longer believe in papal infallibility; we seem to believe in the infallibility of the Nobel, though, as we saw in Chapter 17.

Easier Than You Think: The Problem of Decision Under Skepticism

I have said all along that there is a problem with induction and the Black Swan. In fact, matters are far worse: we may have no less of a problem with phony skepticism.

a. I can't do anything to stop the sun from nonrising tomorrow (no matter how hard I try),
b. I can't do anything about whether or not there is an afterlife,
c. I can't do anything about Martians or demons taking hold of my brain.

But I have plenty of ways to avoid being a sucker. It is not much more difficult than that.

I conclude Part Three by reiterating that my antidote to Black Swans is precisely to be noncommoditized in my thinking. But beyond avoiding being a sucker, this attitude lends itself to a protocol of how to act—not how to think, but how to convert knowledge into action and figure out what knowledge is worth. Let us examine what to do or not do with this in the concluding section of this book.

Part 4

THE END

Chapter Nineteen

HALF AND HALF, OR HOW TO GET EVEN WITH THE BLACK SWAN

The other half—Remember Apelles—When missing a train can be painful

---

It is now time for a few last words.

Half the time I am a hyperskeptic; the other half I hold certainties and can be intransigent about them, with a very stubborn disposition. Of course I am hyperskeptic where others, particularly those I call *bildungsphilisters,* are gullible, and gullible where others seem skeptical. I am skeptical about confirmation—though only when errors are costly—not about disconfirmation. Having plenty of data will not provide confirmation, but a single instance can disconfirm. I am skeptical when I suspect wild randomness, gullible when I believe that randomness is mild.

Half the time I hate Black Swans, the other half I love them. I like the randomness that produces the texture of life, the positive accidents, the success of Apelles the painter, the potential gifts you do not have to pay for. Few understand the beauty in the story of Apelles; in fact, most people exercise their error avoidance by repressing the Apelles in them.

Half the time I am hyperconservative in the conduct of my own affairs; the other half I am hyperaggressive. This may not seem exceptional, except that my conservatism applies to what others call risk taking, and my aggressiveness to areas where others recommend caution.

I worry less about small failures, more about large, potentially terminal ones. I worry far more about the "promising" stock market, particularly the "safe" blue chip stocks, than I do about speculative ventures—the former present invisible risks, the latter offer no surprises since you know how volatile they are and can limit your downside by investing smaller amounts.

I worry less about advertised and sensational risks, more about the more vicious hidden ones. I worry less about terrorism than about diabetes, less about matters people usually worry about because they are obvious worries, and more about matters that lie outside our consciousness and common discourse (I also have to confess that I do not worry a lot—I try to worry about matters I can do something about). I worry less about embarrassment than about missing an opportunity.

In the end this is a trivial decision making rule: I am very aggressive when I can gain exposure to positive Black Swans—when a failure would be of small moment—and very conservative when I am under threat from a negative Black Swan. I am very aggressive when an error in a model can benefit me, and paranoid when the error can hurt. This may not be too interesting except that it is exactly what other people do not do. In finance, for instance, people use flimsy theories to manage their risks and put wild ideas under "rational" scrutiny.

Half the time I am intellectual, the other half I am a no-nonsense practitioner. I am no-nonsense and practical in academic matters, and intellectual when it comes to practice.

Half the time I am shallow, the other half I want to avoid shallowness. I am shallow when it comes to aesthetics; I avoid shallowness in the context of risks and returns. My aestheticism makes me put poetry before prose, Greeks before Romans, dignity before elegance, elegance before culture, culture before erudition, erudition before knowledge, knowledge before intellect, and intellect before truth. But only for matters that are Black Swan free. Our tendency is to be very rational, except when it comes to the Black Swan.

Half the people I know call me irreverent (you have read my comments about your local Platonified professors), half call me fawning (you have seen my slavish devotion to Huet, Bayle, Popper, Poincaré, Montaigne, Hayek, and others).

Half the time I hate Nietzsche, the other half I like his prose.

WHEN MISSING A TRAIN IS PAINLESS

I once received another piece of life-changing advice, which, unlike the advice I got from a friend in Chapter 3, I find applicable, wise, and empirically valid. My classmate in Paris, the novelist-to-be Jean-Olivier Tedesco, pronounced, as he prevented me from running to catch a subway, "I don't run for trains."

Snub your destiny. I have taught myself to resist running to keep on schedule. This may seem a very small piece of advice, but it registered. In refusing to run to catch trains, I have felt the true value of *elegance* and aesthetics in behavior, a sense of being in control of my time, my schedule, and my life. *Missing a train is only painful if you run after it!* Likewise, not matching the idea of success others expect from you is only painful if that's what you are seeking.

You stand *above* the rat race and the pecking order, not *outside* of it, if you do so by choice.

Quitting a high-paying position, if it is *your* decision, will seem a better payoff than the utility of the money involved (this may seem crazy, but I've tried it and it works). This is the first step toward the stoic's throwing a four-letter word at fate. You have far more control over your life if you decide on your criterion by yourself.

Mother Nature has given us some defense mechanisms: as in Aesop's fable, one of these is our ability to consider that the grapes we cannot (or did not) reach are sour. But an aggressively stoic *prior* disdain and rejection of the grapes is even more rewarding. Be aggressive; be the one to resign, if you have the guts.

It is more difficult to be a loser in a game you set up yourself.

In Black Swan terms, this means that you are exposed to the improbable only if you let it control you. You always control what *you* do; so make this your end.

THE END

But all these ideas, all this philosophy of induction, all these problems with knowledge, all these wild opportunities and scary possible losses, everything palls in front of the following metaphysical consideration.

I am sometimes taken aback by how people can have a miserable day or get angry because they feel cheated by a bad meal, cold coffee, a social rebuff, or a rude reception. Recall my discussion in Chapter 8 on the difficulty in seeing the true odds of the events that run your own life. We are quick to forget that just being alive is an extraordinary piece of good luck, a remote event, a chance occurrence of monstrous proportions.

Imagine a speck of dust next to a planet a billion times the size of the earth. The speck of dust represents the odds in favor of your being born; the huge planet would be the odds against it. So stop sweating the small stuff. Don't be like the ingrate who got a castle as a present and worried about the mildew in the bathroom. Stop looking the gift horse in the mouth—remember that you are a Black Swan. And thank you for reading my book.

Epilogue

YEVGENIA'S WHITE SWANS

Yevgenia Krasnova went into the long hibernation that was necessary for producing a new book. She stayed in New York City, where she found it easiest to find tranquillity, alone with her text. It was easiest to concentrate after long periods during which she was surrounded by crowds, hoping to run into Nero so she could make a snide remark to him, perhaps humiliate him, possibly win him back. She canceled her e-mail account, switched to writing longhand, since she found it soothing, and hired a secretary to type her text. She spent eight years writing, erasing, correcting, venting her occasional anger at the secretary, interviewing new secretaries, and quietly rewriting. Her apartment was full of smoke, with papers strewn on every surface. Like all artists she remained dissatisfied with the state of completion of her work, yet she felt that she had gone far deeper than with her first book. She laughed at the public who extolled her earlier work, for she now found it shallow, hurriedly completed, and undistilled.

When the new book, which was aptly called *The Loop,* came out, Yevgenia was wise enough to avoid the press and ignore her reviews, and stayed insulated from the external world. As expected by her publisher, the reviews were laudatory. But, strangely, few were buying. People must be talking about the book without reading it, he thought. Her fans had been waiting for it and talking about it for years. The publisher, who now owned a very large collection of pink glasses and led a flamboyant lifestyle, was presently betting the farm on Yevgenia. He had no other hits and none in sight. He needed to score big to pay for his villa in Carpentras in Provence and his dues on the financial settlement with his estranged wife, as well as to buy a new convertible Jaguar (pink). He had been certain that he had a good shot with Yevgenia's long-awaited book, and he could not figure out why almost everyone called it a masterpiece yet no one was buying it. A year and a half later, *The Loop* was effectively out of print. The publisher, now in severe financial distress, thought he knew the reason: the book was "too f***ing long!"—Yevgenia should have written a shorter one. After a long but soothing lachrymal episode, Yevgenia thought of the characters in the rainy novels of Georges Simenon and Graham Greene. They lived in a state of numbing and secure mediocrity. Second-rateness had charm, Yevgenia thought, and she had always preferred charm over beauty.

So Yevgenia's second book too was a Black Swan.

ACKNOWLEDGMENTS

I derived an unexpected amount of enjoyment in writing this book—in fact, it just wrote itself—and I want the reader to experience the same. I would like to thank the following friends.

My friend and adviser Rolf Dobelli, the novelist, entrepreneur, and voracious reader, kept up with the various versions of this text. I also built up a large debt toward Peter Bevelin, an erudite and pure "thinking doer" with extreme curiosity who spends his waking hours chasing ideas and spotting the papers I am usually looking for; he scrutinized the text. Yechezkel Zilber, a Jerusalem-based idea-starved autodidact who sees the world *ab ovo,* from the egg, asked very tough questions, to the point of making me ashamed of the formal education I received and uncomfortable for not being a true autodidact like him—it is thanks to no-nonsense people that I am grounding my Black Swan idea in academic libertarianism. The scholar Philip Tetlock, who knows more about prediction than anyone since the Delphic times, went through the manuscript and scrutinized my arguments. Phil is so valuable and thorough that he was even more informational with the absence of comments than he was with his comments. I owe a big debt to Danny Kahneman who, in addition to the long conversations on my topics of human nature (and noting with horror that I remembered almost every comment), put me in contact with Phil Tetlock. I thank Maya Bar Hillel for inviting me to address the Society of Judgment and Decision Making at their annual meeting in Toronto in November 2005—thanks to the generosity of the researchers there, and the stimulating discussions, I came back having taken away far more than I gave. Robert Shiller asked me to purge some "irreverent" comments, but the fact that he criticized the aggressiveness of the delivery, but not the content, was quite informational. Mariagiovanna Muso was the first to become conscious of the Black Swan effect on the arts and sent me along the right lines of research in sociology and anthropology. I had long discussions with the literary scholar Mihai Spariosu on Plato, Balzac, ecological intelligence, and cafés in Bucharest. Didier Sornette, always a phone call away, kept e-mailing me papers on various unadvertised, but highly relevant, subjects in statistical physics. Jean-Philippe Bouchaud offered a great deal of help on the problems associated with the statistics of large deviations. Michael Allen wrote a monograph for writers looking to get published, based on the ideas of Chapter 8—I subsequently rewrote Chapter 8 through the eyes of a writer looking at his lot in life. Mark Blyth was always helpful as a sounding board, reader, and adviser. My friends at the DoD, Andy Marshall and Andrew Mays, supplied me with ideas and questions. Paul Solman, a voracious mind, went through the manuscript with severe scrutiny. I owe the term *Extremistan* to Chris Anderson, who found my earlier designation too bookish. Nigel Harvey guided me through the literature on forecasting.

I plied the following scientists with questions: Terry Burnham, Robert Trivers, Robyn Dawes, Peter Ayton, Scott Atran, Dan Goldstein, Alexander Reisz, Art De Vany, Raphael Douady, Piotr Zielonka, Gur Huberman, Elkhonon Goldberg, and Dan Sperber. Ed Thorp, the true living owner of the "Black-Scholes formula" was helpful; I realized, speaking to him, that economists ignore intellectual productions outside their club—regardless how valuable. Lorenzo Perilli was extremely generous with his comments about Menodotus and helped correct a few errors. Duncan Watts allowed me to present the third part of this book at a Columbia University seminar in sociology and collect all manner of comments. David Cowan supplied the graph in the Poincaré discussion, making mine pale by comparison. I also benefited from James Montier's wonderful brief pieces on human nature. Bruno Dupire, as always, provides the best walking conversations.

It does not pay to be the loyal friend of a pushy author too close to his manuscript. Marie-Christine Riachi was given the thankless task of reading chapters in inverse order; I only gave her the incomplete pieces and, of those, only the ones (then) patently lacking in clarity. Jamil Baz received the full text every time but chose to read it backwards. Laurence Zuriff read and commented on every chapter. Philip Halperin, who knows more about risk management than anyone (still) alive, offered wonderful comments and observations. Other victims: Cyrus Pirasteh, Bernard Oppetit, Pascal Boulard, Guy Riviere, Joelle Weiss, Didier Javice, Andreea Munteanu, Andrei Pokrovsky, Philippe Asseily, Farid Karkaby, George Nasr, Alina Stefan, George Martin, Stan Jonas, and Flavia Cymbalista.

I received helpful comments from the voracious intellectual Paul Solman (who went through the manuscript with a microscope). I owe a lot to Phil Rosenczweig, Avishai Margalit, Peter Forbes, Michael Schrage, Driss Ben Brahim, Vinay Pande, Antony Van Couvering, Nicholas Vardy, Brian Hinchcliffe, Aaron Brown, Espen Haug, Neil Chriss, Zvika Afik, Shaiy Pilpel, Paul Kedrosky, Reid Bernstein, Claudia Schmid, Jay Leonard, Tony Glickman, Paul Johnson, Chidem Kurdas (and the NYU Austrian economists), Charles Babbitt, plus so many anonymous persons I have forgotten about* . . .

Ralph Gomory and Jesse Ausubel of the Sloan Foundation run a research funding program called the Known, the Unknown, and the Unknowable. They offered their moral and financial help for the promotion of my ideas—I took the invaluable moral option. I also thank my business partners, coauthors, and intellectual associates: Espen Haug, Mark Spitznagel, Benoît Mandelbrot, Tom Witz, Paul Wilmott, Avital Pilpel, and Emanuel Derman. I also thank John Brockman and Katinka Matson for making this book possible, and Max Brockman for his comments on the draft. I thank Cindy, Sarah, and Alexander for their tolerance. In addition, Alexander helped with the graphs and Sarah worked on the bibliography.

I tried to give my editor, Will Murphy, the impression of being an unbearably stubborn author, only to discover that I was fortunate that he was an equally stubborn editor (but good at hiding it). He protected me from the intrusions of the standardizing editors. They have an uncanny ability to inflict maximal damage by breaking the internal rhythm of one's prose with the minimum of changes. Will M. is also the right kind of party

* I lost his business card, but would like to warmly thank a scientist traveling to Vienna aboard British Airways flight 700 on December 11, 2003, for suggesting the billiard ball illustration in Chapter 11. All I know about him is that he was fifty-two, gray-haired, English-born, wrote poetry on yellow notepads, and was traveling with seven suitcases since he was moving in with his thirty-five-year-old Viennese girlfriend.

animal. I was also flattered that Daniel Menaker took the time to edit my text. I also thank Janet Wygal and Steven Meyers. The staff at Random House was accommodating—but they never got used to my phone pranks (like my trying to pass for Bernard-Henri Lévy). One of the highlights of my writing career was a long lunch with William Goodlad, my editor at Penguin, and Stefan McGrath, the managing director of the group. I suddenly realized that I could not separate the storyteller in me from the scientific thinker; as a matter of fact, the story came first to my mind, rather than as an after-the-fact illustration of the concept.

Part Three of this book inspired my class lectures at the University of Massachusetts at Amherst. I also thank my second home, the Courant Institute of Mathematical Sciences of New York University, for allowing me to lecture for three quarters of a decade.

It is unfortunate that one learns most from people one disagrees with—something Montaigne encouraged half a millennium ago but is rarely practiced. I discovered that it puts your arguments through robust seasoning since you know that these people will identify the slightest crack—and you get information about thc limits of their theories as well as the weaknesses of your own. I tried to be more graceful with my detractors than with my friends—particularly those who were (and stayed) civilized. So, over my career, I learned a few tricks from a series of public debates, correspondence, and discussions with Robert C. Merton, Steve Ross, Myron Scholes, Philippe Jorion, and dozens of others (though, aside from Elie Ayache's critique, the last time I heard something remotely new against my ideas was in 1994). These debates were valuable since I was looking for the extent of the counterarguments to my Black Swan idea and trying to figure out how my detractors think—or what they did not think about. Over the years I have ended up reading more material from those I disagree with than from those whose opinion I share—I read more Samuelson than Hayek, more Merton (the younger) than Merton (the elder), more Hegel than Montaigne, and more Descartes than Sextus. It is the duty of every author to represent the ideas of his adversaries as faithfully as possible.

My greatest accomplishment in life is to have managed to befriend people, such as Elie Ayache and Jim Gatheral, in spite of some intellectual disagreements.

Most of this book was written during a peripatetic period when I freed myself of (almost) all business obligations, routines, and pressures, and went on meditative urban walks in a variety of cities where I gave a series of lectures on the Black Swan idea.* I wrote it largely in cafés—my preference has always been for dilapidated (but elegant) cafés in regular neighborhoods, as unpolluted with persons of commerce as possible. I also spent much time in Heathrow Terminal 4, absorbed in my writing to the point that I forgot about my allergy to the presence of strained businessmen around me.

* It is impossible to go very deep into an idea when you run a business, no matter the number of hours the occupation entails—simply put, unless you are insensitive, the worries and feelings of responsibility occupy precious cognitive space. You may be able to study, meditate, and write if you are an employee, but not when you own a business—unless you are of an irresponsible nature. I thank my partner, Mark Spitznagel, for allowing me—thanks to the clarity of his mind and his highly systematic, highly disciplined, and well engineered approach—to gain exposure to high-impact rare events without my having to get directly involved in business activities.

GLOSSARY

Academic libertarian: someone (like myself) who considers that knowledge is subjected to strict rules but not institutional authority, as the interest of organized knowledge is self-perpetuation, not necessarily truth (as with governments). Academia can suffer from an acute **expert problem** (q.v.), producing cosmetic but fake knowledge, particularly in **narrative disciplines** (q.v.), and can be a main source of Black Swans.

Apelles-style strategy: A strategy of seeking gains by collecting positive accidents from maximizing exposure to "good Black Swans."

Barbell strategy: a method that consists of taking both a defensive attitude and an excessively aggressive one at the same time, by protecting assets from all sources of uncertainty while allocating a small portion for high-risk strategies.

Bildungsphilister: a philistine with cosmetic, nongenuine culture. Nietzsche used this term to refer to the dogma-prone newspaper reader and opera lover with cosmetic exposure to culture and shallow depth. I extend it to the buzzword-using researcher in nonexperimental fields who lacks in imagination, curiosity, erudition, and culture and is closely centered on his ideas, on his "discipline." This prevents him from seeing the conflicts between his ideas and the texture of the world.

Black Swan blindness: the underestimation of the role of the Black Swan, and occasional overestimation of a specific one.

Black Swan ethical problem: Owing to the nonrepeatable aspect of the Black Swan, there is an asymmetry between the rewards of those who prevent and those who cure.

Confirmation error (or Platonic confirmation): You look for instances that confirm your beliefs, your construction (or model)—and find them.

Empty-suit problem (or "expert problem"): Some professionals have no differential abilities from the rest of the population, but for some reason, and against their empirical records, are believed to be experts: clinical psychologists, academic economists, risk "experts," statisticians, political analysts, financial "experts," military analysts, CEOs, et cetera. They dress up their expertise in beautiful language, jargon, mathematics, and often wear expensive suits.

Epilogism: A theory-free method of looking at history by accumulating facts with minimal generalization and being conscious of the side effects of making causal claims.

Epistemic arrogance: Measure the difference between what someone actually knows and how much he thinks he knows. An excess will imply arrogance, a deficit humility. An epistemocrat is someone of epistemic humility, who holds his own knowledge in greatest suspicion.

Epistemic opacity: Randomness is the result of incomplete information at some layer. It is functionally indistinguishable from "true" or "physical" randomness.

Extremistan: the province where the total can be conceivably impacted by a single observation.

Fallacy of silent evidence: Looking at history, we do not see the full story, only the rosier parts of the process.

Fooled by randomness: the general confusion between luck and determinism, which leads to a variety of superstitions with practical consequences, such as the belief that higher earnings in some professions are generated by skills when there is a significant component of luck in them.

Future blindness: our natural inability to take into account the properties of the future—like autism, which prevents one from taking into account the existence of the minds of others.

Locke's madman: someone who makes impeccable and rigorous reasoning from faulty premises—such as Paul Samuelson, Robert Merton the minor, and Gerard Debreu—thus producing phony models of uncertainty that make us vulnerable to Black Swans.

Lottery-ticket fallacy: the naïve analogy equating an investment in collecting positive Black Swans to the accumulation of lottery tickets. Lottery tickets are not scalable.

Ludic fallacy (or uncertainty of the nerd): the manifestation of the Platonic fallacy in the study of uncertainty; basing studies of chance on the narrow world of games and dice. A-Platonic randomness has an additional layer of uncertainty concerning the rules of the game in real life. The bell curve (Gaussian), or GIF (Great Intellectual Fraud), is the application of the ludic fallacy to randomness.

Mandelbrotian Gray Swan: Black Swans that we can somewhat take into account—earthquakes, blockbuster books, stock market crashes—but for which it is not possible to completely figure out the properties and produce precise calculations.

Mediocristan: the province dominated by the mediocre, with few extreme successes or failures. No single observation can meaningfully affect the aggregate. The bell curve is grounded in Mediocristan. There is a qualitative difference between Gaussians and scalable laws, much like gas and water.

Narrative discipline: the discipline that consists in fitting a convincing and well-sounding story to the past. Opposed to experimental discipline.

Narrative fallacy: our need to fit a story or pattern to a series of connected or disconnected facts. The statistical application is data mining.

Nerd knowledge: the belief that what cannot be Platonized and studied does not exist at all, or is not worth considering. There even exists a form of skepticism practiced by the nerd.

Platonic fold: the place where our Platonic representation enters into contact with reality and you can see the side effects of models.

Platonicity: the focus on those pure, well-defined, and easily discernible objects like triangles, or more social notions like friendship or love, at the cost of ignoring those objects of seemingly messier and less tractable structures.

Probability distribution: the model used to calculate the odds of different events, how they are "distributed." When we say that an event is distributed according to the bell curve, we mean that the Gaussian bell curve can help provide probabilities of various occurrences.

Problem of induction: the logical-philosophical extension of the Black Swan problem.

Randomness as incomplete information: simply, what I cannot guess is random because my knowledge about the causes is incomplete, not necessarily because the process has truly unpredictable properties.

Retrospective distortion: examining past events without adjusting for the forward passage of time. It leads to the illusion of posterior predictability.

Reverse-engineering problem: It is easier to predict how an ice cube would melt into a puddle than, looking at a puddle, to guess the shape of the ice cube that may have caused it. This "inverse problem" makes narrative disciplines and accounts (such as histories) suspicious.

Round-trip fallacy: the confusion of absence of evidence of Black Swans (or something else) for evidence of absence of Black Swans (or something else). It affects statisticians and other people who have lost part of their reasoning by solving too many equations.

Scandal of prediction: the poor prediction record in some forecasting entities (particularly narrative disciplines) mixed with verbose commentary and a lack of awareness of their own dire past record.

Scorn of the abstract: favoring contextualized thinking over more abstract, though more relevant, matters. "The death of one child is a tragedy; the death of a million is a statistic."

Statistical regress argument (or the problem of the circularity of statistics): We need data to discover a probability distribution. How do we know if we have enough? From the probability distribution. If it is a Gaussian, then a few points of data will suffice. How do we know it is a Gaussian? From the data. So we need the data to tell us what probability distribution to assume, and we need a probability distribution to tell us how much data we need. This causes a severe regress argument, which is somewhat shamelessly circumvented by resorting to the Gaussian and its kin.

Uncertainty of the deluded: people who tunnel on sources of uncertainty by producing precise sources like the great uncertainty principle, or similar, less consequential matters, to real life; worrying about subatomic particles while forgetting that we can't predict tomorrow's crises.

NOTES

BEHIND THE CURTAIN: ADDITIONAL NOTES, TECHNICAL COMMENTS, REFERENCES, AND READING RECOMMENDATIONS

I separate topics thematically; so general references will mostly be found in the chapter in which they first occur. I prefer to use a logical sequence here rather than stick to chapter division.

PROLOGUE and CHAPTER 1

Black Swan in logic: First, mine is not a problem in logic. The philosophical problem is about the possibility of a Black Swan. Mine is about the *impact*. Also, it may not be too relevant who came up with the metaphor first, but the earliest mention of Black Swan problem I could find is in John Stuart Mill's *A System of Logic*. It was later used by many (including Charles Sanders Peirce) before it became associated with Karl Popper.

Bell curve: When I write *bell curve* I mean the Gaussian bell curve, a.k.a. normal distribution. All curves look like bells, so this is a nickname. Also, when I write *the Gaussian basin* I mean all distributions that are similar and for which the improbable is inconsequential and of low impact (more technically, nonscalable—all moments are finite). Note that the visual presentation of the bell curve in histogram form masks the contribution of the remote event, as such an event will be a point to the far right or far left of the center.

Diamonds: See Eco (2002).

Platonicity: I'm simply referring to incurring the risk of using a wrong form—not that forms don't exist. I am not against essentialisms; I am often skeptical of our reverse engineering and identification of the right form. It is an inverse problem!

Empiricist: If I call myself an empiricist, or an empirical philosopher, it is because I am just suspicious of confirmatory generalizations and hasty theorizing. Do not confuse this with the British empiricist tradition. Also, many statisticians, as we will see with the Makridakis competition, call themselves "empirical" researchers, but are in fact just the opposite—they fit theories to the past.

Mention of Christ: See Flavius Josephus's *The Jewish War.*

Great War and prediction: Ferguson (2006b).

Hindsight bias (retrospective distortion): See Fischhoff (1982b).
Historical fractures: Braudel (1985), p. 169, quotes a little known passage from Gautier. He writes, " 'This long history,' wrote Emile-Félix Gautier, 'lasted a dozen centuries, longer than the entire history of France. Encountering the first Arab sword, the Greek language and thought, all that heritage went up in smoke, as if it never happened.' " For discussions of discontinuity, see also Gurvitch (1957), Braudel (1953), Harris (2004).
Religions spread as bestsellers: Veyne (1971). See also Veyne (2005).
Clustering in political opinions: Pinker (2002).
Categories: Rosch (1973, 1978). See also Umberto Eco's *Kant and the Platypus.*
Ontological uncertainty: Some of the literature discusses my categorization problem as *ontological uncertainty,* meaning there can be uncertainty concerning the entities themselves.
Historiography and philosophy of history: Bloch (1953), Carr (1961), Gaddis (2002), Braudel (1969, 1990), Bourdé and Martin (1989), Certeau (1975), *Muqaddamat* Ibn Khaldoun illustrate the search for causation, which we see already present in Herodotus. For philosophy of history, Aron (1961), Fukuyama (1992). For postmodern views, see Jenkins (1991). I show in Part Two how historiographers are unaware of the epistemological difference between forward and backward processes (i.e., between projection and reverse engineering).
Information and markets: See Shiller (1981, 1989), DeLong et al. (1991), and Cutler et al. (1989). The bulk of market moves does not have a "reason," just a contrived explanation.
Of descriptive value for crashes: See Galbraith (1997), Shiller (2000), and Kindleberger (2001).

CHAPTER 3

Movies: See De Vany (2002). See also Salganik et al. (2006) for the contagion in music buying.
Religion and domains of contagion: See Boyer (2001).
Wisdom (madness) of crowds: Collectively, we can both get wiser or far more foolish. We may collectively have intuitions for Mediocristan-related matters, such as the weight of an ox (see Surowiecki, 2004), but my conjecture is that we fail in more complicated predictions (economic variables for which crowds incur pathologies—two heads are worse than one). For decision errors and groups, see Sniezek and Buckley (1993). Classic: Charles Mackay's *Extraordinary Popular Delusions and the Madness of Crowds.*
Increase in the severity of events: Zajdenweber (2000).
Modern life: The nineteenth-century novelist Émile Zola welcomed the arrival of the market for culture in the late 1800s, of which he seemed to be one of the first beneficiaries. He predicted that the writers' and artists' ability to exploit the commercial system freed them from a dependence on patrons' whims. Alas, this was accompanied with more severe concentration—very few people benefited from the system. Lahire (2006) shows how most writers, throughout history, have starved. Remarkably, we have ample data from France about the literary tradition.

CHAPTER 4

Titanic: The quote is from Dave Ingram's presentation at the Enterprise Risk Management Symposium in Chicago on May 2, 2005. For more on LTCM, see Lowenstein (2000), Dunbar (1999).
Hume's exposition: Hume (1748, 2000).
Sextus Empriricus: "It is easy, I think, to reject the method of induction (επαγωγη). For since by way of it they want to make universals convincing on the basis of particu-

lars, they will do this surveying all the particulars or some of them. But if some, the induction will be infirm, it being that some of the particulars omitted in the induction should be contrary to the universal; and if all, they will labor at an impossible task, since the particulars and infinite are indeterminate. Thus in either case it results, I think, that induction totters." *Outline of Pyrrhonism,* Book II, p. 204.

Bayle: The *Dictionnaire historique et critique* is long (twelve volumes, close to 6,000 pages) and heavy (40 pounds), yet it was an intellectual bestseller in its day, before being supplanted by the *philosophes.* It can be downloaded from the French Bibliothèque Nationale at www.bn.fr.

Hume's inspiration from Bayle: See Popkin (1951, 1955). Any reading of Bishop Huet (further down) would reveal the similarities with Hume.

Pre-Bayle thinkers: *Dissertation sur la recherche de la vérité,* Simon Foucher, from around 1673. It is a delight to read. It makes the heuristics and biases tradition look like the continuation of the pre-Enlightenment prescientific revolution atmosphere.

Bishop Huet and the problem of induction: "Things cannot be known with perfect certainty because their causes are infinite," wrote Pierre-Daniel Huet in his *Philosophical Treatise on the Weaknesses of the Human Mind.* Huet, former bishop of Avranches, wrote this under the name Théocrite de Pluvignac, Seigneur de la Roche, Gentilhomme de Périgord. The chapter has another exact presentation of what became later known as "Hume's problem." That was in 1690, when the future David Home (later Hume) was minus twenty-two, so of no possible influence on Monseigneur Huet.

Brochard's work: I first encountered the mention of Brochard's work (1888) in Nietzsche's *Ecce Homo,* in a comment where he also describes the skeptics as straight talkers. "An excellent study by Victor Brochard, *Les sceptiques grecs,* in which my Laertiana are also employed. The skeptics! the only *honourable* type among the two and five fold ambiguous philosopher crowd!" *More* trivia: Brochard taught Proust (see Kristeva, 1998).

Brochard seems to have understood Popper's problem (a few decades before Popper's birth). He presents the views of the negative empiricism of Menodotus of Nicomedia in similar terms to what we would call today "Popperian" empiricism. I wonder if Popper knew anything about Menodotus. He does not seem to quote him anywhere. Brochard published his doctoral thesis, *De l'erreur,* in 1878 at the University of Paris, on the subject of error—wonderfully modern.

Epilogism: We know very little about Menodotus except for attacks on his beliefs by his detractor Galen in the extant Latin version of the *Outline of Empiricism* (*Subfiguratio empirica),* hard to translate:

> *Memoriam et sensum et vocans* epilogismum *hoc tertium, multotiens autem et preter memoriam nihil aliud ponens quam* epilogismum. (In addition to perception and recollection, the third method is *epilogism sensum,* as the practitioner has, besides memory, nothing other than *epilogism senses;* Perilli's correction.

But there is hope. Perilli (2004) reports that, according to a letter by the translator Is-haq Bin Hunain, there may be a "transcription" of Menodotus's work in Arabic somewhere for a scholar to find.

Pascal: Pascal too had an idea of the confirmation problem and the asymmetry of inference. In his preface to the *Traité du vide,* Pascal writes (and I translate):

> In the judgment they made that nature did not tolerate a vacuum, they only meant nature in the state in which they knew it, since, so claim so in general, it would not be sufficient to witness it in a hundred different encounters, nor in a thousand, not in any other number no matter how large, since it would be a single case that would deny the general definition, and if one was contrary, a single one . . .

Hume's biographer: Mossner (1970). For a history of skepticism, Victor Cousin's lectures *Leçons d'histoire de la philosophie à la Sorbonne* (1828) and Hippolyte Taine's *Les philosophes classiques,* 9th edition (1868, 1905). Popkin (2003) is a modern account. Also see Heckman (2003) and Bevan (1913). I have seen nothing in the modern philosophy of probability linking it to skeptical inquiry.

Sextus: See Popkin (2003), Sextus, House (1980), Bayle, Huet, Annas and Barnes (1985), and Julia Anna and Barnes's introduction in Sextus Empiricus (2000). Favier (1906) is hard to find; the only copy I located, thanks to Gur Huberman's efforts, was rotten—it seems that it has not been consulted in the past hundred years.

Menodotus of Nicomedia and the marriage between empiricism and skepticism: According to Brochard (1887), Menodotus is responsible for the mixing of empiricism and Pyrrhonism. See also Favier (1906). See skepticism about this idea in Dye (2004), and Perilli (2004).

Function not structure; empirical tripod: There are three sources, and three only, for experience to rely upon: observation, history (i.e., recorded observation), and judgment by analogy.

Algazel: See his *Tahafut al falasifah,* which is rebutted by Averroës, a.k.a. Ibn-Rushd, in *Tahafut Attahafut.*

Religious skeptics: There is also a medieval Jewish tradition, with the Arabic-speaking poet Yehuda Halevi. See Floridi (2002).

Algazel and the ultimate/proximate causation: ". . . their determining, from the sole observation, of the nature of the necessary relationship between the cause and the effect, as if one could not witness the effect without the attributed cause of the cause without the same effect." (*Tahafut*)

At the core of Algazel's idea is the notion that if you drink because you are thirsty, thirst should not be seen as a *direct* cause. There may be a greater scheme being played out; in fact, there *is,* but it can only be understood by those familiar with evolutionary thinking. See Tinbergen (1963, 1968) for a modern account of the proximate. In a way, Algazel builds on Aristotle to attack him. In his *Physics,* Aristotle had already seen the distinction between the different layers of cause (formal, efficient, final, and material).

Modern discussions on causality: See Reichenbach (1938), Granger (1999), and Pearl (2000).

Children and natural induction: See Gelman and Coley (1990), Gelman and Hirschfeld (1999), and Sloman (1993).

Natural induction: See Hespos (2006), Clark and Boyer (2006), Inagaki and Hatano (2006), Reboul (2006). See summary of earlier works in Plotkin (1998).

CHAPTERS 5–7

"Economists": What I mean by "economists" are most members of the mainstream, neoclassical economics and finance establishment in universities—not fringe groups such as the Austrian or the Post-Keynesian schools.

Small numbers: Tversky and Kahneman (1971), Rabin (2000).

Domain specificity: Williams and Connolly (2006). We can see it in the usually overinterpreted Wason Selection Test: Wason (1960, 1968). See also Shaklee and Fischhoff (1982), Barron Beaty, and Hearshly (1988). Kahneman's "They knew better" in Gilovich et al. (2002).

Updike: The blurb is from Jaynes (1976).

Brain hemispheric specialization: Gazzaniga and LeDoux (1978), Gazzaniga et al. (2005). Furthermore, Wolford, Miller, and Gazzaniga (2000) show probability matching by the left brain. When you supply the right brain with, say, a lever that produces desirable goods 60% of the time, and another lever 40%, the right brain will correctly push the first lever as the optimal policy. If, on the other hand, you supply the left brain with the same options, it will push the first lever 60 percent of the time and the other one 40—it will refuse to accept randomness. Goldberg (2005) argues that the specialty is along different lines: left-brain damage does not bear severe effects in children, unlike right-brain lesions, while this is the reverse for the elderly. I thank Elkhonon Goldberg for referring me to Snyder's work; Snyder (2001). The experiment is from Snyder et al. (2003).

Sock selection and retrofit explanation: The experiment of the socks is presented in Carter (1999); the original paper appears to be Nisbett and Wilson (1977). See also Montier (2007).

Astebro: Astebro (2003). See "Searching for the Invisible Man," *The Economist,* March 9, 2006. To see how the overconfidence of entrepreneurs can explain the high failure rate, see Camerer (1995).

Dopamine: Brugger and Graves (1997), among many other papers. See also Mohr et al. (2003) on dopamine asymmetry.

Entropy and information: I am purposely avoiding the notion of entropy because the way it is conventionally phrased makes it ill-adapted to the type of randomness we experience in real life. Tsallis entropy works better with fat tails.

Notes on George Perec: Eco (1994).

Narrativity and illusion of understanding: Wilson, Gilbert, and Centerbar (2003): "Helplessness theory has demonstrated that if people feel that they cannot control or predict their environments, they are at risk for severe motivational and cognitive deficits, such as depression." For the writing down of a diary, see Wilson (2002) or Wegner (2002).

E. M. Forster's example: reference in Margalit (2002).

National character: Terracciano et al. (2005) and Robins (2005) for the extent of individual variations. The illusion of nationality trait, which I usually call the "nationality heuristic," does connect to the halo effect: see Rosenzweig (2006) and Cialdini (2001). See Anderson (1983) for the ontology of nationality.

Consistency bias: What psychologists call the consistency bias is the effect of revising memories in such a way to make sense with respect to subsequent information. See Schacter (2001).

Memory not like storage on a computer: Rose (2003), Nader and LeDoux (1999).

The myth of repressed memory: Loftus and Ketcham (2004).

Chess players and disconfirmation: Cowley and Byrne (2004).

Quine's problem: Davidson (1983) argues in favor of local, but against total, skepticism.

Narrativity: Note that my discussion is not existential here, but merely practical, so my idea is to look at narrativity as an informational compression, nothing more involved philosophically (like whether a self is sequential or not). There is a literature on the "narrative self"—Bruner (2002) or whether it is necessary—see Strawson (1994) and his attack in Strawson (2004). The debate: Schechtman (1997), Taylor (1999), Phelan (2005). Synthesis in Turner (1996).

"Postmodernists" and the desirability of narratives: See McCloskey (1990) and Frankfurter and McGoun (1996).

Narrativity of sayings and proverbs: Psychologists have long examined the gullibility of people in social settings when faced with well-sounding proverbs. For instance, experiments have been made since the 1960s where people are asked whether they believe that a proverb is right, while another cohort is presented with the opposite meaning. For a presentation of the hilarious results, see Myers (2002).

Science as a narrative: Indeed scientific papers can succeed by the same narrativity bias that "makes a story." You need to get attention. Bushman and Wells (2001).

Discovering probabilities: Barron and Erev (2003) show how probabilities are underestimated when they are not explicitly presented. Also personal communication with Barron.

Risk and probability: See Slovic, Fischhoff, and Lichtenstein (1976), Slovic et al. (1977), and Slovic (1987). For risk as analysis and risk as feeling theory, see Slovic et al. (2002, 2003), and Taleb (2004c). See Bar-Hillel and Wagenaar (1991).

Link between narrative fallacy and clinical knowledge: Dawes (1999) has a message for economists: see here his work on interviews and the concoction of a narrative. See also Dawes (2001) on the retrospective effect.

Two systems of reasoning: See Sloman (1996, 2002), and the summary in Kahneman and Frederick (2002). Kahneman's Nobel lecture sums it all up; it can be found at www.nobel.se. See also Stanovich and West (2000).

Risk and emotions: Given the growing recent interest in the emotional role in behavior, there has been a growing literature on the role of emotions in both risk bearing and risk avoidance: the "risk as feeling" theory. See Loewenstein et al. (2001) and Slovic et al. (2003a). For a survey see Slovic et al. (2003b) and see also Slovic (1987). For a discussion of the "affect heuristic," see Finucane et al. (2000). For modularity, see Bates (1994).

Emotions and cognition: For the effect of emotions on cognition, see LeDoux (2002). For risk, see Bechara et al. (1994).

Availability heuristic (how easily things come to mind): See Tversky and Kahneman (1973).

Real incidence of catastrophes: For an insightful discussion, see Albouy (2002), Zajdenweber (2000), or Sunstein (2002).

Terrorism exploitation of the sensational: See the essay in Taleb (2004c).

General books on psychology of decision making (heuristics and biases): Baron (2000) is simply the most comprehensive on the subject. Kunda (1999) is a summary from the standpoint of social psychology (sadly, the author died prematurely); shorter: Plous (1993). Also Dawes (1988) and Dawes (2001). Note that a chunk of the original papers are happily compiled in Kahneman et al. (1982), Kahneman and Tversky (2000), Gilovich et al. (2002), and Slovic (2001a and 2001b). See also Myers (2002) for an account on intuition and Gigerenzer et al. (2000) for an ecological presentation of the subject. The most complete account in economics and finance is Montier (2007), where his beautiful summary pieces that fed me for the last four years are compiled—not being an academic, he gets straight to the point. See also Camerer, Loewenstein, and Rabin (2004) for a selection of technical papers. A recommended review article on clinical "expert" knowledge is Dawes (2001).

More general psychology of decision presentations: Klein (1998) proposes an alternative model of intuition. See Cialdini (2001) for social manipulation. A more specialized work, Camerer (2003), focuses on game theory.

General review essays and comprehensive books in cognitive science: Newell and Simon (1972), Varela (1988), Fodor (1983), Marr (1982), Eysenck and Keane (2000), Lakoff and Johnson (1980). The *MIT Encyclopedia of Cognitive Science* has review articles by main thinkers.

Evolutionary theory and domains of adaptation: See the original Wilson (2000), Kreps and Davies (1993), and Burnham (1997, 2003). Very readable: Burnham and Phelan (2000). The compilation of Robert Trivers's work is in Trivers (2002). See also Wrangham (1999) on wars.

Politics: "The Political Brain: A Recent Brain-imaging Study Shows That Our Political Predilections Are a Product of Unconscious Confirmation Bias," by Michael Shermer, *Scientific American*, September 26, 2006.

Neurobiology of decision making: For a general understanding of our knowledge about the brain's architecture: Gazzaniga et al. (2002). Gazzaniga (2005) provides literary summaries of some of the topics. More popular: Carter (1999). Also recommended: Ratey (2001), Ramachandran (2003), Ramachandran and Blakeslee (1998), Carter (1999, 2002), Conlan (1999), the very readable Lewis, Amini, and Lannon (2000), and Goleman (1995). See Glimcher (2002) for probability and the brain. For the emotional brain, the three books by Damasio (1994, 2000, 2003), in addition to LeDoux (1998) and the more detailed LeDoux (2002), are the classics. See also the shorter Evans (2002). For the role of vision in aesthetics, but also in interpretation, Zeki (1999).

General works on memory: In psychology, Schacter (2001) is a review work of the memory biases with links to the hindsight effects. In neurobiology, see Rose (2003) and Squire and Kandel (2000). A general textbook on memory (in empirical psychology) is Baddeley (1997).

Intellectual colonies and social life: See the account in Collins (1998) of the "lineages" of philosophers (although I don't think he was aware enough of the Casanova problem to take into account the bias making the works of solo philosophers less likely to survive). For an illustration of the aggressiveness of groups, see Uglow (2003).

Hyman Minsky's work: Minsky (1982).

Asymmetry: Prospect theory (Kahneman and Tversky [1979] and Tversky and Kahneman [1992]) accounts for the asymmetry between bad and good random events, but it also shows that the negative domain is convex while the positive domain is concave, meaning that a loss of 100 is less painful than 100 losses of 1 but that a gain of 100 is also far less pleasurable than 100 times a gain of 1.

Neural correlates of the asymmetry: See Davidson's work in Goleman (2003), Lane et al. (1997), and Gehring and Willoughby (2002). Csikszentmihalyi (1993, 1998) further explains the attractiveness of steady payoffs with his theory of "flow."

Deferred rewards and its neural correlates: McLure et al. (2004) show the brain activation in the cortex upon making a decision to defer, providing insight on the limbic impulse behind immediacy and the cortical activity in delaying. See also Loewenstein et al. (1992), Elster (1998), Berridge (2005). For the neurology of preferences in Capuchin monkeys, Chen et al. (2005).

Bleed or blowup: Gladwell (2002) and Taleb (2004c). Why bleed is painful can be explained by dull stress; Sapolsky et al. (2003) and Sapolsky (1998). For how companies like steady returns, Degeorge and Zeckhauser (1999). Poetics of hope: Mihailescu (2006).

Discontinuities and jumps: Classified by René Thom as constituting seven classes; Thom (1980).

Evolution and small probabilities: Consider also the naïve evolutionary thinking positing the "optimality" of selection. The founder of sociobiology, the great E. O. Wilson, does not agree with such optimality when it comes to rare events. In Wilson (2002), he writes:

> The human brain evidently evolved to commit itself emotionally only to a small piece of geography, a limited band of kinsmen, and two or three generations into the future. To look neither far ahead nor far afield is elemental in a Darwinian sense. *We are innately inclined to ignore any distant possibility not yet requiring examination. It is, people say, just good common sense.* Why do they think in this shortsighted way?
>
> The reason is simple: it is a hardwired part of our Paleolithic heritage. For hundreds of millennia, those who worked for short-term gain within a small circle of relatives and friends lived longer and left more offspring—even when their collective striving caused their chiefdoms and empires to crumble around them. The long view that might have saved their distant descendants required a vision and extended altruism instinctively difficult to marshal.

See also Miller (2000): *"Evolution has no foresight. It lacks the long-term vision of drug company management. A species can't raise venture capital to pay its bills while its research team . . . This makes it hard to explain innovations."*

Note that neither author considered my age argument.

CHAPTER 8

Silent evidence bears the name *wrong reference class* in the nasty field of philosophy of probability, *anthropic bias* in physics, and *survivorship bias* in statistics (economists present the interesting attribute of having rediscovered it a few times while being severely fooled by it).

Confirmation: Bacon says in *On Truth*, "No pleasure is comparable to the standing upon the vantage ground of truth (a hill not to be commanded and where the air is always clear and serene), and to see the errors, and wanderings, and mists, and tempests, in the vale below." This easily shows how great intentions can lead to the confirmation fallacy.

Bacon did not understand the empiricists: He was looking for the golden mean. Again, from *On Truth:*

> There are three sources of error and three species of false philosophy; the sophistic, the empiric and the superstitious. . . . Aristotle affords the most eminent instance of the first; for he corrupted natural philosophy by logic—thus he formed the world of categories. . . . Nor is much stress to be laid on his frequent recourse to experiment in his books on animals, his problems and other treatises, for he had already decided, without having properly consulted experience as the basis of his decisions and axioms. . . . The empiric school produces dogmas of a more deformed and monstrous nature than the sophistic or theoretic school; not being founded in the light of common notions (which however poor and superstitious, is yet in a manner universal and of general tendency), but in the confined obscurity of a few experiments.

Bacon's misconception may be the reason it took us a while to understand that they treated history (and experiments) as mere and vague "guidance," i.e., epilogy.

Publishing: Allen (2005), Klebanoff (2002), Epstein (2001), de Bellaigue (2004), and Blake (1999). For a funny list of rejections, see Bernard (2002) and White (1982). Michael Korda's memoir, Korda (2000), adds some color to the business. These books are anecdotal, but we will see later that books follow steep scale-invariant structures with the implication of a severe role for randomness.

Anthropic bias: See the wonderful and comprehensive discussion in Bostrom (2002). In physics, see Barrow and Tipler (1986) and Rees (2004). Sornette (2004) has Gott's derivation of survival as a power law. In finance, Sullivan et al. (1999) discuss survivorship bias. See also Taleb (2004a). Studies that ignore the bias and state inappropriate conclusions: Stanley and Danko (1996) and the more foolish Stanley (2000).

Manuscripts and the Phoenicians: For survival and science, see Cisne (2005). Note that the article takes into account physical survival (like fossil), not cultural, which implies a selection bias. Courtesy Peter Bevelin.

Stigler's law of eponymy: Stigler (2002).

French book statistics: *Lire,* April 2005.

Why dispersion matters: More technically, the distribution of the extremum (i.e., the maximum or minimum) of a random variable depends more on the variance of the process than on its mean. Someone whose weight tends to fluctuate a lot is more likely to show you a picture of himself very thin than someone else whose weight is on average lower but remains constant. The mean (read skills) sometimes plays a very, very small role.

Fossil record: I thank the reader Frederick Colbourne for his comments on this subject. The literature calls it the "pull of the recent," but has difficulty estimating the effects, owing to disagreements. See Jablonski et al. (2003).

Undiscovered public knowledge: Here is another manifestation of silent evidence: you can actually do lab work sitting in an armchair, just by linking bits and pieces of research by people who labor apart from one another and miss on connections. Using bibliographic analysis, it is possible to find links between published information that had not been known previously by researchers. I "discovered" the vindication of the armchair in Fuller (2005). For other interesting discoveries, see Spasser (1997) and Swanson (1986a, 1986b, 1987).

Crime: The definition of economic "crime" is something that comes in hindsight. Regulations, once enacted, do not run retrospectively, so many activities causing excess are never sanctioned (e.g., bribery).

Bastiat: See Bastiat (1862–1864).

Casanova: I thank the reader Milo Jones for pointing out to me the exact number of volumes. See Masters (1969).

Reference point problem: Taking into account background information requires a form of thinking in *conditional* terms that, oddly, many scientists (especially the better ones) are incapable of handling. The difference between the two odds is called, simply, conditional probability. We are computing the probability of surviving *conditional* on our being in the sample itself. Simply put, you cannot compute probabilities if your survival is part of the condition of the realization of the process.

Plagues: See McNeill (1976).

CHAPTER 9

Intelligence and Nobel: Simonton (1999). If IQ scores correlate, they do so very weakly with subsequent success.

"Uncertainty": Knight (1923). My definition of such risk (Taleb, 2007c) is that it is a normative situation, where we can be certain about probabilities, i.e., no metaprobabilities. Whereas, if randomness and risk result from epistemic opacity, the difficulty in seeing causes, then necessarily the distinction is bunk. Any reader of Cicero would recognize it as his probability; see epistemic opacity in his *De Divinatione,* Liber primus, LVI, 127:

> Qui enim teneat causas rerum futurarum, idem necesse est omnia teneat quae futura sint. Quod cum nemo facere nisi deus possit, relinquendum est homini, ut signis quibusdam consequentia declarantibus futura praesentiat.

"He who knows the causes will understand the future, except that, given that nobody outside God possesses such faculty . . ."

Philosophy and epistemology of probability: Laplace. *Treatise,* Keynes (1920), de Finetti (1931), Kyburg (1983), Levi (1970), Ayer, Hacking (1990, 2001), Gillies (2000), von Mises (1928), von Plato (1994), Carnap (1950), Cohen (1989), Popper (1971), Eatwell et al. (1987), and Gigerenzer et al. (1989).

History of statistical knowledge and methods: I found no intelligent work in the history of statistics, i.e., one that does not fall prey to the ludic fallacy or Gaussianism. For a conventional account, see Bernstein (1996) and David (1962).

General books on probability and information theory: Cover and Thomas (1991); less technical but excellent, Bayer (2003). For a probabilistic view of information theory: the posthumous Jaynes (2003) is the only mathematical book other than de Finetti's work that I can recommend to the general reader, owing to his Bayesian approach and his allergy for the formalism of the idiot savant.

Poker: It escapes the ludic fallacy; see Taleb (2006a).

Plato's normative approach to left and right hands: See McManus (2002).

Nietzsche's *bildungsphilister:* See van Tongeren (2002) and Hicks and Rosenberg (2003). Note that because of the confirmation bias academics will tell you that intellectuals "lack rigor," and will bring examples of those who do, not those who don't.

Economics books that deal with uncertainty: Carter et al. (1962), Shackle (1961, 1973), Hayek (1994). Hirshleifer and Riley (1992) fits uncertainty into neoclassical economics.

Incomputability: For earthquakes, see Freedman and Stark (2003) (courtesy of Gur Huberman).

Academia and philistinism: There is a round-trip fallacy; if academia means rigor (which I doubt, since what I saw called "peer reviewing" is too often a masquerade), nonacademic does not imply nonrigorous. Why do I doubt the "rigor"? By the confirmation bias they show you their contributions yet in spite of the high number of laboring academics, a relatively minute fraction of our results come from them. A disproportionately high number of contributions come from freelance researchers and those dissingly called amateurs: Darwin, Freud, Marx, Mandelbrot, even the early Einstein. Influence on the part of an academic is usually accidental. This even held in the Middle Ages and the Renaissance, see Le Goff (1985). Also, the Enlightenment figures (Voltaire, Rousseau, d'Holbach, Diderot, Montesquieu) were all nonacademics at a time when academia was large.

CHAPTER 10

Overconfidence: Albert and Raiffa (1982) (though apparently the paper languished for a decade before formal publication). Lichtenstein and Fischhoff (1977) showed that overconfidence can be influenced by item difficulty; it typically diminishes and turns into underconfidence in easy items (compare with Armelius [1979]). Plenty of papers since have tried to pin down the conditions of calibration failures or robustness (be they task training, ecological aspects of the domain, level of education, or nationality): Dawes (1980), Koriat, Lichtenstein, and Fischhoff (1980), Mayseless and Kruglanski (1987), Dunning et al. (1990), Ayton and McClelland (1997), Gervais and Odean (1999), Griffin and Varey (1996), Juslin (1991, 1993, 1994), Juslin and Olsson (1997), Kadane and Lichtenstein (1982), May (1986), McClelland and Bolger (1994), Pfeifer (1994), Russo and Schoernaker (1992), Klayman et al. (1999). Note the decrease (unexpectedly) in overconfidence under group decisions: see Sniezek and Henry (1989)—and solutions in Plous (1995). I am suspicious here of the Mediocristan/Extremistan distinction and the unevenness of the variables. Alas, I found no paper making this distinction. There are also solutions in Stoll (1996), Arkes et al. (1987). For overconfidence in finance, see Thorley (1999) and Barber and Odean (1999). For cross-boundaries effects, Yates et al. (1996, 1998), Angele et al. (1982). For simultaneous overconfidence and underconfidence, see Erev, Wallsten, and Budescu (1994).

Frequency vs. probability—the ecological problem: Hoffrage and Gigerenzer (1998) think that overconfidence is less significant when the problem is expressed in frequencies as opposed to probabilities. In fact, there has been a debate about the difference between "ecology" and laboratory; see Gigerenzer et al. (2000), Gigerenzer and Richter (1990), and Gigerenzer (1991). We are "fast and frugal" (Gigerenzer and Goldstein [1996]). As far as the Black Swan is concerned, these problems of ecology do not arise: we do not live in an environment in which we are supplied with frequencies or, more generally, for which we are fit. Also in ecology, Spariosu (2004) for the ludic aspect, Cosmides and Tooby (1990). Leary (1987) for Brunswikian ideas, as well as Brunswik (1952).

Lack of awareness of ignorance: "In short, the same knowledge that underlies the ability to produce correct judgment is also the knowledge that underlies the ability to recognize correct judgment. To lack the former is to be deficient in the latter." From Kruger and Dunning (1999).

Expert problem in isolation: I see the expert problem as indistinguishable from Matthew effects and Extremism fat tails (more later), yet I found no such link in the literatures of sociology and psychology.

Clinical knowledge and its problems: See Meehl (1954) and Dawes, Faust, and Meehl (1989). Most entertaining is the essay "Why I Do Not Attend Case Conferences" in Meehl (1973). See also Wagenaar and Keren (1985, 1986).

Financial analysts, herding, and forecasting: See Guedj and Bouchaud (2006), Abarbanell and Bernard (1992), Chen et al. (2002), De Bondt and Thaler (1990), Easterwood and Nutt (1999), Friesen and Weller (2002), Foster (1977), Hong and Kubik (2003), Jacob et al. (1999), Lim (2001), Liu (1998), Maines and Hand (1996), Mendenhall (1991), Mikhail et al. (1997, 1999), Zitzewitz (2001), and El-Galfy and Forbes (2005). For a comparison with weather forecasters (unfavorable): Tyszka and Zielonka (2002).

Economists and forecasting: Tetlock (2005), Makridakis and Hibon (2000), Makridakis et al. (1982), Makridakis et al. (1993), Gripaios (1994), Armstrong (1978, 1981); and rebuttals by McNees (1978), Tashman (2000), Blake et al. (1986), Onkal et al. (2003), Gillespie (1979), Baron (2004), Batchelor (1990, 2001), Dominitz and Grether (1999). Lamont (2002) looks for reputational factors: established forecasters get worse as they produce more radical forecasts to get attention—consistent with Tetlock's hedgehog effect. Ahiya and Doi (2001) look for herd behavior in Japan. See McNees (1995), Remus et al. (1997), O'Neill and Desai (2005), Bewley and Fiebig (2002), Angner (2006), Bénassy-Quéré (2002); Brender and Pisani (2001) look at the Bloomberg consensus; De Bondt and Kappler (2004) claim evidence of weak persistence in fifty-two years of data, but I saw the slides in a presentation, never the paper, which after two years might never materialize. Overconfidence, Braun and Yaniv (1992). See Hahn (1993) for a general intellectual discussion. More general, Clemen (1986, 1989). For Game theory, Green (2005).

Many operators, such as James Montier, and many newspapers and magazines (such as *The Economist*), run casual tests of prediction. Cumulatively, they must be taken seriously since they cover more variables.

Popular culture: In 1931, Edward Angly exposed forecasts made by President Hoover in a book titled *Oh Yeah?* Another hilarious book is Cerf and Navasky (1998), where, incidentally, I got the pre-1973 oil-estimation story.

Effects of information: The major paper is Bruner and Potter (1964). I thank Danny Kahneman for discussions and pointing out this paper to me. See also Montier (2007), Oskamp (1965), and Benartzi (2001). These biases become ambiguous information (Griffin and Tversky [1992]). For how they fail to disappear with expertise and training, see Kahneman and Tversky (1982) and Tversky and Kahneman (1982). See Kunda (1990) for how preference-consistent information is taken at face value, while preference-inconsistent information is processed critically.

Planning fallacy: Kahneman and Tversky (1979) and Buehler, Griffin, and Ross (2002). The planning fallacy shows a consistent bias in people's planning ability, even with matters of a repeatable nature—though it is more exaggerated with nonrepeatable events.

Wars: Trivers (2002).

Are there incentives to delay?: Flyvbjerg et al. (2002).

Oskamp: Oskamp (1965) and Montier (2007).

Task characteristics and effect on decision making: Shanteau (1992).

***Epistēmē* vs. *Technē*:** This distinction harks back to Aristotle, but it recurs then dies? down—it most recently recurs in accounts such as tacit knowledge in "know how." See Ryle (1949), Polanyi (1958/1974), and Mokyr (2002).

Catherine the Great: The number of lovers comes from Rounding (2006).

Life expectancy: www.annuityadvantage.com/lifeexpectancy.htm. For projects, I have used a probability of exceeding with a power-law exponent of 3/2: $f= Kx^{3/2}$. Thus the conditional expectation of x, knowing that x exceeds a

$$E[x|x>a]=\frac{\int_a^\infty x f(x)dx}{\int_a^\infty f(x)dx}.$$

CHAPTERS 11–13

Serendipity: See Koestler (1959) and Rees (2004). Rees also has powerful ideas on forecastability. See also Popper's comments in Popper (2002), and Waller (2002a), Cannon

(1940), Mach (1896) (cited in Simonton [1999]), and Merton and Barber (2004). See Simonton (2004) for a synthesis. For serendipity in medicine and anesthesiology, see Vale et al. (2005).

"Renaissance man": See www.bell-labs.com/project/feature/archives/cosmology/.

Laser: As usual, there are controversies as to who "invented" the technology. After a successful discovery, precursors are rapidly found, owing to the retrospective distortion. Charles Townsend won the Nobel prize, but was sued by his student Gordon Gould, who held that he did the actual work (see *The Economist,* June 9, 2005).

Darwin/Wallace: Quammen (2006).

Popper's attack on historicism: See Popper (2002). Note that I am reinterpreting Popper's idea in a modern manner here, using my own experiences and knowledge, not commenting on comments about Popper's work—with the consequent lack of fidelity to his message. In other words, these are not directly Popper's arguments, but largely mine phrased in a Popperian framework. The conditional expectation of an unconditional expectation is an unconditional expectation.

Forecast for the future a hundred years earlier: Bellamy (1891) illustrates our mental projections of the future. However, some stories might be exaggerated: "A Patently False Patent Myth still! Did a patent official really once resign because he thought nothing was left to invent? Once such myths start they take on a life of their own." *Skeptical Inquirer,* May–June, 2003.

Observation by Peirce: Olsson (2006), Peirce (1955).

Predicting and explaining: See Thom (1993).

Poincaré: The three body problem can be found in Barrow-Green (1996), Rollet (2005), and Galison (2003). On Einstein, Pais (1982). More recent revelations in Hladik (2004).

Billiard balls: Berry (1978) and Pisarenko and Sornette (2004).

Very general discussion on "complexity": Benkirane (2002), Scheps (1996), and Ruelle (1991). For limits, Barrow (1998).

Hayek: See www.nobel.se. See Hayek (1945, 1994). Is it that mechanisms do not correct themselves from railing by influential people, but either by mortality of the operators, or something even more severe, by being put out of business? Alas, because of contagion, there seems to be little logic to how matters improve; luck plays a part in how soft sciences evolve. See Ormerod (2006) for network effects in "intellectuals and socialism" and the power-law distribution in influence owing to the scale-free aspect of the connections—and the consequential arbitrariness. Hayek seems to have been a prisoner of Weber's old differentiation between *Natur-Wissenschaften* and *Geistes Wissenschaften*—but thankfully not Popper.

Insularity of economists: Pieters and Baumgartner (2002). One good aspect of the insularity of economists is that they can insult me all they want without any consequence: it appears that only economists read other economists (so they can write papers for other economists to read). For a more general case, see Wallerstein (1999). Note that Braudel fought "economic history." It was history.

Economics as religion: Nelson (2001) and Keen (2001). For methodology, see Blaug (1992). For high priests and lowly philosophers, see Boettke, Coyne, and Leeson (2006). Note that the works of Gary Becker and the Platonists of the Chicago School are all marred by the confirmation bias: Becker is quick to show you situations in which people are moved by economic incentives, but does not show you cases (vastly more numerous) in which people don't care about such materialistic incentives.

The smartest book I've seen in economics is Gave et al. (2005) since it transcends the constructed categories in academic economic discourse (one of the authors is the journalist Anatole Kaletsky).

General theory: This fact has not deterred "general theorists." One hotshot of the Platonifying variety explained to me during a long plane ride from Geneva to New York that the ideas of Kahneman and his colleagues must be rejected because they do not allow us to develop a general equilibrium theory, producing "time-inconsistent preferences." For a minute I thought he was joking: he blamed the psychologists' ideas and human incoherence for interfering with his ability to build his Platonic model.

Samuelson: For his optimization, see Samuelson (1983). Also Stiglitz (1994).

Plato's dogma on body symmetry: "Athenian Stranger to Cleinias: In that the right and left hand are supposed to be by nature differently suited for our various uses of them; whereas no difference is found in the use of the feet and the lower limbs; but in the use of the hands we are, as it were, maimed by the folly of nurses and mothers; for although our several limbs are by nature balanced, we create a difference in them by bad habit," in Plato's *Laws*. See McManus (2002).

Drug companies: Other such firms, I was told, are run by commercial persons who tell researchers where they find a "market need" and ask them to "invent" drugs and cures accordingly—which accords with the methods of the dangerously misleading Wall Street security analysts. They formulate projections as if they know what they are going to find.

Models of the returns on innovations: Sornette and Zajdenweber (1999) and Silverberg and Verspagen (2005).

Evolution on a short leash: Dennet (2003) and Stanovich and West (2000).

Montaigne: We don't get much from the biographies of a personal essayist; some information in Frame (1965) and Zweig (1960).

Projectibility and the grue paradox: See Goodman (1955). See also an application (or perhaps misapplication) in King and Zheng (2005).

Constructionism: See Berger and Luckmann (1966) and Hacking (1999).

Certification vs, true skills or knowledge: See Donhardt (2004). There is also a franchise protection. Mathematics may not be so necessary a tool for economics, except to protect the franchise of those economists who know math. In my father's days, the selection process for the mandarins was made using their abilities in Latin (or Greek). So the class of students groomed for the top was grounded in the classics and knew some interesting subjects. They were also trained in Cicero's highly probabilistic view of things—and selected on erudition, which carries small side effects. If anything it allows you to handle fuzzy matters. My generation was selected according to mathematical skills. You made it based on an engineering mentality; this produced mandarins with mathematical, highly structured, logical minds, and, accordingly, they will select their peers based on such criteria. So the papers in economics and social science gravitated toward the highly mathematical and protected their franchise by putting high mathematical barriers to entry. You could also smoke the general public who is unable to put a check on you. Another effect of this franchise protection is that it might have encouraged putting "at the top" those idiot-savant-like researchers who lacked in erudition, hence were insular, parochial, and closed to other disciplines.

Freedom and determinism: a speculative idea in Penrose (1989) where only the quantum effects (with the perceived indeterminacy there) can justify consciousness.

Projectibility: uniqueness assuming least squares or MAD.

Chaos theory and the backward/forward confusion: Laurent Firode's *Happenstance,* a.k.a. *Le battement d'ailes du papillon / The Beating of a Butterfly's Wings* (2000).

Autism and perception of randomness: See Williams et al. (2002).

Forecasting and misforecasting errors in hedonic states: Wilson, Meyers, and Gilbert (2001), Wilson, Gilbert, and Centerbar (2003), and Wilson et al. (2005). They call it "emotional evanescence."

Forecasting and consciousness: See the idea of "aboutness" in Dennett (1995, 2003) and Humphrey (1992). However, Gilbert (2006) believes that we are not the only animal that forecasts—which is wrong as it turned out. Suddendorf (2006) and Dally, Emery, and Clayton (2006) show that animals too forecast!

Russell's comment on Pascal's wager: Ayer (1988) reports this as a private communication.

History: Carr (1961), Hexter (1979), and Gaddis (2002). But I have trouble with historians throughout, because they often mistake the forward and the backward processes. Mark Buchanan's *Ubiquity* and the quite confused discussion by Niall Ferguson in *Nature*. Neither of them seem to realize the problem of calibration with power laws. See also Ferguson, *Why Did the Great War?*, to gauge the extent of the forward-backward problems.

For the traditional nomological tendency, i.e., the attempt to go beyond cause into a general theory, see *Muqaddamah* by Ibn Khaldoun. See also Hegel's *Philosophy of History*.

Emotion and cognition: Zajonc (1980, 1984).

Catastrophe insurance: Froot (2001) claims that insurance for remote events is overpriced. How he determined this remains unclear (perhaps by backfitting or bootstraps), but reinsurance companies have not been making a penny selling "overpriced" insurance.

Postmodernists: Postmodernists do not seem to be aware of the differences between narrative and prediction.

Luck and serendipity in medicine: Vale et al. (2005). In history, see Cooper (2004). See also Ruffié (1977). More general, see Roberts (1989).

Affective forecasting: See Gilbert (1991), Gilbert et al. (1993), and Montier (2007).

CHAPTERS 14–17

This section will also serve another purpose. Whenever I talk about the Black Swan, people tend to supply me with anecdotes. But these anecdotes are just corroborative: you need to show that *in the aggregate* the world is dominated by Black Swan events. To me, the rejection of nonscalable randomness is sufficient to establish the role and significance of Black Swans.

Matthew effects: See Merton (1968, 1973a, 1988). Martial, in his *Epigrams*: "*Semper pauper eris, si pauper es, Aemiliane./Dantur opes nullis (nunc) nisi divitibus.*" (Epigr. V 81). See also Zuckerman (1997, 1998).

Cumulative advantage and its consequences on social fairness: review in DiPrete et al. (2006). See also Brookes-Gun and Duncan (1994), Broughton and Mills (1980), Dannefer (2003), Donhardt (2004), Hannon (2003), and Huber (1998). For how it may explain precocity, see Elman and O'Rand (2004).

Concentration and fairness in intellectual careers: Cole and Cole (1973), Cole (1970), Conley (1999), Faia (1975), Seglen (1992), Redner (1998), Lotka (1926), Fox and Kochanowski (2004), and Huber (2002).

Winner take all: Rosen (1981), Frank (1994), Frank and Cook (1995), and Attewell (2001).

Arts: Bourdieu (1996), Taleb (2004e).

Wars: War is concentrated in an Extremistan manner: Lewis Fry Richardson noted last century the uneveness in the distribution of casualties (Richardson [1960]).

Modern wars: Arkush and Allen (2006). In the study of the Maori, the pattern of fighting with clubs was sustainable for many centuries—modern tools cause 20,000 to 50,000 deaths a year. We are simply not made for technical warfare. For an anecdotal and causative account of the history of a war, see Ferguson (2006).

S&P 500: See Rosenzweig (2006).

The long tail: Anderson (2006).

Cognitive diversity: See Page (2007). For the effect of the Internet on schools, see Han et al. (2006).

Cascades: See Schelling (1971, 1978) and Watts (2002). For information cascades in economics, see Bikhchandani, Hirshleifer, and Welch (1992) and Shiller (1995). See also Surowiecki (2004).

Fairness: Some researchers, like Frank (1999), see arbitrary and random success by others as no different from pollution, which necessitates the enactment of a tax. De Vany, Taleb, and Spitznagel (2004) propose a market-based solution to the problem of allocation through the process of voluntary self-insurance and derivative products. Shiller (2003) proposes cross-country insurance.

The mathematics of preferential attachment: This argument pitted Mandelbrot against the cognitive scientist Herbert Simon, who formalized Zipf's ideas in a 1955 paper (Simon [1955]), which then became known as the Zipf-Simon model. Hey, you need to allow for people to fall from favor!

Concentration: Price (1970). Simon's "Zipf derivation," Simon (1955). More general bibliometrics, see Price (1976) and Glänzel (2003).

Creative destruction revisited: See Schumpeter (1942).

Networks: Barabási and Albert (1999), Albert and Barabási (2000), Strogatz (2001, 2003), Callaway et al. (2000), Newman et al. (2000), Newman, Watts, and Strogatz (2000), Newman (2001), Watts and Strogatz (1998), Watts (2002, 2003), and Amaral et al. (2000). It supposedly started with Milgram (1967). See also Barbour and Reinert (2000), Barthélémy and Amaral (1999). See Boots and Sasaki (1999) for infections. For extensions, see Bhalla and Iyengar (1999). Resilence, Cohen et al. (2000), Barabási and Bonabeau (2003), Barabási (2002), and Banavar et al. (2000). Power laws and the Web, Adamic and Huberman (1999) and Adamic (1999). Statistics of the Internet: Huberman (2001), Willinger et al. (2004), and Faloutsos, Faloutsos, and Faloutsos (1999). For DNA, see Vogelstein et al. (2000).

Self-organized criticality: Bak (1996).

Pioneers of fat tails: For wealth, Pareto (1896), Yule (1925, 1944). Less of a pioneer Zipf (1932, 1949). For linguistics, see Mandelbrot (1952).

Pareto: See Bouvier (1999).

Endogenous vs. exogenous: Sornette et al. (2004).

Sperber's work: Sperber (1996a, 1996b, 1997).

Regression: If you hear the phrase *least square regression*, you should be suspicious about the claims being made. As it assumes that your errors wash out rather rapidly, it underestimates the total possible error, and thus overestimates what knowledge one can derive from the data.

The notion of central limit: very misunderstood: it takes a long time to reach the central limit—so as we do not live in the asymptote, we've got problems. All various random variables (as we started in the example of Chapter 16 with a +1 or –1, which is called a Bernouilli draw) under summation (we did sum up the wins of the 40 tosses) become Gaussian. Summation is key here, since we are considering the results of adding up the 40 steps, which is where the Gaussian, under the first and second central assumptions becomes what is called a "distribution." (A distribution tells you how you are likely to have your outcomes spread out, or distributed.) However, they may get there at different speeds. This is called the central limit theorem: if you add random variables coming from these individual tame jumps, it will lead to the Gaussian.

Where does the central limit not work? If you do not have these central assumptions, but have jumps of random size instead, then we would not get the Gaussian. Furthermore, we sometimes converge very slowly to the Gaussian. For preasymptotics and scalability, Mandelbrot and Taleb (2007a), Bouchaud and Potters (2003). For the problem of working outside asymptotes, Taleb (2007).

Aureas mediocritas: historical perspective, in Naya and Pouey-Mounou (2005) aptly called *Éloge de la médiocrité*.

Reification (hypostatization): Lukacz, in Bewes (2002).

Catastrophes: Posner (2004).

Concentration and modern economic life: Zajdenweber (2000).

Choices of society structure and compressed outcomes: The classical paper is Rawls (1971), though Frohlich, Oppenheimer, and Eavy (1987a, 1987b), as well as Lissowski, Tyszka, and Okrasa (1991), contradict the notion of the desirability of Rawl's veil (though by experiment). People prefer maximum average income subjected to a floor constraint on some form of equality for the poor, inequality for the rich type of environment.

Gaussian contagion: Quételet in Stigler (1986). Francis Galton (as quoted in Ian Hacking's *The Taming of Chance*): "I know of scarcely anything so apt to impress the imagination as the wonderful form of cosmic order expressed by 'the law of error.' "

"Finite variance" nonsense: Associated with CLT is an assumption called "finite variance" that is rather technical: none of these building-block steps can take an infinite value if you square them or multiply them by themselves. They need to be bounded at some number. We simplified here by making them all one single step, or finite standard deviation. But the problem is that some fractal payoffs may have finite variance, but still not take us there rapidly. See Bouchaud and Potters (2003).

Lognormal: There is an intermediate variety that is called the lognormal, emphasized by one Gibrat (see Sutton [1997]) early in the twentieth century as an attempt to explain the distribution of wealth. In this framework, it is not quite that the wealthy get wealthier, in a pure preferential attachment situation, but that if your wealth is at 100 you will vary by 1, but when your wealth is at 1,000, you will vary by 10. The relative changes in your wealth are Gaussian. So the lognormal superficially resembles the fractal, in the sense that it may tolerate some large deviations, but it is dangerous because these rapidly taper off at the end. The introduction of the lognormal was a very bad compromise, but a way to conceal the flaws of the Gaussian.

Extinctions: Sterelny (2001). For extinctions from abrupt fractures, see Courtillot (1995) and Courtillot and Gaudemer (1996). Jumps: Eldredge and Gould.

FRACTALS, POWER LAWS, and SCALE-FREE DISTRIBUTIONS

Definition: Technically, $P_{>x} = K\, x^{-\alpha}$ where α is supposed to be the power-law exponent. It is said to be scale free, in the sense that it does not have a characteristic scale: relative deviation of $\frac{P_{>x}}{P_{>nx}}$ does not depend on x, but on n—for *x* "large enough." Now, in the other class of distribution, the one that I can intuitively describe as nonscalable, with the typical shape p(x) = Exp[-a x], the scale will be *a*.

Problem of "how large": Now the problem that is usually misunderstood. This scalability might stop somewhere, but I do not know where, so I might consider it infinite. The statements *very large and I don't know how large* and *infinitely large* are epistemologically substitutable. There might be a point at which the distributions flip. This will show once we look at them more graphically.

Log P>x = -α Log X +C^t for a scalable. When we do a log-log plot (i.e., plot P>x and x on a logarithmic scale), as in Figures 15 and 16, we should see a straight line.

Fractals and power laws: Mandelbrot (1975, 1982). Schroeder (1991) is imperative. John Chipman's unpublished manuscript *The Paretian Heritage* (Chipman [2006]) is the best review piece I've seen. See also Mitzenmacher (2003).

"To come very near true theory and to grasp its precise application are two very different things as the history of science teaches us. Everything of importance has been said before by somebody who did not discover it." Whitehead (1925).

Fractals in poetry: For the quote on Dickinson, see Fulton (1998).

Lacunarity: Brockman (2005). In the arts, Mandelbrot (1982).

Fractals in medicine: "New Tool to Diagnose and Treat Breast Cancer," *Newswise*, July 18, 2006.

General reference books in statistical physics: The most complete (in relation to fat tails) is Sornette (2004). See also Voit (2001) or the far deeper Bouchaud and Potters (2002) for financial prices and econophysics. For "complexity" theory, technical books: Bocarra (2004), Strogatz (1994), the popular Ruelle (1991), and also Prigogine (1996).

Fitting processes: For the philosophy of the problem, Taleb and Pilpel (2004). See also Pisarenko and Sornette (2004), Sornette et al. (2004), and Sornette and Ide (2001).

Poisson jump: Sometimes people propose a Gaussian distribution with a small probability of a "Poisson" jump. This may be fine, but how do you know how large the jump is going to be? Past data might not tell you how large the jump is.

FIGURE 15: TYPICAL DISTRIBUTION WITH POWER-LAW TAILS (HERE A STUDENT T)

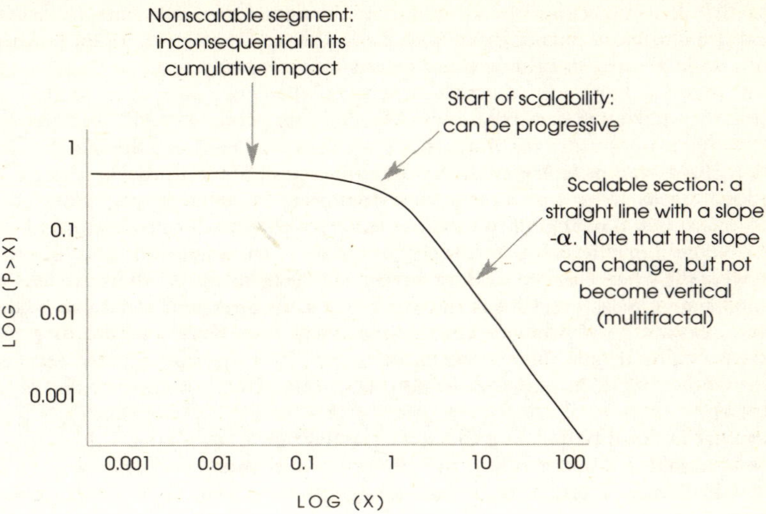

FIGURE 16

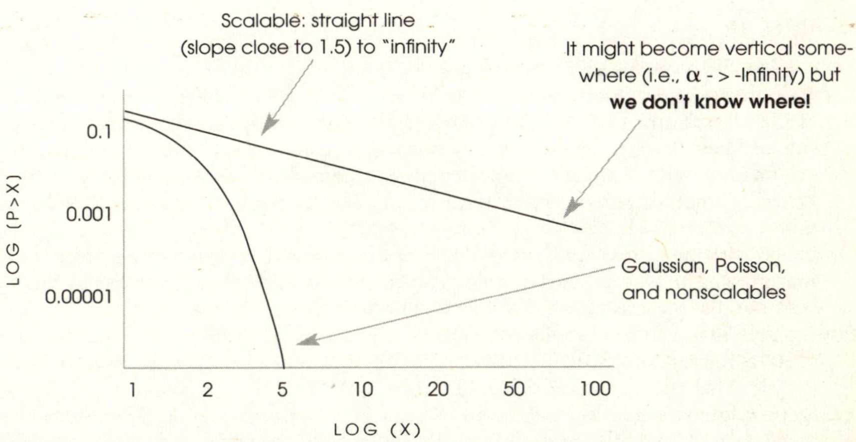

The two exhaustive domains of attraction: vertical or straight line with slopes either negative infinity or constant negative α. Note that since probabilities need to add up to 1 (even in France) there cannot be other alternatives to the two basins, which is why I narrow it down to these two exclusively.

My ideas are made very simple with this clean cut polarization—added to the problem of not knowing which basin we are in owing to the scarcity of data on the far right.

Small sample effect: Weron (2001). Officer (1972) is quite ignorant of the point.

Recursivity of statistics: Taleb and Pilpel (2004), Blyth et al. (2005).

Biology: Modern molecular biology pioneers Salvador Luria and Max Delbrück witnessed a clustering phenomenon with the occasional occurrence of extremely large mutants in a bacterial colony, larger than all other bacteria.

Thermodynamics: Entropy maximization without the constraints of a second moment leads to a Levy-stable distribution—Mandelbrot's thesis of 1952 (see Mandelbrot [1997a]). Tsallis's more sophisticated view of entropy leads to a Student T.

Imitation chains and pathologies: An informational cascade is a process where a purely rational agent elects a particular choice ignoring his own private information (or judgment) to follow that of others. You run, I follow you, because you may be aware of a danger I may be missing. It is efficient to do what others do instead of having to reinvent the wheel every time. But this copying the behavior of others can lead to imitation chains. Soon everyone is running in the same direction, and it can be for spurious reasons. This behavior causes stock market bubbles and the formation of massive cultural fads. Bikhchandani et al. (1992). In psychology, see Hansen and Donoghue (1977). In biology/selection, Dugatkin (2001), Kirpatrick and Dugatkin (1994).

Self-organized criticality: Bak and Chen (1991), Bak (1996).

Economic variables: Bundt and Murphy (2006). Most economic variables seem to follow a "stable" distribution. They include foreign exchange, the GDP, the money supply, interest rates (long and short term), and industrial production.

Statisticians not accepting scalability: Flawed reasoning mistaking for sampling error in the tails for a boundedness: Perline (2005), for instance, does not understand the difference between absence of evidence and evidence of absence.

Time series and memory: You can have "fractal memory," i.e., the effect of past events on the present has an impact that has a "tail." It decays as power-law, not exponentially.

Marmott's work: Marmott (2004).

CHAPTER 18

Economists: Weintraub (2002), Szenberg (1992).

Portfolio theory and modern finance: Markowitz (1952, 1959), Huang and Litzenberger (1988) and Sharpe (1994, 1996). What is called the Sharpe ratio is meaningless outside of Mediocristan. The contents of Steve Ross's book (Ross [2004]) on "neoclassical finance" are completely canceled if you consider Extremistan in spite of the "elegant" mathematics and the beautiful top-down theories. "Anecdote" of Merton minor in Merton (1992).

Obsession with measurement: Crosby (1997) is often shown to me as convincing evidence that measuring was a great accomplishment not knowing that it applied to Mediocristan and Mediocristan only. Bernstein (1996) makes the same error.

Power laws in finance: Mandelbrot (1963), Gabaix et al. (2003), and Stanley et al. (2000). Kaizoji and Kaizoji (2004), Véhel and Walter (2002). Land prices: Kaizoji (2003). Magisterial: Bouchaud and Potters (2003).

Equity premium puzzle: If you accept fat tails, there is no equity premium puzzle. Benartzi and Thaler (1995) offer a psychological explanation, not realizing that variance is not the measure. So do many others.

Covered writes: a sucker's game as you cut your upside—conditional on the upside being breached, the stock should rally a lot more than intuitively accepted. For a representative mistake, see Board et al. (2000).

Nobel family: "Nobel Descendant Slams Economics Prize," *The Local*, September 28, 2005, Stockholm.

Double bubble: The problem of derivatives is that if the underlying security has mild fat tails and follows a mild power law (i.e., a tail exponent of three or higher), the derivative will produce far fatter tails (if the payoff is in squares, then the tail exponent of the derivatives portfolio will be half that of the primitive). This makes the Black-Scholes-Merton equation twice as unfit!

Poisson busting: The best way to figure out the problems of the Poisson as a substitute for a scalable is to calibrate a Poisson and compute the errors out of sample. The same applies to methods such as GARCH—they fare well in sample, but horribly, horribly outside (even a trailing three-month past historical volatility or mean deviation will outperform a GARCH of higher orders).

Why the Nobel: Derman and Taleb (2005), Haug (2007).

Claude Bernard and experimental medicine: *"Empiricism pour le présent, avec direction a aspiration scientifique pour l'avenir."* From Claude Bernard, *Principe de la médecine expérimentale*. See also Fagot-Largeault (2002) and Ruffié (1977). Modern evidence-based medicine: Ierodiakonou and Vandenbroucke (1993) and Vandenbroucke (1996) discuss a stochastic approach to medicine.

CHAPTER 19

Popper quote: From *Conjectures and Refutations,* pages 95–97.

The lottery paradox: This is one example of scholars not understanding the high-impact rare event. There is a well-known philosophical conundrum called the "lottery paradox," originally posed by the logician Henry Kyburg (see Rescher [2001] and Clark [2002]), which goes as follows: "I do not believe that any ticket will win the lottery, but I do believe that all tickets will win the lottery." To me (and a regular person) this statement does not seem to have anything strange in it. Yet for an academic philosopher trained in classical logic, this is a paradox. But it is only so if one tries to squeeze probability statements into commonly used logic that dates from Aristotle and is *all or nothing*. An *all or nothing* acceptance and rejection ("I believe" or "I do not believe") is inadequate with the highly improbable. We need shades of belief, degrees of faith you have in a statement other than 100% or 0%.

One final philosophical consideration. For my friend the options trader and Talmudic scholar Rabbi Tony Glickman: life is convex and to be seen as a series of derivatives. Simply put, when you cut the negative exposure, you limit your vulnerability to unknowledge, Taleb (2005).

BIBLIOGRAPHY

Abarbanell, Jeffery S., and Victor L. Bernard, 1992, "Test of Analysts' Overreaction/Underreaction of Earnings Information as an Explanation for Anomalous Stock Price Behavior." *Journal of Finance* 47: 1181–1207.

Aczel, Amir D, 2004, *Chance: A Guide to Gambling, Love, the Stock Market, and Just About Everything Else.* New York: Thunder's Mouth Press.

Adamic, Lada, 1999, "The Small World Web." *Lecture Notes in Computational Science* 1696: 443–452.

Adamic, Lada, and Bernardo A. Huberman, 1999, "The Nature of Markets in the World Wide Web." *Quarterly Journal of Electronic Commerce* 1: 5–12.

Albert, R., and A.-L. Barabási, 2000, "Topology of Evolving Networks: Local Events and Universality." *Physical Review Letters* 85: 5234–5237.

Albert, R., H. Jeong, and A.-L. Barabási, 2000, "Error and Attack Tolerance of Complex Networks." *Nature* 406: 378–382.

Albouy, François-Xavier, 2002, *Le temps des catastrophes.* Paris: Descartes & Cie.

Al-Ghazali, 1989, "Mikhtarat Min Ahthar Al-Ghazali." In Saliba, Jamil, *Tarikh Al Falsafa Al Arabiah.* Beirut: Al Sharikah Al Ahlamiah Lilk-itab.

Allen, Mark S., 2006, "Transformations in Maori Warfare: Toa, Pa, and Pu." In Elizabeth N. Arkush and Mark W. Allen, 2006.

Allen, Michael, 2003, *The Truth About Writing.* Wiltshire: Kingsfield Publications.

———, 2005, *On the Survival of Rats in the Slushpile: Essays and Criticism.* Wiltshire: Kingsfield Publications.

Allport, D. A., 1975, "The State of Cognitive Psychology." *Quarterly Journal of Experimental Psychology* 27: 141–152.

Allwood, C. M., and H. Montgomery, 1987, "Response Selection Strategies and Realism of Confidence Judgments." *Organizational Behavior and Human Decision Processes* 39: 365–383.

Alpert, M., and H. Raiffa, 1982, "A Progress Report on the Training of Probability Assessors." In D. Kahneman, P. Slovic, and A. Tversky, eds., 1982.

Amaral, L. A. N., A. Scala, M. Barthélémy, and H. E. Stanley, 2000, "Classes of Behavior of Small-world Networks." *Proceedings of the National Academy of Science* 97: 11149–11152.

Anderson, Benedict, 1983, *Imagined Communities.* New York: Verso.

Anderson, Chris, 2006, *The Long Tail.* New York: Hyperion.

Anderson, N. H., 1986, "A Cognitive Theory of Judgment and Decision." In B. Brehmer, H. Jungermann, P. Lourens, and G. Sevón, eds., *New Directions in Research on Decision Making*. Amsterdam: North-Holland.

Angele, U., B. Beer-Binder, R. Berger, C. Bussmann, H. Kleinbölting, and B. Mansard, 1982, *Über- und Unterschätzung des eigenen Wissens in Abhängigkeit von Geschlecht und Bildungsstand* (*Overestimation and Underestimation of One's Knowledge as a Function of Sex and Education*). Unpublished manuscript, University of Konstanz, Federal Republic of Germany.

Angner, Erik, 2006, "Economists as Experts: Overconfidence in Theory and Practice." *Journal of Economic Methodology* 13(1): 1–24.

Annas, Julia, and Julian Barnes, 1985, *Modes of Skepticism*. Cambridge: Cambridge University Press.

Arkes, H. R., C. Christensen, C. Lai, and C. Blumer, 1987, "Two Methods of Reducing Overconfidence." *Organizational Behavior and Human Decision Processes* 39: 133–144.

Arkes, H. R., and K. R. Hammond, 1986, *Judgment and Decision Making: An Interdisciplinary Reader*. Cambridge: Cambridge University Press.

Arkush, Elizabeth N., and Mark W. Allen, eds., 2006, *The Archaeology of Warfare: Prehistories of Raiding and Conquest*. Gainesville: University of Florida Press.

Armelius, B., and K. Armelius, 1974, "The Use of Redundancy in Multiple-cue Judgments: Data from a Suppressor–variable task. *American Journal of Psychology* 87: 385–392.

Armelius, K., 1979, "Task Predictability and Performance as Determinants of Confidence in Multiple-cue Judgments." *Scandinavian Journal of Psychology* 20: 19–25.

Armstrong, J. Scott, 1978, "Are Econometricians Useful? Folklore Versus Fact." *Journal of Business* 51(4): 549–564.

———, 1981, "How Expert Are the Experts?" *Inc.*, Dec. 1981: 15–16.

Aron, Raymond, 1961, *Dimensions de la conscience historique*. Paris: Agora.

Arrow, Kenneth, 1987, "Economic Theory and the Postulate of Rationality." In J. Eatwell, M. Milgate, and P. Newman, eds., 1987, 2: 69–74.

Arthur, Brian W., 1994, *Increasing Returns and Path Dependence in the Economy*. Ann Arbor: University of Michigan Press.

Astebro, Thomas, 2003, "The Return to Independent Invention: Evidence of Unrealistic Optimism, Risk Seeking or Skewness Loving?" *Economic Journal* 113(484): 226–239.

Ashiya, Masahiro, and Takero Doi, 2001, "Herd Behavior of Japanese Economists." *Journal of Economic Behavior and Organization* 46: 343–346.

Attewell, P., 2001, "The Winner-take-all High School: Organizational Adaptations to Educational Stratification." *Sociology of Education* 74: 267–295.

Ayache, E., 2004a, "The Back of Beyond," *Wilmott* (Spring): 26–29.

———. 2004b, "A Beginning, in the End," *Wilmott* (Winter): 6–11.

Ayer, A. J., 1958, *The Problem of Knowledge*. London: Penguin Books.

———, 1972, *Probability and Evidence*. New York: Columbia University Press.

———, 1988, *Voltaire*. London: Faber and Faber.

Ayton, P., and A. G. R. McClelland, 1997, "How Real Is Overconfidence?" *Journal of Behavioral Decision Making* 10: 153–285.

Baddeley, Alan, 1997, *Human Memory: Theory and Practice*. London: Psychology Press.

Bak, Per, 1996, *How Nature Works*. New York: Copernicus.

Bak, P., and K. Chen, 1991, "Self-organized criticality." *Scientific American* 264: 46–53.

Ball, Philip, 2004, *Critical Mass: How One Thing Leads to Another*. London: Arrow Books.

———, 2006, "Econophysics: Culture Crash." *Nature* 441: 686–688.

Banavar, J. R., F. Colaiori, A. Flammini, A. Maritan, and A. Rinaldo, 2000, "A Topology of the Fittest Transportation Network." *Physical Review Letters* 84: 4745–4748.

Barabási, Albert-László, 2002, *Linked: The New Science of Networks*. Boston: Perseus Publishing.

Barabási, Albert-László, and Réka Albert, 1999, "Emergence of Scaling in Random Networks." *Science* 286: 509–512.

Barabási, Albert-László, Réka Albert, and H. Jeong, 1999, "Mean-field Theory for Scale-free Random Networks." *Physica A* 272: 173–197.

Barabási, Albert-László, and Eric Bonabeau, 2003, "Scale-free Networks." *Scientific American* 288(5): 50–59.

Baranski, J. V., and W. M. Petrusic, 1994, "The Calibration and Resolution of Confidence in Perceptual Judgments." *Perception and Psychophysics* 55: 412–428.

Barber, B. M., and T. Odean, 1999, "Trading Is Hazardous to Your Wealth: The Common Stock Investment Performance of Individual Investors." Working Paper.

Barbour, A. D., and G. Reinert, 2000, "Small worlds." Preprint cond-mat/0006001 at http://xxx.lanl.gov.

Bar-Hillel, M., and W. A. Wagenaar, 1991, "The perception of randomness." *Advances in Applied Mathematics* 12(4): 428–454.

Baron, Jonathan, 2000, *Thinking and Deciding,* 3rd ed. New York: Cambridge University Press.

Barron, G., and I. Erev, 2003, "Small Feedback-based Decisions and Their Limited Correspondence to Description-based Decisions." *Journal of Behavioral Decision Making* 16: 215–233.

Barrow, John D., 1998, *Impossibility: The Limits of Science and the Science of Limits.* London: Vintage.

Barrow, John D., and Frank J. Tipler, 1986, *The Anthropic Cosmological Principle.* Oxford: Oxford University Press.

Barrow-Green, June, 1996, *Poincaré and the Three Body Problem. History of Mathematics,* Vol. 11, American Mathematical Society.

Barthélémy, M., and L. A. N. Amaral, 1999, "Small-world Networks: Evidence for a Crossover Picture." *Physical Review Letters* 82: 3180–3183.

Bastiat, Frédéric, 1862–1864, *Oeuvres complètes de Frédéric Bastiat,* 6 vols. Paris: Guillaumin.

Batchelor, R. A., 1990, "All Forecasters Are Equal." *Journal of Business and Economic Statistics* 8(1): 143–144.

———, 2001, "How Useful Are the Forecasts of Intergovernmental Agencies? The IMF and OECD Versus the Consensus." *Applied Economics* 33(2): 225–235.

Bates, Elisabeth, 1994, "Modularity, Domain Specificity, and the Development of Language." In D. C. Gajdusek, G. M. McKhann, and C. L. Bolis, eds., *Evolution and Neurology of Language: Discussions in Neuroscience* 10: 1–2, 136–149.

Bauman, A. O., R. B. Deber, and G. G. Thompson, 1991, "Overconfidence Among Physicians and Nurses: The 'micro certainty, macro certainty' phenomenon." *Social Science and Medicine* 32: 167–174.

Bayer, Hans Christian, 2003, *Information: The New Language of Science.* London: Orion Books, Ltd.

Bechara, A., A. R. Damasio, H. Damasio, and S. W. Anderson, 1994, "Insensitivity to Future Consequences Following Damage to Human Prefrontal Cortex." *Cognition* 50: 1–3, 7–15.

Becker, Lawrence C., 1998, *A New Stoicism.* Princeton, N.J.: Princeton University Press.

Bellamy, Edward, 1891, *Cent ans après, ou l'an 2000,* trad. de l'anglais par Paul Rey; avec une préf. par M. Théodore Reinach. Paris: E. Dentu.

Benartzi, Shlomo, 2001. "Excessive Extrapolation and the Allocation of 401(k) Accounts to Company Stock," *Journal of Finance* 56(5): 1,747–1,764

Benartzi, Shlomo, and Richard Thaler, 1995, "Myopic Loss Aversion and the Equity Premium Puzzle." *Quarterly Journal of Economics* 110(1): 73–92.

Bénassy-Quéré, Agnès, 2002, *"Euro/dollar: tout le monde peut se tromper." La Lettre du CEPII* 215.

Benkirane, R., 2002, *La complexité, vertiges et promesses: 18 histoires de sciences.* Paris: Le Pommier.

Berger, Peter L., and Thomas Luckmann, 1966, *The Social Construction of Reality: A Treatise in the Sociology of Knowledge.* New York: Anchor Books.

Bernard, André, 2002, *Rotten Rejections: The Letters That Publisher Wish They'd Never Sent.* London: Chrysalis Books.

Bernard, Claude, 1878, *La science expérimentale.* Paris: J.-B. Baillière.

Bernoulli, Daniel, 1954, "Exposition of a New Theory on the Measurement of Risk." *Econometrica* 22(1): 23–36.

Bernstein, Peter L., 1996, *Against the Gods: The Remarkable Story of Risk.* New York: Wiley.

Berridge, Kent C., 2003, "Irrational Pursuits: Hyper-incentives from a Visceral Brain." In I. Brocas and J. Carillo, eds., 2003.

Berry, M., 1978, "Regular and Irregular Motion, in Topics in Nonlinear Mechanics," ed. S. Jorna, *American Institute of Physics Conference Proceedings* No. 46, 16–120.

Bevan, Edwyn, 1913, *Stoics and Sceptics.* Chicago: Ares Publishers, Inc.

Bewes, Timothy, 2002, *Reification: or The Anxiety of Late Capitalism.* London: Verso.

Bewley, Ronald A., and Denzil G. Fiebig, 2002, "On the Herding Instinct of Interest Rate Forecasters." *Empirical Economics* 27(3): 403–425.

Bhalla, U. S., and R. Iyengar, 1999, "Emergent Properties of Networks of Biological Signalling Pathways. *Science* 283: 381–387.

Bharat, Barot, 2004, "How Accurate are the Swedish Forecasters on GDP-Growth, CPI-Inflation and Unemployment?, 1993–2001." *Brussels Economic Review/Cahiers Economiques de Bruxelles* 47, 2 Editions du DULBEA, Université libre de Bruxelles, 249–278.

Bikhchandani, Sushil, David Hirshleifer, and Ivo Welch, 1992, "A Theory of Fads, Fashion, Custom, and Cultural Change as Informational Cascades." *Journal of Political Economy* 100 (5): 992–1026.

Binmore, K., 1999, "Why Experiment in Economics?" *Economic Journal* 109(453): 16–24.

Birnbaum, M. H., 1983, "Base Rates in Bayesian Inference: Signal Detection Analysis of the Cab Problem." *American Journal of Psychology* 96(1): 85–94.

Björkman, M., 1987, "A Note on Cue Probability Learning: What Conditioning Data Reveal About Cue Contrast." *Scandinavian Journal of Psychology* 28: 226–232.

———, 1994, "Internal Cue Theory: Calibration and Resolution of Confidence in General Knowledge." *Organizational Behavior and Human Decision Processes* 58: 386–405.

Bjorkman, M., P. Juslin, and A. Winman, 1993, "Realism of Confidence in Sensory Discrimination: The Underconfidence Phenomenon." *Perception and Psychophysics* 54: 75–81.

Blake, Carole, 1999, *From Pitch to Publication.* London: Pan.

Blake, David, Michael Beenstock, and Valerie Brasse, 1986, "The Performance of UK Exchange Rate Forecasters." *Economic Journal* 96(384): 986–999.

Blaug, Mark, 1992, *The Methodology of Economics,* 2nd ed. Cambridge: Cambridge University Press.

Bloch, Marc, 1953, *The Historian's Craft.* New York: Vintage Books.

Blyth, M. R. Abdelal, and Cr. Parsons, 2005, *Constructivist Political Economy.* Preprint, forthcoming, 2006: Oxford University Press.

Board, J., C. Sutcliffe, and E. Patrinos, 2000, "Performance of Covered Calls." *European Journal of Finance* 6(1): 1–17.

Bocarra, Nino, 2004, *Modeling Complex Systems.* Heidelberg: Springer.

Boettke, Peter J., Christopher J. Coyne, and Peter T. Leeson, 2006, "High Priests and Lowly Philosophers: The Battle for the Soul of Economics," a forthcoming article in the *Case Western Law Review.*

Boots, M., and A. Sasaki, 1999, " 'Small worlds' and the Evolution of Virulence: Infection Occurs Locally and at a Distance," *Proceedings of the Royal Society of London* B266: 1933–1938.

Bostrom, Nick, 2002, *Anthropic Bias: Observation Selection Effects in Science and Philosophy.* London: Routledge.

Bouchaud, J.-P., and M. Potters, 2003, *Theory of Financial Risks and Derivatives Pricing: From Statistical Physics to Risk Management*, 2nd ed. Cambridge: Cambridge University Press.

Bourdé, Guy, and Hervé Martin, 1989, *Les écoles historiques.* Paris: Éditions du Seuil.

Bourdieu, Pierre, 1992, *Les règles de l'art.* Paris: Éditions du Seuil.

———, 1996, *Sur la télévision suivi de l'emprise du journalisme.* Paris: Raison d'Agir.

———, 2000, *Esquisse d'une théorie de la pratique.* Paris: Éditions de Seuil.

Bouvier, Alban, ed., 1999, *Pareto aujourd'hui.* Paris: Presses Universitaires de France.

Boyer, Pascal, 2001, *Religion Explained: The Evolutionary Origins of Religious Thought.* New York: Basic Books.

Braudel, Fernand, 1953, "Georges Gurvitch ou la discontinuité du social." *Annales E.S.C.* 8: 347–361.

———, 1969, *Écrits sur l'histoire.* Paris: Flammarion.

———, 1985, *La Méditerranée: L'espace et l'histoire.* Paris: Flammarion.

———, 1990, *Écrits sur l'histoire II.* Paris: Flammarion.

Braun, P. A., and I. Yaniv, 1992, "A Case Study of Expert Judgment: Economists' Probabilities Versus Base-rate Model Forecasts." *Journal of Behavioral Decision Making* 5: 217–231.

Brehmer, B., and C. R. B. Joyce, eds., 1988, *Human Judgment: The SJT View.* Amsterdam: North-Holland.

Brender, A., and F. Pisani, 2001, *Les Marchés et la croissance.* Economica.

Brenner, L. A., D. J. Koehler, V. Liberman, and A. Tversky, 1996, "Overconfidence in Probability and Frequency Judgments: A Critical Examination." *Organizational Behavior and Human Decision Processes* 65: 212–219.

Brocas, I., and J. Carillo, eds., 2003, *The Psychology of Economic Decisions,* Vol. 1: *Rationality and Well-being.* Oxford: Oxford University Press.

Brochard, Victor, 1878, *De l'erreur.* Paris: Université de Paris.

———, 1888, *Les sceptiques grecs.* Paris: Imprimerie Nationale.

Brock, W. A., and P. J. F. De Lima, 1995, "Nonlinear Time Series, Complexity Theory, and Finance." University of Wisconsin, Madison—Working Papers 9523.

Brock, W. A., D. A. Hsieh, and B. LeBaron, 1991, *Nonlinear Dynamics, Chaos, and Instability: Statistical Theory and Economic Evidence.* Cambridge, Mass.: The MIT Press.

Brockman, John, 2005, Discussion with Benoît Mandelbrot, www.edge.org.

Brookes-Gunn, J., and G. Duncan, 1994, *Consequences of Growing Up Poor.* New York: Russell Sage.

Broughton, W., and E. W. Mills, 1980, "Resource Inequality and Accumulative Advantage: Stratification in the Ministry." *Social Forces* 58: 1289–1301.

Brugger, P., and R. E. Graves, 1997, "Right Hemispatial Inattention and Magical Ideation." *European Archive of Psychiatry and Clinical Neuroscience* 247(1): 55–57.

Bruner, Jerome, 1994, "The 'Remembered' Self." In Ulric Neisser and Robyn Fivush, eds., *The Remembering Self: Construction and Accuracy in the Self-Narrative.* Cambridge: Cambridge University Press.

———, 2002, *Making Stories: Law, Literature, Life.* New York: Farrar, Straus & Giroux.

Bruner, Jerome S., and Mary C. Potter, 1964, "Interference in Visual Recognition" *Science* 144(3617): 424–425.

Brunswik, E., 1952, *The Conceptual Framework of Psychology.* Chicago: The University of Chicago Press.

———, 1955, "Representative Design and Probabilistic Theory in a Functional Psychology." *Psychological Review* 62: 193–217.

Buchanan, Mark, 2001, *Ubiquity: Why Catastrophes Happen.* New York: Three Rivers Press.

———, 2002, *Nexus: Small Worlds and the Groundbreaking Theory of Networks.* New York: W. W. Norton and Company.

Budescu, D. V., I. Erev, and T. S. Wallsten, 1997, "On the Importance of Random Error in the Study of Probability Judgment. Part I: New Theoretical Developments." *Journal of Behavioral Decision Making* 10: 157–171.

Buehler, R., D. Griffin, and M. Ross, 2002, "Inside the Planning Fallacy: The Causes and Consequences of Optimistic Time Predictions." In T. Gilovich, D. Griffin, and D. Kahneman, eds., 2002.

Bundt, Thomas, and Robert P. Murphy, 2006, "Are Changes in Macroeconomic Variables Normally Distributed? Testing an Assumption of Neoclassical Economics." Preprint, NYU Economics Department.

Burnham, Terence C., 1997, *Essays on Genetic Evolution and Economics.* New York: Dissertation.com.

———, 2003, "Caveman Economics." Preprint, Harvard Business School.

Burnham, T., and J. Phelan, 2000, *Mean Genes.* Boston: Perseus Publishing.

Bushman, B. J., and G. L. Wells, 2001, "Narrative Impressions of Literature: The Availability Bias and the Corrective Properties of Meta-analytic Approaches." *Personality and Social Psychology Bulletin* 27: 1123–1130.

Callaway, D. S., M. E. J. Newman, S. H. Strogatz, and D. J. Watts, 2000, "Network Robustness and Fragility: Percolation on Random Graphs." *Physical Review Letters* 85: 5468–5471.

Camerer, C., 1995, "Individual Decision Making." In John H. Kagel and Alvin E. Roth, eds., *The Handbook of Experimental Economics*. Princeton, N.J.: Princeton University Press.

———, 2003, *Behavioral Game Theory: Experiments in Strategic Interaction*. Princeton, N.J.: Princeton University Press.

Camerer, Colin F., George Loewenstein, and D. Prelec, 2003, "Neuroeconomics: How Neuroscience Can Inform Economics." Caltech Working Paper.

Camerer, Colin F., George Loewenstein, and Matthew Rabin, 2004, *Advances in Behavioral Economics*. Princeton, N.J.: Princeton University Press.

Cannon, Walter B., 1940, "The Role of Chance in Discovery." *Scientific Monthly* 50: 204–209.

Carnap, R., 1950, *The Logical Foundations of Probability*. Chicago: The University of Chicago Press.

———, 1966, *Philosophical Foundations of Physics*. New York: Basic Books.

Carr, Edward Hallett, 1961, *What Is History?* New York: Vintage Books.

Carter, C. F., G. P. Meredith, and G. L. S. Shackle, 1962, *Uncertainty and Business Decisions*. Liverpool: Liverpool University Press.

Carter, Rita, 1999, *Mapping the Mind*. Berkeley: University of California Press.

———, 2002, *Exploring Consciousness*. Berkeley: University of California Press.

Casanova, Giovanni Giacomo, 1880, *Mémoires de J. Casanova de Seingalt*. Paris: Garnier Frères.

Casscells, W., A. Schoenberger, and T. Grayboys, 1978, "Interpretation by Physicians of Clinical Laboratory Results." *New England Journal of Medicine* 299: 999–1000.

Cerf, Christopher, and Victor Navasky, 1998, *The Expert Speaks: The Definitive Compendium of Authoritative Misinformation*. New York: Villard Books.

Certeau, Michel de, 1975, *L'Ecriture de l'histoire*. Paris: Gallimard.

Chamley, Christophe P., 2004, *Rational Herds: Economic Models of Social Learning*. Cambridge: Cambridge University Press.

Chancellor, Edward, 1999, *Devil Take the Hindmost: A History of Financial Speculation*. New York: Farrar, Straus & Giroux.

Chartier, Roger, 1996, *Culture et société. L'ordre des livres, XVIe–XVIIIe*. Paris: Albin Michel.

Chen, Keith, Venkat Lakshminarayanan, and Laurie Santos, 2005, "The Evolution of Our Preferences: Evidence from Capuchin Monkey Trading Behavior." Cowles Foundation Discussion Paper No. 1524.

Chen, Qi, Jennifer Francis, and Wei Jiang, 2002, "Investor Learning About Analyst Predictive Ability." Working Paper, Duke University.

Cherniak, C., 1994, "Component Placement Optimization in the Brain." *Journal of Neuroscience* 14: 2418–2427.

Chipman, John, 2006, "The Paretian Heritage." Working Paper, University of Minnesota.

Cialdini, Robert B., 2001, *Influence: Science and Practice*. Boston: Allyn and Bacon.

Cisne, John L., 2005, "Medieval Manuscripts' 'Demography' and Classic Texts' Extinction." *Science* 307(5713): 1305–1307.

Clark, Barrett, and Pascal Boyer, 2006, "*Causal Inferences: Evolutionary Domains and Neural Systems.*" Interdisciplines Conference on Causality, see www.interdiscplines.org.

Clark, Michael, 2002, *Paradoxes from A to Z*. London: Routledge.

Clemen, R. T., 1986, "Calibration and the Aggregation of Probabilities." *Management Science* 32: 312–314.

———, 1989, "Combining Forecasts: A Review and Annotated Bibliography." *International Journal of Forecasting* 5: 559–609.

Cohen, L. J., 1989, *The Philosophy of Induction and Probability*. Oxford: Clarendon Press.

Cohen, R., K. Erez, D. ben-Avraham, and S. Havlin, 2000, "Resilience of the Internet to Random Breakdowns." *Physical Review Letters* 85: 4626–4628.

Cole, J. R., and S. Cole, 1973, *Social Stratification in Science*. Chicago: The University of Chicago Press.

Cole, J. R., and B. Singer, 1991, "A Theory of Limited Differences: Explaining the Productivity Puzzle in Science." In J. C. H. Zuckerman and J. Bauer, eds., *The Outer Circle: Women in the Scientific Community*. New York: W. W. Norton and Company.

Cole, Peter, 2002, *Access to Philosophy: The Theory of Knowledge*. London: Hodder and Stoughton.

Cole, S., 1970, "Professional Standing and the Reception of Scientific Discoveries." *American Journal of Sociology* 76: 286–306.

Cole, S., J. C. Cole, and G. A. Simon, 1981, "Chance and Consensus in Peer Review." *Science* 214: 881–886.

Collins, Randall, 1998, *The Sociology of Philosophies: A Global Theory of Intellectual Change.* Cambridge, Mass.: The Belknap Press of Harvard University Press.

Conley, D., 1999, *Being Black, Living in the Red: Race, Wealth and Social Policy in America.* Los Angeles: University of California Press.

Cooper, John M., 2004, *Knowledge, Nature, and the Good,* Chapter 1: "Method and Science in on Ancient Medicine." Princeton, N.J.: Princeton University Press.

Cootner, Paul H., 1964, *The Random Character of Stock Market Prices.* London: Risk Books.

Cosmides, L., and J. Tooby, 1990, "Is the Mind a Frequentist?" Paper presented at the 31st annual meeting of the Psychonomics Society, New Orleans, La.

———, 1992, "Cognitive Adaptations for Social Exchange." In Jerome H. Barkow, Leda Cosmides, and John Tooby, eds., *The Adapted Mind.* Oxford: Oxford University Press.

———, 1996, "Are Humans Good Intuitive Statisticians After All? Rethinking Some Conclusions from the Literature on Judgment and Uncertainty." *Cognition* 58(1): 187–276.

Courtillot, V., 1995, *La vie en catastrophes.* Paris: Fayard.

Courtillot, V., and Y. Gaudemer, 1996, "Effects of Mass-Extinctions on Biodiversity." *Nature* 381: 146–147.

Cousin, Victor, 1820, *Cours d'histoire de la philosophie morale au dix-huitième siècle.* Paris: Ladrange.

Cover, T. M., and J. A. Thomas, 1991, *Elements of Information Theory.* New York: Wiley.

Cowley, Michelle, and Ruth M. J. Byrne, 2004, "Chess Master's Hypothesis Testing." In Kenneth Forbus, Dedre Gentner, and Terry Regier, eds., *Proceedings of 26th Annual Conference of the Cognitive Science Society, CogSci 2004,* Mahwah, N.J.: Lawrence Erlbaum.

Crosby, Alfred W., 1997, *The Measure of Reality: Quantification and Western Society, 1250–1600.* Cambridge: Cambridge University Press.

Csikszentmihalyi, Mihaly, 1993, *Flow: The Psychology of Optimal Experience.* New York: Perennial Press.

———, 1998, *Finding Flow: The Psychology of Engagement with Everyday Life.* New York: Basic Books.

Cutler, David, James Poterba, and Lawrence Summers, 1989, "What Moves Stock Prices?" *Journal of Portfolio Management* 15: 4–12.

Dally J. M., N. J. Emery, and N. S. Clayton, 2006, "Food-Catching Western Scrub-Jays Keep Track of Who Was Watching When." *Science* 312 (5780): 1,662–1,665.

Damasio, Antonio, 1994, *Descartes' Error: Emotion, Reason, and the Human Brain.* New York: Avon Books.

———, 2000, *The Feeling of What Happens: Body and Emotion in the Making of Consciousness.* New York: Harvest Books.

———, 2003, *Looking for Spinoza: Joy, Sorrow and the Feeling Brain.* New York: Harcourt.

Dannefer, D., 1987, "Aging as Intracohort Differentiation: Accentuation, the Matthew Effect and the Life Course." *Sociological Forum* 2: 211–236.

———, 2003, "Cumulative Advantage/Disadvantage and the Life Course: Cross-fertilizing Age and Social Science." *Journal of Gerontology Series B: Psychological Sciences and Social Sciences* 58: 327–337.

Darwin, Charles, 1859, *On Natural Selection.* London: Penguin Books, Great Ideas.

Daston, L. J., 1988, *Classical Probability in the Enlightenment.* Princeton, N.J.: Princeton University Press.

David, Florence Nightingale, 1962, *Games, Gods, and Gambling: A History of Probability and Statistical Ideas.* Oxford: Oxford University Press.

Dawes, Robyn M., 1980, "Confidence in Intellectual Judgments vs. Confidence in Perceptual Judgments." In E. D. Lantermann and H. Feger, eds., *Similarity and Choice: Papers in Honor of Clyde Coombs.* Bern, Switzerland: Huber.

———,1988, *Rational Choice in an Uncertain World.* New York: Harcourt.

———, 1989, "Measurement Models for Rating and Comparing Risks: The Context of AIDS." *Conference Proceedings Health Services Research Methodology: A Focus on AIDS,* September 1989.

———, 1999, "A Message from Psychologists to Economists: Mere Predictability Doesn't Matter Like It Should, Without a Good Story Appended to It." *Journal of Economic Behavior and Organization.* 39: 29–40.

———, 2001a, "Clinical Versus Actuarial Judgment." *International Encyclopedia of the Social and Behavioral Sciences* 2048–2051.

———, 2001b, *Everyday Irrationality: How Pseudo-Scientists, Lunatics, and the Rest of Us Systematically Fail to Think Rationally.* Oxford: Westview Press.

———, 2002, "The Ethics of Using or Not Using Statistical Prediction Rules in Psychological Practice and Related Consulting Activities." *Philosophy of Science* 69: 178–184.

Dawes, Robyn M., D. Faust, and P. E. Meehl, 1989, "Clinical Versus Actuarial Judgment." *Science* 243: 1668–1674.

Dawes, Robyn M., R. Fildes, M. Lawrence, and K. Ord, 1994, "The Past and the Future of Forecasting Research." *International Journal of Forecasting* 10: 151–159.

Dawes, Robyn M., and T. L. Smith, 1985, "Attitude and Opinion Measurement." In G. Lindzey and E. Aronson, *The Handbook of Social Psychology,* Vol. 1. Hillsdale, N.J.: Lawrence Erlbaum.

de Bellaigue, Eric, de., 2004, *British Book Publishing as a Business Since the 1960s.* London: The British Library.

De Bondt, Werner, and Andreas Kappler, 2004, "Luck, Skill, and Bias in Economists' Forecasts." Working Paper, Driehaus Center for Behavioral Finance, DePaul University.

De Bondt, Werner F. M., and Richard M. Thaler, 1990, "Do Security Analysts Overreact?" *American Economic Review* 80: 52–57.

Debreu, Gerard, 1959, *Theorie de la valeur,* Dunod, tr. *Theory of Value.* New York: Wiley.

de Finetti, Bruno, 1931, 1989, "Probabilism." *Erkenntnis* 31: 169–223.

———, 1975, 1995, *Filosophia della probabilita.* Milan: Il Saggiatore.

Degeorge, François, Jayendu Patel, and Richard Zeckhauser, 1999, "Earnings Management to Exceed Thresholds." *Journal of Business* 72(1): 1–33.

DeLong, Bradford, Andrei Shleifer, Lawrence Summers, and Robert J. Waldmann, 1991. "The Survival of Noise Traders in Financial Markets." *Journal of Business* 64(1): 1–20.

Dennett, Daniel C., 1995, *Darwin's Dangerous Idea: Evolution and the Meanings of Life.* New York: Simon & Schuster.

———, 2003, *Freedom Evolves.* New York: Penguin Books.

Derman, E., and N. N. Taleb, 2005, "The Illusions of Dynamic Replication." *Quantitative Finance* 5: 323–326.

De Vany, Arthur, 2002, *Hollywood Economics: Chaos in the Movie Industry.* London: Routledge.

De Vany, Arthur, Nassim Nicholas Taleb, and Mark Spitznagel, 2004, "Can We Shield Artists from Wild Uncertainty?" presented at the Fort Lauderdale Film Festival Scholar's Workshop, June 2004.

DiPrete, Thomas A., and Greg Eirich, 2006, "Cumulative Advantage as a Mechanism for Inequality: A Review of Theoretical and Empirical Developments." *Annual Review of Sociology* 32: 271–297.

Dominitz, Jeff, and David Grether, 1999, "I Know What You Did Last Quarter: Economic Forecasts of Professional Forecasters." Working Paper, Caltech.

Donhardt, Gary L., 2004, "In Search of the Effects of Academic Achievement in Postgraduation Earnings." *Research in Higher Education* 45(3): 271–284.

Dugatkin, Lee Alan, 2001, *The Imitation Factor: Evolution Beyond the Gene.* New York: Simon & Schuster.

Dunbar, Nicholas, 1999, *Inventing Money: The Story of Long-Term Capital Management and the Legends Behind It.* Chichester, England: John Wiley & Sons, Ltd.

Dunning, D., D. W. Griffin, J. Milojkovic, and L. Ross, 1990, "The Overconfidence Effect in Social Prediction." *Journal of Personality and Social Psychology* 58: 568–581.

Dye, Guillaume, 2004, A review of Lorenzo Perilli's *Menodoto di Nicomedia,* Munich and Leipzig: K. G. Saur, in *Bryn Mawr Classical Review,* December 20.

Easterwood, John C., and Stacey R. Nutt, 1999, "Inefficiency in Analysts' Earnings Forecasts: Systematic Misreaction or Systematic Optimism?" *Journal of Finance* 54: 1777–1797.

Eatwell, J., M. Milgate, and P. Newman, eds., 1987, *The New Palgrave: A Dictionary of Economics*. London: Macmillan.

Eco, Umberto, 1992, *How to Travel with a Salmon and Other Essays*. San Diego: Harcourt.

———, 1994, *Six Walks in the Fictional Woods*. Cambridge, Mass.: Harvard University Press.

———, 2000, *Kant and the Platypus: Essays on Language and Cognition*. New York: Harvest Books.

———, 2002, *On Literature*. Orlando: Harcourt Books.

———, 2003, *Mouse or Rat? Translation as Negotiation*. London: Orion Books.

Einhorn, H. J., and R. M. Hogarth, 1981, "Behavioral Decision Theory: Processes of Judgment and Choice." *Annual Review of Psychology* 32: 53–88.

Ekeland, Ivar, 1990, *Mathematics of the Unexpected*. Chicago: The University of Chicago Press.

Eldredge, Niles, and Stephen Jay Gould, 1972, "Punctuated Equilibria: An Alternative to Phyletic Gradualism." *Models in Paleobiology*, ed., T.J.M. Schopf. New York: Freeman.

El-Galfy, A. M., and W. P. Forbes, 2005, "An Evaluation of U.S. Security Analysts Forecasts, 1983–1999." Working Paper.

Elman, C., and A. M. O'Rand, 2004, "The Race Is to the Swift: Socioeconomic Origins, Adult Education, and Wage Attainment." *American Journal of Sociology* 110: 123–160.

Empiricus, Sextus, 1997, *Esquisses pyrrhoniennes*. Paris: Éditions du Seuil.

———, 2002, *Contre les professeurs*. Paris: Éditions du Seuil.

Epstein, Jason, 2001, *Book Business*. London: W. W. Norton.

Erev, I., T. S. Wallsten, and D. V. Budescu, 1994, "Simultaneous Over- and Underconfidence: The Role of Error in Judgment Processes." *Psychological Review* 101: 519–528.

Estoup, J. B., 1916, *Gammes Stenographique*. Paris: Institut Stenographique de France.

Evans, Dylan, 2002, *Emotions: The Science of Sentiment*. Oxford: Oxford University Press.

Eysenck, M. W., and M. T. Keane, 2000, *Cognitive Psychology*, 4th ed. London: Psychology Press.

Fagot-Largeault, Anne, 2002, *Philosophie des sciences biologiques et medicales*. Paris: College de France.

Faia, M., 1975, "Productivity Among Scientists: A Replication and Elaboration." *American Sociological Review* 40: 825–829.

Faloutsos, M., P. Faloutsos, and C. Faloutsos, 1999, "On Power-law Relationships of the Internet Topology." *Computer Communications Review* 29: 251–262.

Favier, A., 1906, *Un médecin grec du deuxième siècle ap. J.-C., précurseur de la méthode expérimentale moderne: Ménodote de Nicomédie*. Paris: Jules Roisset.

Ferguson, Niall, 2005, *1914: Why the World Went to War*. London: Penguin.

———, 2006a, *The War of the World: History's Age of Hatred*. London: Allen Lane.

———, 2006b, "Political Risk and the International Bond Market Between the 1848 Revolution and the Outbreak of the First World War." *Economic History Review* 59(1): 70–112.

Ferraro, K. F., and J. A. Kelley-Moore, 2003, "Cumulative Disadvantage and Health: Long-term Consequences of Obesity?" *American Sociological Review* 68: 707–729.

Feyerabend, Paul, 1987, *Farewell to Reason*. London: Verso.

Finucane, M. L., A. Alhakami, P. Slovic, and S. M. Johnson, 2000, "The Affect a Heuristic in Judgments of Risks and Benefits." *Journal of Behavioral Decision Making* 13: 1–17.

Fischhoff, Baruch, 1982a, "Debiasing." In D. Kahneman, P. Slovic, and A. Tversky, eds., *Judgment Under Uncertainty: Heuristics and Biases*. Cambridge: Cambridge University Press.

———, 1982b, "For Those Condemned to Study the Past: Heuristics and Biases in Hindsight." In D. Kahneman, P. Slovic, and A. Tversky, *Judgment Under Uncertainty: Heuristics and Biases*. Cambridge: Cambridge University Press.

Fischhoff, B., and D. MacGregor, 1983, "Judged Lethality: How Much People Seem to Know Depends on How They Are Asked." *Risk Analysis* 3: 229–236.

Fischhoff, Baruch, Paul Slovic, and Sarah Lichtenstein, 1977, "Knowing with Certainty: The Appropriateness of Extreme Confidence." *Journal of Experimental Psychology* 3(4): 552–564.

Floridi, Luciano, 2002, *The Transmission and Recovery of Pyrrhonism*. Oxford: Oxford University Press.

Flyvbjerg, Bent, Mette Skamris Holm, and Søren Buhl, 2002, "Underestimating Costs in Public

Works Projects—Error or Lie." *American Journal of Planning* 68(3), http://home.planet.nl/~viss1197/japaflyvbjerg.pdf.

Fodor, Jerry A., 1983, *The Modularity of Mind: An Essay on Faculty Psychology.* Cambridge, Mass.: The MIT Press.

Foster, George, 1977, "Quarterly Accounting Data: Time-series Properties and Predictive Ability Results." *Accounting Review* 52: 1–21.

Fox, M. A., and P. Kochanowski, 2004, "Models of Superstardom: An Application of the Lotka and Yule Distributions." *Popular Music and Society* 27: 507–522.

Frame, Donald M., 1965, *Montaigne: A Biography.* New York: Harcourt Brace and World.

Frank, Jerome D., 1935, "Some Psychological Determinants of the Level of Aspiration." *American Journal of Psychology* 47: 285–293.

Frank, Robert, 1994, "Talent and the Winner-Take-All Society." A review of Derek Bok's *The Cost of Talent: How Executives and Professionals Are Paid and How It Affects America,* New York: The Free Press, 1993, in *The American Prospect* 5(17), www.prospect.org/print/V5/17/frank-r.html.

Frank, Robert H., 1985, *Choosing the Right Pond: Human Behavior and the Quest for Status.* Oxford: Oxford University Press.

Frank, Robert H., and P. J. Cook, 1995, *The Winner-Take-All Society: Why the Few at the Top Get So Much More Than the Rest of Us.* New York: The Free Press.

Frankfurter, G. M., and E. G. McGoun, 1996, *Toward Finance with Meaning: The Methodology of Finance: What It Is and What It Can Be.* Greenwich, Conn.: JAI Press.

Freedman, D. A., and P. B. Stark, 2003, "What Is the Chance of an Earthquake?" Technical Report 611 of the Department of Statistics, University of California, Berkeley, September 2001, revised January 2003.

Friesen, Geoffrey, and Paul A. Weller, 2002, "Quantifying Cognitive Biases in Analyst Earnings Forecasts." Working Paper, University of Iowa.

Frohlich, N., J. A. Oppenheimer, and C. L. Eavy, 1987a, "Laboratory Results on Rawls's Distributive Justice." *British Journal of Political Science* 17: 1–21.

———, 1987b, "Choices of Principles of Distributive Justice in Experimental Groups." *American Journal of Political Science* 31(3): 606–636.

Froot, K. A., 2001, "The Market for Catastrophe Risk: A Clinical Examination," *Journal of Financial Economics* 60(2–3): 529–571.

Fukuyama, Francis, 1992, *The End of History and the Last Man.* New York: The Free Press.

Fuller, Steve, 2005, *The Intellectual.* London: Icon Books.

Fulton, Alice, 1998, "Fractal Amplifications: Writing in Three Dimensions." *Thumbscrew* 12 (winter).

Gabaix, X., P. Gopikrishnan, V. Plerou, and H. E. Stanley, 2003, "A Theory of Power-law Distributions in Financial Market Fluctuations." *Nature* 423: 267–270.

Gaddis, John Lewis, 2002, *The Landscape of History: How Historians Map the Past.* Oxford: Oxford University Press.

Galbraith, John Kenneth, 1997, *The Great Crash 1929.* New York: Mariner Books.

Galison, Peter, 2003, *Einstein's Clocks, Poincaré's Maps: Empires of Time.* New York: W. W. Norton and Company.

Gave, Charles, Anatole Kaletsky, and Louis-Vincent Gave, 2005, *Our Brave New World.* London: GaveKal Research.

Gazzaniga, M. S., R. Ivry, and G. R. Mangun, 2002, *Cognitive Neuroscience: The Biology of the Mind,* 2nd ed. New York: W. W. Norton and Company.

Gazzaniga, Michael, and Joseph LeDoux, 1978, *The Integrated Mind.* Plenum Press.

Gazzaniga, Michael S., 2005, *The Ethical Brain.* New York: Dana Press.

Gehring, W. J., and A. R.Willoughby, 2002,"The Medial Frontal Cortex and the Rapid Processing of Monetary Gains and Losses." *Science* 295: 2279–2282.

Gelman, S. A., 1988, "The Development of Induction Within Natural Kind and Artifact Categories." *Cognitive Psychology* 20: 65–95.

Gelman, S. A., and J. D. Coley, 1990, "The Importance of Knowing a Dodo Is a Bird: Categories and Inferences in Two-year-old Children." *Developmental Psychology* 26: 796–804.

Gelman, S. A., and L. A. Hirschfeld, 1999, "How Biological Is Essentialism?" In D. L. Medin and S. Atran, eds., *Folkbiology.* Cambridge, Mass.: The MIT Press.

Gelman, S. A., and E. M. Markman, 1986, "Categories and Induction in Young Children." *Cognition* 23: 183–209.

Gervais, Simon, and Terrance Odean, 1999, "Learning to Be Overconfident." Working Paper, University of Pennsylvania.

Gigerenzer, G., P. M. Todd, and the ABC Research Group, 2000, *Simple Heuristics That Make Us Smart.* Oxford: Oxford University Press.

Gigerenzer, Gerd, 1984, "External Validity of Laboratory Experiments: The Frequency-Validity Relationship." *American Journal of Psychology* 97: 185–195.

———, 1987, "Survival of the Fittest Probabilist: Brunswik, Thurstone, and the Two Disciplines of Psychology." In L. Krüger, G. Gigerenzer, and M. S. Morgan, eds., *The Probabilistic Revolution,* Vol. 2: *Ideas in the Sciences.* Cambridge, Mass.: The MIT Press.

———, 1991, "From Tools to Theories: A Heuristic of Discovery in Cognitive Psychology." *Psychological Review* 98(2): 254–267.

Gigerenzer, G., J. Czerlinski, and L. Martignon, 2002, "How Good Are Fast and Frugal Heuristics?" In T. Gilovich, D. Griffin, and D. Kahneman, eds., 2002.

Gigerenzer, G., and D. G. Goldstein, 1996, "Reasoning the Fast and Frugal Way: Models of Bounded Rationality." *Psychological Review* 103: 650–669.

Gigerenzer, Gerd, W. Hell, and H. Blank, 1988, "Presentation and Content: The Use of Base Rates as a Continuous Variable." *Journal of Experimental Psychology: Human Perception and Performance* 14: 513–525.

Gigerenzer, G., U. Hoffrage, and H. Kleinbolting, 1991, "Probabilistic Mental Models: A Brunswikian Theory of Confidence." *Psychological Review* 98: 506–528.

Gigerenzer, G., and H. R. Richter, 1990, "Context Effects and Their Interaction with Development: Area Judgments." *Cognitive Development* 5: 235–264.

Gigerenzer, G., Z. Swijtink, T. Porter, L. J. Daston, J. Beatty, and L. Krüger, 1989, *The Empire of Chance: How Probability Changed Science and Everyday Life.* Cambridge: Cambridge University Press.

Gilbert, D., E. Pinel, T. D. Wilson, S. Blumberg, and T. Weatley, 2002, "Durability Bias in Affective Forecasting." In T. Gilovich, D. Griffin, and D. Kahneman, eds., 2002.

Gilbert, Daniel, 2006, *Stumbling on Happiness.* New York: Knopf.

Gilbert, Daniel T., 1991, "How Mental Systems Believe." *American Psychologist* 46: 107–119.

Gilbert, Daniel T., Romin W. Tafarodi, and Patrick S. Malone, 1993, "You Can't Not Believe Everything You Read." *Journal of Personality and Social Psychology* 65: 221–233.

Gillespie, John V., 1979, Review of William Ascher's *Forecasting: An Appraisal for Policy-Makers and Planners* in *The American Political Science Review* 73(2): 554–555.

Gillies, Donald, 2000, *Philosophical Theories of Probability.* London: Routledge.

Gilovich, T., D. Griffin, and D. Kahneman, eds., 2002, *Heuristics and Biases: The Psychology of Intuitive Judgment.* Cambridge: Cambridge University Press.

Gladwell, Malcolm, 1996, "The Tipping Point: Why Is the City Suddenly So Much Safer—Could It Be That Crime Really Is an Epidemic?" *The New Yorker,* June 3.

———, 2000, *The Tipping Point: How Little Things Can Make a Big Difference.* New York: Little, Brown.

———, 2002, "Blowing Up: How Nassim Taleb Turned the Inevitability of Disaster into an Investment Strategy." *The New Yorker,* April 22 and 29.

Glänzel, W., 2003, *Bibliometrics as a Research Field: A Course on the Theory and Application of Bibliometric Indicators.* Preprint.

Gleik, James, 1987, *Chaos: Making a New Science.* London: Abacus.

Glimcher, Paul, 2002, *Decisions, Uncertainty, and the Brain: The Science of Neuroeconomics.* Cambridge, Mass.: The MIT Press.

Goldberg, Elkhonon, 2001, *The Executive Brain: Frontal Lobes and the Civilized Mind.* Oxford: Oxford University Press.

———, 2005, *The Wisdom Paradox: How Your Mind Can Grow Stronger as Your Brain Grows Older.* New York: Gotham.

Goleman, Daniel, 1995, *Emotional Intelligence: Why It Could Matter More Than IQ*. New York: Bantam Books.

———, 2003, *Destructive Emotions, How Can We Overcome Them? A Scientific Dialogue with the Dalai Lama*. New York: Bantam.

Goodman, N., 1955, *Fact, Fiction, and Forecast*. Cambridge, Mass.: Harvard University Press.

———, 1972, "Seven Strictures on Similarity." In N. Goodman, ed., *Problems and Projects*. New York: Bobbs-Merrill.

Gopnik, A., 2004, C. Glymour, D. M. Sobel, L. E. Schulz, T. Kushnir, and D. Danks, D., press, "A Theory of Causal Learning in Children: Causal Maps and Bayes Nets." *Psychological Review* 111: 3–32.

Granger, Clive W. J., 1999, *Empirical Modeling in Economics: Specification and Evaluation*. Cambridge: Cambridge University Press.

Gray, John, 2002, *Straw Dogs: Thoughts on Humans and Other Animals*. London: Granta Books.

Green, Jack, 1962, *Fire the Bastards!* New York: Dalkey Archive Press.

Green, K. C. 2005, "Game Theory, Simulated Interaction, and Unaided Judgement for Forecasting Decisions in Conflicts: Further Evidence." *International Journal of Forecasting* 21: 463–472.

Griffin, D. W., and A. Tversky, 1992, "The Weighing of Evidence and the Determinants of Confidence." *Cognitive Psychology* 24: 411–435.

Griffin, D. W., and C. A. Varey, 1996, "Towards a Consensus on Overconfidence." *Organizational Behavior and Human Decision Processes* 65: 227–231.

Gripaios, Peter, 1994, "The Use and Abuse of Economic Forecasts." *Management Decision* 32(6): 61–64.

Guedj, Olivier, and Jean-Philippe Bouchaud, 2006, "Experts' Earning Forecasts: Bias, Herding and Gossamer Information," forthcoming.

Guglielmo, Cavallo, and Roger Chartier, 1997, *Histoire de la lecture dans le monde occidental*. Paris: Éditions du Seuil.

Gurvitch, Georges, 1957, "Continuité et discontinuité en histoire et sociologie." *Annales E.S.C.*: 73–84.

———, 1966, *The Social Framework of Knowledge*. New York: Harper Torchbooks.

Hacking, Ian, 1965, *Logic of Statistical Inference*. Cambridge: Cambridge University Press.

———, 1983, *Representing and Intervening: Introductory Topics in the Philosophy of Natural Science*. Cambridge: Cambridge University Press.

———, 1990, *The Taming of Chance*. Cambridge: Cambridge University Press.

———, 1999, *The Social Construction of What?* Cambridge, Mass.: Harvard University Press.

———, 2001, *An Introduction to Probability and Inductive Logic*. Cambridge: Cambridge University Press.

Hahn, Frank, 1993, "Predicting the Economy." In Leo Howe and Alan Wain, eds., 1993.

Hannon, L., 2003, "Poverty, Delinquency, and Educational Attainment: Cumulative Disadvantage or Disadvantage Saturation?" *Sociological Inquiry* 73: 575–594.

Hansen, R. D., and J. M. Donoghue, 1977, "The Power of Consensus: Information Derived from One's Own and Others' Behavior." *Journal of Personality and Social Psychology* 35: 294–302.

Hardy, G. H., 1940, *A Mathematician's Apology*. Cambridge: Cambridge University Press.

Harris, Olivia, 2004, "Braudel: Historical Time and the Horror of Discontinuity." *History Workshop Journal* 57: 161–174.

Harvey, N., 1997, "Confidence in Judgment." *Trends in Cognitive Science* 1: 78–82.

Hasher, L., and R. T. Zacks, 1979, "Automatic and Effortful Processes in Memory." *Journal of Experimental Psychology: General* 108: 356–388.

Haug, Espen, 2007, *Derivatives: Models on Models*. New York: Wiley.

Hausman, Daniel M., ed., 1994, *The Philosophy of Economics: An Anthology*, 2nd ed. New York: Cambridge University Press.

Hayek, F. A., 1945, "The Use of Knowledge in Society." *American Economic Review* 35(4): 519–530.

———, 1994, *The Road to Serfdom*. Chicago: The University of Chicago Press.

Hecht, Jennifer Michael, 2003, *Doubt: A History*. New York: Harper Collins.

Hempel, C., 1965, *Aspects of Scientific Explanation*. New York: The Free Press.

Henderson, Bill, and André Bernard, eds., *Rotten Reviews and Rejections*. Wainscott, N.Y.: Pushcart.

Hespos, Susan, 2006, "Physical Causality in Human Infants." Interdisciplines Conference on Causality, www.interdisciplines.org.

Hexter, J. H., 1979, *On Historians, Reappraisals of Some of the Masters of Modern History*. Cambridge, Mass.: Harvard University Press.

Hicks, Steven V., and Alan Rosenberg, 2003, "The 'Philosopher of the Future' as the Figure of Disruptive Wisdom." *Journal of Nietzsche Studies* 25: 1–34.

Hilton, Denis, 2003, "Psychology and the Financial Markets: Applications to Understanding and Remedying Irrational Decision-making." In I. Brocas and J. Carillo, eds., 2003.

Hintzman, D. L., G. Nozawa, and M. Irmscher, 1982, "Frequency as a Nonpropositional Attribute of Memory." *Journal of Verbal Learning and Verbal Behavior* 21: 127–141.

Hirshleifer, J., and J. G. Riley, 1992, *The Analytics of Uncertainty and Information*. Cambridge: Cambridge University Press.

Hladik, Jean, 2004, *Comment le jeune et ambitieux Einstein s'est approprié la relativité restreinte de Poincaré*. Paris: Ellipses.

Hoffrage, U., and G. Gigerenzer, 1998, "Using Natural Frequencies to Improve Diagnostic Inferences." *Academic Medicine* 73(5): 538–540.

Hong, Harrison, and Jeffrey Kubik, 2003, "Analyzing the Analysts: Career Concerns and Biased Earnings Forecasts." *Journal of Finance* 58(1): 313–351.

Hopfield, J. J., 1994, "Neurons, Dynamics, and Computation." *Physics Today* 47: 40–46.

Horkheimer, Max, and Theodor W. Adorno, 2002, *Dialectic of Enlightenment: Philosophical Fragments*. Stanford: Stanford University Press.

House, D. K., 1980, "The Life of Sextus Empiricus." *The Classical Quarterly, New Series* 30(1): 227–238.

Howe, Leo, and Alan Wain, eds., 1993, *Predicting the Future*. Cambridge: Cambridge University Press.

Hsee, C. K., and Y. R. Rottenstreich, 2004, "Music, Pandas and Muggers: On the Affective Psychology of Value." *Journal of Experimental Psychology*, forthcoming.

Hsieh, David A., 1991, "Chaos and Nonlinear Dynamics: Application to Financial Markets." *Journal of Finance* 46(5): 1839–1877.

Huang, C. F., and R. H. Litzenberger, 1988, *Foundations for Financial Economics*. New York/Amsterdam/London: North-Holland.

Huber, J. C., 1998, "Cumulative Advantage and Success-Breeds-Success: The Value of Time Pattern Analysis." *Journal of the American Society for Information Science and Technology* 49: 471–476.

———, 2002, "A New Model That Generates Lotka's Law." *Journal of the American Society for Information Science and Technology* 53: 209–219.

Huberman, Bernardo A., 2001, *The Laws of the Web: Patterns in the Ecology of Information*. Cambridge, Mass.: The MIT Press.

Hume, David, 1748, 2000, *A Treatise of Human Nature: Being an Attempt to Introduce the Experimental Method of Reasoning into Moral Subjects*. Oxford: Oxford University Press.

Humphrey, Nicholas, 1992, *A History of the Mind: Evolution and the Birth of Consciousness*. New York: Copernicus.

Husserl, Edmund, 1954, *The Crisis of European Sciences and Transcendental Phenomenology*. Evanston, Ill.: Northwestern University Press.

Ierodiakonou, K., and J. P. Vandenbroucke, 1993, "Medicine as a Stochastic Art." *Lancet* 341: 542–543.

Inagaki, Kayoko, and Giyoo Hatano, 2006, "Do Young Children Possess Distinct Causalities for the Three Core Domains of Thought?" Interdisciplines Conference on Causality, www.interdisciplines.org.

Jablonski, D., K. Roy, J. W. Valentine, R. M. Price, and P. S. Anderson, 2003, "The Impact of the Pull of the Recent on the History of Marine Diversity." *Science* 300(5622): 1133–1135.

Jacob, John, Thomas Lys, and Margaret Neale, 1999, "Expertise in Forecasting Performance of Security Analysts." *Journal of Accounting and Economics* 28: 51–82.

Jaynes, E. T., 2003, *Probability Theory: The Logic of Science.* Cambridge: Cambridge University Press.

Jaynes, Julian, 1976, *The Origin of Consciousness in the Breakdown of the Bicameral Mind.* New York: Mariner Books.

Jenkins, Keith, 1991, *Re-Thinking History.* London: Routledge.

Jeong, H., B. Tombor, R. Albert, Z. N. Oltavi, and A.-L. Barabási, 2000, "The Large-scale Organization of Metabolic Networks." *Nature* 407: 651–654.

Joung, Wendy, Beryl Hesketh, and Andrew Neal, 2006, "Using 'War Stories' to Train for Adaptive Performance: Is It Better to Learn from Error or Success?" *Applied Psychology: An International Review* 55(2): 282–302.

Juslin, P., 1991, *Well-calibrated General Knowledge: An Ecological Inductive Approach to Realism of Confidence.* Manuscript submitted for publication. Uppsala, Sweden.

———, 1993, "An Explanation of the Hard-Easy Effect in Studies of Realism of Confidence in One's General Knowledge." *European Journal of Cognitive Psychology* 5:55–71.

———, 1994, "The Overconfidence Phenomenon as a Consequence of Informal Experimenter-guided Selection of Almanac Items." *Organizational Behavior and Human Decision Processes* 57: 226–246.

Juslin, P., and H. Olsson, 1997, "Thurstonian and Brunswikian Origins of Uncertainty in Judgment: A Sampling Model of Confidence in Sensory Discrimination." *Psychological Review* 104: 344–366.

Juslin, P., H. Olsson, and M. Björkman, 1997, "Brunswikian and Thurstonian Origins of Bias in Probability Assessment: On the Interpretation of Stochastic Components of Judgment." *Journal of Behavioral Decision Making* 10: 189–209.

Juslin, P., H. Olsson, and A. Winman, 1998, "The Calibration Issue: Theoretical Comments on Suantak, Bolger, and Ferrell." *Organizational Behavior and Human Decision Processes* 73: 3–26.

Kadane, J. B., and S. Lichtenstein, 1982, "A Subjectivist View of Calibration." Report No. 82–86, Eugene, Ore.: Decision Research.

Kahneman, D., 2003, "Why People Take Risks." In *Gestire la vulnerabilità e l'incertezza; un incontro internazionale fra studiosi e capi di impresa.* Rome: Italian Institute of Risk Studies.

Kahneman, D., E. Diener, and N. Schwarz, eds., 1999, *Well-being: The Foundations of Hedonic Psychology.* New York: Russell Sage Foundation.

Kahneman, D., and S. Frederick, 2002, "Representativeness Revisited: Attribute Substitution in Intuitive Judgment." In T. Gilovich, D. Griffin, and D. Kahneman, eds., 2002.

Kahneman, D., J. L. Knetsch, and R. H. Thaler, 1986, "Rational Choice and the Framing of Decisions." *Journal of Business* 59(4): 251–278.

Kahneman, D., and D. Lovallo, 1993, "Timid Choices and Bold Forecasts: A Cognitive Perspective on Risk-taking." *Management Science* 39: 17–31.

Kahneman, D., and A. Tversky, 1972, "Subjective Probability: A Judgment of Representativeness." *Cognitive Psychology* 3: 430–454.

———, 1973, "On the Psychology of Prediction." *Psychological Review* 80: 237–251.

———, 1979, "Prospect Theory: An Analysis of Decision Under Risk." *Econometrica* 46(2): 171–185.

———, 1982, "On the Study of Statistical Intuitions." In D. Kahneman, P. Slovic, and A. Tversky, eds., *Judgment Under Uncertainty: Heuristics and Biases.* Cambridge: Cambridge University Press.

———, 1996, "On the Reality of Cognitive Illusions." *Psychological Review* 103: 582–591.

———, eds., 2000, *Choices, Values, and Frames.* Cambridge: Cambridge University Press.

———, 1991, "Anomalies: The Endowment Effect, Loss Aversion, and Status Quo Bias." In D. Kahneman and A. Tversky, eds., 2000.

Kaizoji, Taisei, 2003, "Scaling Behavior in Land Markets." *Physica A: Statistical Mechanics and Its Applications* 326(1–2): 256–264.

Kaizoji, Taisei, and Michiyo Kaizoji, 2004, "Power Law for Ensembles of Stock Prices." *Physica*

A: Statistical Mechanics and Its Applications 344(1–2), *Applications of Physics in Financial Analysis* 4 (APFA4) (December 1): 240–243.

Katz, J. Sylvan, 1999, "The Self-similar Science System." *Research Policy* 28(5): 501–517.

Keen, Steve, 2001, *Debunking Economics: The Naked Emperor of the Social Classes.* London: Pluto Press.

Kemp, C., and J. B. Tenenbaum, 2003, "Theory-based Induction." *Proceedings of the Twenty-fifth Annual Conference of the Cognitive Science Society,* Boston, Mass.

Keren, G., 1988, "On the Ability of Assessing Non-verdical Perceptions: Some Calibration Studies." *Acta Psychologica* 67: 95–119.

———, 1991, "Calibration and Probability Judgments: Conceptual and Methodological Issues." *Acta Psychologica* 77: 217–273.

Keynes, John Maynard, 1920, *Treatise on Probability.* London: Macmillan.

———, 1937, "The General Theory." *Quarterly Journal of Economics* 51: 209–233.

Kidd, John B., 1970, "The Utilization of Subjective Probabilities in Production Planning." *Acta Psychologica* 34(2/3): 338–347.

Kim, E. Han, Adair Morse, and Luigi Zingales, 2006, "Are Elite Universities Losing Their Competitive Edge?" NBER Working Paper 12245.

Kindleberger, Charles P., 2001, *Manias, Panics, and Crashes.* New York: Wiley.

King, Gary, and Langche Zeng, 2005, "When Can History Be Our Guide? The Pitfalls of Counterfactual Inference." Working Paper, Harvard University.

Kirkpatrick, Mark, and Lee Alan Dugatkin, 1994, "Sexual Selection and the Evolutionary Effects of Copying Mate Choice." *Behavioral Evolutionary Sociobiology* 34: 443–449.

Klayman, J., 1995, "Varieties of Confirmation Bias." In J. Busemeyer, R. Hastie, and D. L. Medin, eds., *Decision Making from a Cognitive Perspective. The Psychology of Learning and Motivation* 32: 83–136. New York: Academic Press.

Klayman, J., and Y.-W. Ha, 1987, "Confirmation, Disconfirmation, and Information in Hypothesis Testing." *Psychological Review* 94: 211–228.

Klayman, Joshua, Jack B. Soll, Claudia Gonzalez-Vallejo, and Sema Barlas, 1999, "Overconfidence: It Depends on How, What, and Whom You Ask." *Organizational Behavior and Human Decision Processes* 79(3): 216–247.

Klebanoff, Arthur, 2002, *The Agent.* London: Texere.

Klein, Gary, 1998, *Sources of Power: How People Make Decisions.* Cambridge: The MIT Press.

Knight, Frank, 1921, 1965, *Risk, Uncertainty and Profit.* New York: Harper and Row.

Koehler, J. J., B. J. Gibbs, and R. M. Hogarth, 1994, "Shattering the Illusion of Control: Multi-shot Versus Single-shot Gambles." *Journal of Behavioral Decision Making* 7: 183–191.

Koestler, Arthur, 1959, *The Sleepwalkers: A History of Man's Changing Vision of the Universe.* London: Penguin.

Korda, Michael, 2000, *Another Life: A Memoir of Other People.* New York: Random House.

Koriat, A., S. Lichtenstein, and B. Fischhoff, 1980, "Reasons for Confidence." *Journal of Experimental Psychology: Human Learning and Memory* 6: 107–118.

Kreps, J., and N. B. Davies, 1993, *An Introduction to Behavioral Ecology,* 3rd ed. Oxford: Blackwell Scientific Publications.

Kristeva, Julia, 1998, *Time and Sense.* New York: Columbia University Press.

Kruger, J., and D. Dunning, 1999, "Unskilled and Unaware of It: How Difficulties in Recognizing One's Own Incompetence Lead to Inflated Self-Assessments." *Journal of Personality and Social Psychology* 77(6): 1121–1134.

Kunda, Ziva, 1990, "The Case for Motivated Reasoning." *Psychological Bulletin* 108: 480–498.

———, 1999, *Social Cognition: Making Sense of People.* Cambridge: The MIT Press.

Kurz, Mordecai, 1997,"Endogenous Uncertainty: A Unified View of Market Volatility." Working Paper: Stanford University Press.

Kyburg, Henry E., Jr., 1983, *Epistemology and Inference.* Minneapolis: University of Minnesota Press.

Lad, F., 1984, "The Calibration Question." *British Journal of the Philosophy of Science* 35: 213–221.

Lahire, Bernard, 2006, *La condition littéraire.* Paris: Editions La Découverte.

Lakoff, George, and Mark Johnson, 1980, *Metaphors We Live By.* Chicago: The University of Chicago Press.

Lamont, Owen A., 2002, "Macroeconomic Forecasts and Microeconomic Forecasters." *Journal of Economic Behavior and Organization* 48(3): 265–280.

Lane, R. D., E. M. Reiman, M. M. Bradley, P. J. Lang, G. L. Ahern, R. J. Davidson, and G. E. Schwartz, 1997, "Neuroanatomical correlates of pleasant and unpleasant emotion." *Neuropsychologia* 35(11): 1437–1444.

Langer, E. J., 1975, "The Illusion of Control." *Journal of Personality and Social Psychology* 32: 311–328.

Larrick, R. P., 1993, "Motivational Factors in Decision Theories: The Role of Self-Protection." *Psychological Bulletin* 113: 440–450.

Leary, D. E., 1987, "From Act Psychology to Probabilistic Functionalism: The Place of Egon Brunswik in the History of Psychology." In M. G. Ash and W. R. Woodward, eds., *Psychology in Twentieth-century Thought and Society.* Cambridge: Cambridge University Press.

LeDoux, Joseph, 1998, *The Emotional Brain: The Mysterious Underpinnings of Emotional Life.* New York: Simon & Schuster.

———, 2002, *Synaptic Self: How Our Brains Become Who We Are.* New York: Viking.

Le Goff, Jacques, 1985, *Les intellectuels au moyen age.* Paris: Points Histoire.

Levi, Isaac, 1970, *Gambling with Truth.* Cambridge, Mass.: The MIT Press.

Lichtenstein, Sarah, and Baruch Fischhoff, 1977, "Do Those Who Know More Also Know More About How Much They Know? The Calibration of Probability Judgments." *Organizational Behavior and Human Performance* 20: 159–183.

Lichtenstein, Sarah, and Baruch Fischhoff, 1981, "The Effects of Gender and Instructions on Calibration." *Decision Research Report* 81–5. Eugene, Ore.: Decision Research.

Lichtenstein, Sarah, Baruch Fischhoff, and Lawrence Phillips, 1982, "Calibration of Probabilities: The State of the Art to 1980." In D. Kahneman, P. Slovic, and A. Tversky, eds., *Judgment Under Uncertainty: Heuristics and Biases.* Cambridge: Cambridge University Press.

Lim, T., 2001, "Rationality and Analysts' Forecast Bias." *Journal of Finance* 56(1): 369–385.

Lissowski, Grzegorz, Tadeusz Tyszka, and Wlodzimierz Okrasa, 1991, "Principles of Distributive Justice: Experiments in Poland and America." *Journal of Conflict Resolution* 35(1): 98–119.

Liu, Jing, 1998, "Post-Earnings Announcement Drift and Analysts' Forecasts." Working Paper, UCLA.

Loewenstein, G. F., E. U. Weber, C. K. Hsee, and E. S. Welch, 2001, "Risk as Feelings." *Psychological Bulletin* 127: 267–286.

Loewenstein, George, 1992, "The Fall and Rise of Psychological Explanations in the Economics of Intertemporal Choice." In George Loewenstein and Jon Elster, eds., *Choice over Time.* New York: Russell Sage Foundation.

Loftus, Elizabeth F., and Katherine Ketcham, 1994, *The Myth of Repressed Memory: False Memories and Allegations and Sexual Abuse.* New York: St. Martin's Press.

Lotka, Alfred J., 1926, "The Frequency Distribution of Scientific Productivity." *Journal of the Washington Academy of Sciences* 16(12): 317–323.

Lowenstein, R., 2000, *When Genius Failed: The Rise and Fall of Long-Term Capital Management.* New York: Random House.

Lucas, Robert E., 1978, "Asset Prices in an Exchange Economy." *Econometrica* 46: 1429–1445.

Luce, R. D., and H. Raiffa, 1957, *Games and Decisions: Introduction and Critical Survey.* New York: Wiley.

Mach, E., 1896, "On the Part Played by Accident in Invention and Discovery." *Monist* 6: 161–175.

Machina, M. J., and M. Rothschild, 1987, "Risk." In J. Eatwell, M. Milgate, and P. Newman, eds., 1987.

Magee, Bryan, 1985, *Philosophy and the Real World: An Introduction to Karl Popper.* La Salle, Ill.: Open Court Books.

———, 1997, *Confessions of a Philosopher.* London: Weidenfeld & Nicolson.

Maines, L. A., and J. R. Hand, 1996, "Individuals' Perceptions and Misperceptions of Time-series Properties of Quarterly Earnings." *Accounting Review* 71: 317–336.

Makridakis, S., A. Andersen, R. Carbone, R. Fildes, M. Hibon, R. Lewandowski, J. Newton, R. Parzen, and R. Winkler, 1982, "The Accuracy of Extrapolation (Time Series) Methods: Results of a Forecasting Competition." *Journal of Forecasting* 1: 111–153.

Makridakis, S., C. Chatfield, M. Hibon, M. Lawrence, T. Mills, K. Ord, and L. F. Simmons, 1993, "The M2–Competition: A Real-Time Judgmentally Based Forecasting Study" (with commentary). *International Journal of Forecasting* 5: 29.

Makridakis, S., and M. Hibon, 2000, "The M3-Competition: Results, Conclusions and Implications." *International Journal of Forecasting* 16: 451–476.

Mandelbrot, B., 1963, "The Variation of Certain Speculative Prices." *Journal of Business* 36(4): 394–419.

Mandelbrot, Benoît, 1965, "Information Theory and Psycholinguistics." In B. Wolman and E. Nagel, eds., *Scientific Psychology: Principles and Approaches*. New York: Basic Books.

———, 1975, *Les objets fractals: forme, hasard, et dimension*. Paris: Flammarion.

———, 1982, *The Fractal Geometry of Nature*. New York: W. H. Freeman and Company.

———, 1997a, *Fractales, hasard et finance*. Paris: Flammarion.

———, 1997b, *Fractals and Scaling in Finance: Discontinuity, Concentration, Risk*. New York: Springer-Verlag.

Mandelbrot, Benoît, and Nassim Nicholas Taleb, 2006a, "A Focus on the Exceptions That Prove the Rule." In *Mastering Uncertainty: Financial Times Series*.

———, 2006b, "Matematica della sagessa." *Il Sole 24 Ore*, October 9.

———, 2007a, "Random Jump Not Random Walk." Manuscript.

———, 2007b, "Mild vs. Wild Randomness: Focusing on Risks that Matter." Forthcoming in Frank Diebold, Neil Doherty, and Richard Herring, eds., *The Known, the Unknown and the Unknowable in Financial Institutions*. Princeton, N.J.: Princeton University Press.

Mandler, J. M., and L. McDonough, 1998, "Studies in Inductive Inference in Infancy." *Cognitive Psychology* 37: 60–96.

Margalit, Avishai, 2002, *The Ethics of Memory*. Cambridge, Mass.: Harvard University Press.

Markowitz, Harry, 1952, "Portfolio Selection." *Journal of Finance* (March): 77–91.

———, 1959, *Portfolio Selection: Efficient Diversification of Investments*, 2nd ed. New York: Wiley.

Marmott, Michael, 2004, *The Status Syndrome: How Social Standing Affects Our Health and Longevity*. London: Bloomsbury.

Marr, D., 1982, *Vision*. New York: W. H. Freeman and Company.

Masters, John, 1969, *Casanova*. New York: Bernard Geis Associates.

May, R. M., 1973, *Stability and Complexity in Model Ecosystems*. Princeton, N.J.: Princeton University Press.

May, R. S., 1986, "Overconfidence as a Result of Incomplete and Wrong Knowledge." In R. W. Scholz, ed., *Current Issues in West German Decision Research*. Frankfurt am Main, Germany: Lang.

Mayseless, O., and A. W. Kruglanski, 1987, "What Makes You So Sure? Effects of Epistemic Motivations on Judgmental Confidence. *Organizational Behavior and Human Decision Processes* 39: 162–183.

McClelland, A. G. R., and F. Bolger, 1994, "The Calibration of Subjective Probabilities: Theories and Models, 1980–1994." In G. Wright and P. Ayton, eds., *Subjective Probability*. Chichester, England: Wiley.

McCloskey, Deirdre, 1990, *If You're So Smart: The Narrative of Economic Expertise*. Chicago: The University of Chicago Press.

———, 1992, "The Art of Forecasting: From Ancient to Modern Times." *Cato Journal* 12(1): 23–43.

McClure, Samuel M., David I. Laibson, George F. Loewenstein, and Jonathan D. Cohen, 2004, "Separate Neural Systems Value Immediate and Delayed Monetary Rewards." *Science* 306(5695): 503–507.

McManus, Chris, 2002, *Right Hand, Left Hand*. London: Orion Books.

McNees, Stephen K., 1978, "Rebuttal of Armstrong." *Journal of Business* 51(4): 573–577.

———, 1995, "An Assessment of the 'Official' Economic Forecasts." *New England Economic Review* (July/August): 13–23.

McNeill, William H., 1976, *Plagues and Peoples.* New York: Anchor Books.

Medawar, Peter, 1996, *The Strange Case of the Spotted Mice and Other Classic Essays on Science.* Oxford: Oxford University Press.

Meehl, Paul E., 1954, *Clinical Versus Statistical Predictions: A Theoretical Analysis and Revision of the Literature.* Minneapolis: University of Minnesota Press.

———, 1973, "Why I Do Not Attend in Case Conferences." In *Psychodiagnosis: Selected Papers,* 225–302. Minneapolis: University of Minnesota Press.

Mendenhall, Richard R., 1991, "Evidence of Possible Underweighting of Earnings-related Information." *Journal of Accounting Research* 29: 170–178.

Merton, R. K., 1968. "The Matthew Effect in Science." *Science* 159: 56–63.

———, 1973a, "The Matthew Effect in Science." In N. Storer, ed., *The Sociology of Science.* Chicago: The University of Chicago Press.

———, 1973b, "The Normative Structure of Science." In N. Storer, ed., *The Sociology of Science.* Chicago: The University of Chicago Press.

———, 1988, "The Matthew Effect II: Cumulative Advantage and the Symbolism of Intellectual Property." *Isis* 79: 606–623.

Merton, Robert C., 1972, "An Analytic Derivation of the Efficient Portfolio Frontier." *Journal of Financial and Quantitative Analysis* 7(4): 1851–1872.

———, 1992, *Continuous-Time Finance,* 2nd ed. Cambridge, England: Blackwell.

Merton, Robert K., and Elinor Barber, 2004, *The Travels and Adventures of Serendipity.* Princeton, N.J.: Princeton University Press.

Mihailescu, Calin, 2006, *Lotophysics.* Preprint, University of Western Ontario.

Mikhail, M. B., B. R. Walther, and R. H. Willis, 1999, "Does Forecast Accuracy Matter to Security Analysts?" *The Accounting Review* 74(2): 185–200.

Mikhail, Michael B., Beverly R. Walther, and Richard H. Willis, 1997, "Do Security Analysts Improve Their Performance with Experience?" *Journal of Accounting Research* 35: 131–157.

Milgram, S., 1967, "The Small World Problem." *Psychology Today* 2: 60–67.

Mill, John Stuart, 1860, *A System of Logic Ratiocinative and Inductive, Being a Connected View of the Principle of Evidence and the Methods of Scientific Investigation,* 3rd ed. London: John W. Parker, West Strand.

Miller, Dale T., and Michael Ross, 1975, "Self-Serving Biases in Attribution of Causality: Fact or Fiction?" *Psychological Bulletin* 82(2): 213–225.

Miller, Geoffrey F., 2000, *The Mating Mind: How Sexual Choice Shaped the Evolution of Human Nature.* New York: Doubleday.

Minsky, H., 1982, *Can It Happen Again? Essays on Instability and Finance.* Armonk, N.Y.: M. E. Sharpe.

Mitzenmacher, Michael, 2003, "A Brief History of Generative Models for Power Law and Lognormal Distributions." *Internet Mathematics* 1(2): 226–251.

Mohr, C., T. Landis, H. S. Bracha, and P. Brugger, 2003, "Opposite Turning Behavior in Right-handers and Non-right-handers Suggests a Link Between Handedness and Cerebral Dopamine Asymmetries." *Behavioral Neuroscience* 117(6): 1448–1452.

Mokyr, Joel, 2002, *The Gifts of Athena.* Princeton, N.J.: Princeton University Press.

Montier, James, 2007, *Applied Behavioural Finance.* Chichester, England: Wiley.

Moon, Francis C., 1992, *Chaotic and Fractal Dynamics.* New York: Wiley.

Mossner, E. C., 1970, *The Life of David Hume.* Oxford: Clarendon Press.

Murphy, A. H., and R. Winkler, 1984, "Probability Forecasting in Meteorology." *Journal of the American Statistical Association* 79: 489–500.

Myers, David G., 2002, *Intuition: Its Powers and Perils.* New Haven, Conn.: Yale University Press.

Nader, K., and J. E. LeDoux, 1999, "The Dopaminergic Modulation of Fear: Quinpirole Impairs the Recall of Emotional Memories in Rats." *Behavioral Neuroscience* 113(1): 152–165.

Naya, Emmanuel, and Anne-Pascale Pouey-Mounou, 2005, *Éloge de la médiocrité.* Paris: Éditions Rue d'ulm.

Nelson, Lynn Hankinson, and Jack Nelson, 2000, *On Quine.* Belmont, Calif.: Wadsworth.

Nelson, Robert H., 2001, *Economics as a Religion: From Samuelson to Chicago and Beyond.* University Park, Penn.: The Pennsylvania State University Press.

Newell, A., and H. A. Simon, 1972, *Human Problem Solving.* Englewood Cliffs, N.J.: Prentice-Hall.

Newman, M., 2003, "The Structure and Function of Complex Networks." *SIAM Review* 45: 167–256.

Newman, M. E. J., 2000, "Models of the Small World: A Review. *Journal of Statistical Physics* 101: 819–841.

———, 2001, "The Structure of Scientific Collaboration Networks." *Proceedings of the National Academy of Science* 98: 404–409.

———, 2005, "Power Laws, Pareto Distributions, and Zipf's Law." *Complexity Digest* 2005.02: 1–27.

Newman, M. E. J., C. Moore, and D. J. Watts, 2000, "Mean-field Solution of the Small-World Network Model." *Physical Review Letters* 84: 3201–3204.

Newman, M. E. J., D. J. Watts, and S. H. Strogatz, 2000, "Random Graphs with Arbitrary Degree Distribution and Their Applications." Preprint cond-mat/0007235 at http://xxx.lanl.gov.

Neyman, J., 1977, "Frequentist Probability and Frequentist Statistics." *Synthese* 36: 97–131.

Nietzsche, Friedrich, 1979, *Ecce Homo.* London: Penguin Books.

Nisbett, R. E., D. H. Krantz, D. H. Jepson, and Z. Kunda, 1983, "The Use of Statistical Heuristics in Everyday Inductive Reasoning." *Psychological Review* 90: 339–363.

Nisbett, Richard E., and Timothy D. Wilson, 1977, "Telling More Than We Can Know: Verbal Reports on Mental Processes." *Psychological Bulletin* 84(3): 231–259.

Nussbaum, Martha C., 1986, *The Fragility of Goodness: Luck and Ethics in Greek Tragedy and Philosophy.* Cambridge: Cambridge University Press.

O'Connor, M., and M. Lawrence, 1989, "An Examination of the Accuracy of Judgment Confidence Intervals in Time Series Forecasting." *International Journal of Forecasting* 8: 141–155.

O'Neill, Brian C., and Mausami Desai, 2005, "Accuracy of Past Projections of U.S. Energy Consumption." *Energy Policy* 33: 979–993.

Oberauer K., O. Wilhelm, and R. R. Diaz, 1999, "Bayesian Rationality for the Wason Selection Task? A Test of Optimal Data Selection Theory." *Thinking and Reasoning* 5(2): 115–144.

Odean, Terrance, 1998a, "Are Investors Reluctant to Realize Their Losses?" *Journal of Finance* 53(5): 1775–1798.

———, 1998b. "Volume, Volatility, Price and Profit When All Traders Are Above Average." *Journal of Finance* 53(6): 1887–1934.

Officer, R. R., 1972, "The Distribution of Stock Returns." *Journal of the American Statistical Association* 340(67): 807–812.

Olsson, Erik J., 2006, *Knowledge and Inquiry: Essays on the Pragmatism of Isaac Levi.* Cambridge Studies in Probability, Induction and Decision Theory Series. Cambridge: Cambridge University Press.

Onkal, D., J. F. Yates, C. Simga-Mugan, and S. Oztin, 2003, "Professional and Amateur Judgment Accuracy: The Case of Foreign Exchange Rates." *Organizational Behavior and Human Decision Processes* 91: 169–185.

Ormerod, Paul, 2005, *Why Most Things Fail.* New York: Pantheon Books.

———, 2006, "Hayek, 'The Intellectuals and Socialism,' and Weighted Scale-free Networks." *Economic Affairs* 26: 1–41.

Oskamp, Stuart, 1965, "Overconfidence in Case-Study Judgments." *Journal of Consulting Psychology* 29(3): 261–265.

Paese, P. W., and J. A. Sniezek, 1991, "Influences on the Appropriateness of Confidence in Judgment: Practice, Effort, Information, and Decision Making." *Organizational Behavior and Human Decision Processes* 48: 100–130.

Page, Scott, 2007, *The Difference: How the Power of Diversity Can Create Better Groups, Firms, Schools, and Societies.* Princeton, N.J.: Princeton University Press.

Pais, Abraham, 1982, *Subtle Is the Lord.* New York: Oxford University Press.

Pareto, Vilfredo, 1896, *Cours d'économie politique.* Geneva: Droz.

Park, David, 2005, *The Grand Contraption: The World as Myth, Number, and Chance*. Princeton, N.J.: Princeton University Press.

Paulos, John Allen, 1988, *Innumeracy*. New York: Hill & Wang.

———, 2003, *A Mathematician Plays the Stock Market*. Boston: Basic Books.

Pearl, J., 2000, *Causality: Models, Reasoning, and Inference*. New York: Cambridge University Press.

Peirce, Charles Sanders, 1923, 1998, *Chance, Love and Logic: Philosophical Essays*. Lincoln: University of Nebraska Press.

———, 1955, *Philosophical Writings of Peirce*, edited by J. Buchler. New York: Dover.

Penrose, Roger, 1989, *The Emperor's New Mind*. New York: Penguin.

Pérez, C. J., A. Corral, A. Diáz-Guilera, K. Christensen, and A. Arenas, 1996, "On Self-organized Criticality and Synchronization in Lattice Models of Coupled Dynamical Systems." *International Journal of Modern Physics B* 10: 1111–1151.

Perilli, Lorenzo, 2004, *Menodoto di Nicomedia: Contributo a una storia galeniana della medicina empirica*. Munich, Leipzig: K. G. Saur.

Perline, R., 2005, "Strong, Weak, and False Inverse Power Laws." *Statistical Science* 20(1): 68–88.

Pfeifer, P. E., 1994, "Are We Overconfident in the Belief That Probability Forecasters Are Overconfident?" *Organizational Behavior and Human Decision Processes* 58(2): 203–213.

Phelan, James, 2005, "Who's Here? Thoughts on Narrative Identity and Narrative Imperialism." *Narrative* 13: 205–211.

Piattelli-Palmarini, Massimo, 1994, *Inevitable Illusions: How Mistakes of Reason Rule Our Minds*. New York: Wiley.

Pieters, Rik, and Hans Baumgartner, 2002. "Who Talks to Whom? Intra- and Interdisciplinary Communication of Economics Journals." *Journal of Economic Literature* 40(2): 483–509.

Pinker, Steven, 1997, *How the Mind Works*. New York: W. W. Norton and Company.

———, 2002, *The Blank Slate: The Modern Denial of Human Nature*. New York: Viking.

Pisarenko, V., and D. Sornette, 2004, "On Statistical Methods of Parameter Estimation for Deterministically Chaotic Time-Series." *Physical Review E* 69: 036122.

Plotkin, Henry, 1998, *Evolution in Mind: An Introduction to Evolutionary Psychology*. London: Penguin.

Plous, S., 1993. *The Psychology of Judgment and Decision Making*. New York: McGraw-Hill.

———, 1995, "A Comparison of Strategies for Reducing Interval Overconfidence in Group Judgments." *Journal of Applied Psychology* 80: 443–454.

Polanyi, Michael, 1958/1974, *Personal Knowledge: Towards a Post-Critical Philosophy*. Chicago: The University of Chicago Press.

Popkin, Richard H., 1951, "David Hume: His Pyrrhonism and His Critique of Pyrrhonism." *The Philosophical Quarterly* 1(5): 385–407.

———, 1955, "The Skeptical Precursors of David Hume." *Philosophy and Phenomenological Research* 16(1): 61–71.

———, 2003, *The History of Scepticism: From Savonarola to Bayle*. Oxford: Oxford University Press.

Popper, Karl R., 1971, *The Open Society and Its Enemies*, 5th ed. Princeton, N.J.: Princeton University Press.

———, 1992, *Conjectures and Refutations: The Growth of Scientific Knowledge*, 5th ed. London: Routledge.

———, 1994, *The Myth of the Framework*. London: Routledge.

———, 2002a, *The Logic of Scientific Discovery*, 15th ed. London: Routledge.

———, 2002b, *The Poverty of Historicism*. London: Routledge.

Posner, Richard A., 2004, *Catastrophe: Risk and Response*. Oxford: Oxford University Press.

Price, Derek J. de Solla, 1965, "Networks of Scientific Papers." *Science* 149: 510–515.

———, 1970, "Citation Measures of Hard Science, Soft Science, Technology, and Non-science." In C. E. Nelson and D. K. Pollak, eds., *Communication Among Scientists and Engineers*. Lexington, Mass.: Heat.

———, 1976, "A General Theory of Bibliometric and Other Cumulative Advantage Processes." *Journal of the American Society of Information Sciences* 27: 292–306.

Prigogine, Ilya, 1996, *The End of Certainty: Time, Chaos, and the New Laws of Nature.* New York: The Free Press.

Quammen, David, 2006, *The Reluctant Mr. Darwin.* New York: W. W. Norton and Company.

Quine, W. V., 1951, "Two Dogmas of Empiricism." *The Philosophical Review* 60: 20–43.

———, 1970, "Natural Kinds." In N. Rescher, ed., *Essays in Honor of Carl G. Hempel.* Dordrecht: D. Reidel.

Rabin, M., 1998, "Psychology and Economics." *Journal of Economic Literature* 36: 11–46.

Rabin, M., and R. H. Thaler, 2001, "Anomalies: Risk Aversion." *Journal of Economic Perspectives* 15(1): 219–232.

Rabin, Matthew, 2000, "Inference by Believers in the Law of Small Numbers." Working Paper, Economics Department, University of California, Berkeley, http://repositories.cdlib.org/iber/econ/.

Ramachandran, V. S., 2003, *The Emerging Mind.* London: Portfolio.

Ramachandran, V. S., and S. Blakeslee, 1998, *Phantoms in the Brain.* New York: Morrow.

Rancière, Jacques, 1997, *Les mots de l'histoire. Essai de poétique du savoir.* Paris: Éditions du Seuil.

Ratey, John J., 2001, *A User's Guide to the Brain: Perception, Attention and the Four Theaters of the Brain.* New York: Pantheon.

Rawls, John, 1971, *A Theory of Justice.* Cambridge, Mass.: Harvard University Press.

Reboul, Anne, 2006, "Similarities and Differences Between Human and Nonhuman Causal Cognition." Interdisciplines Conference on Causality, www.interdisciplines.org.

Redner, S., 1998, "How Popular Is Your Paper? An Empirical Study of the Citation Distribution." *European Physical Journal B* 4: 131–134.

Rees, Martin, 2004, *Our Final Century: Will Civilization Survive the Twenty-first Century?* London: Arrow Books.

Reichenbach, H., 1938, *Experience and prediction.* Chicago: The University of Chicago Press.

Remus, W., M. Oapos Connor, and K. Griggs, 1997, "Does Feedback Improve the Accuracy of Recurrent Judgmental Forecasts?" Proceedings of the Thirtieth Hawaii International Conference on System Sciences, January 7–10: 5–6.

Rescher, Nicholas, 1995, *Luck: The Brilliant Randomness of Everyday Life.* New York: Farrar, Straus & Giroux.

———, 2001, *Paradoxes: Their Roots, Range, and Resolution.* Chicago: Open Court Books.

Richardson, L. F., 1960, *Statistics of Deadly Quarrels.* Pacific Grove, Calif.: Boxwood Press.

Rips, L., 2001, "Necessity and Natural Categories." *Psychological Bulletin* 127: 827–852.

Roberts, Royston M., 1989, *Serendipity: Accidental Discoveries in Science.* New York: Wiley.

Robins, Richard W., 2005, "Pscyhology: The Nature of Personality: Genes, Culture, and National Character." *Science* 310: 62–63.

Rollet, Laurent, 2005, *Un mathématicien au Panthéon? Autour de la mort de Henri Poincaré.* Laboratoire de Philosophie et d'Histoire des Sciences—Archives Henri-Poincaré, Université Nancy 2.

Ronis, D. L., and J. F. Yates, 1987, "Components of Probability Judgment Accuracy: Individual Consistency and Effects of Subject Matter and Assessment Method." *Organizational Behavior and Human Decision Processes* 40: 193–218.

Rosch, E., 1978, "Principles of Categorization." In E. Rosch and B. B. Lloyd, eds., *Cognition and Categorization.* Hillsdale, N.J.: Lawrence Erlbaum.

Rosch, E. H., 1973, "Natural Categories." *Cognitive Psychology* 4: 328–350.

Rose, Steven, 2003, *The Making of Memory: From Molecules to Mind,* revised ed. New York: Vintage.

Rosen, S., 1981, "The Economics of Superstars." *American Economic Review* 71: 845–858.

Rosenzweig, Phil, 2006, *The Halo Effect and Other Business Delusions: Why Experts Are So Often Wrong and What Wise Managers Must Know.* New York: The Free Press.

Ross, Stephen A., 2004, *Neoclassical Finance.* Princeton, N.J.: Princeton University Press.

Rounding, Virginia, 2006, *Catherine the Great: Love, Sex and Power.* London: Hutchinson.

Ruelle, David, 1991, *Hasard et chaos*. Paris: Odile Jacob.

Ruffié, Jacques, 1977, *De la biologie à la culture*. Paris: Flammarion.

Russell, Bertrand, 1912, *The Problems of Philosophy*. New York: Oxford University Press.

———, 1993, *My Philosophical Development*. London: Routledge.

———, 1996, *Sceptical Essays*. London: Routledge.

Russo, J. Edward, and Paul J. H. Schoernaker, 1992, "Managing Overconfidence." *Sloan Management Review* 33(2): 7–17.

Ryle, Gilbert, 1949, *The Concept of Mind*. Chicago: The University of Chicago Press.

Salganik, Matthew J., Peter S. Dodds, and Duncan J. Watts, 2006, "Experimental Study of Inequality and Unpredictability in an Artificial Cultural Market." *Science* 311: 854–856.

Samuelson, Paul A., 1983, *Foundations of Economic Analysis*. Cambridge, Mass.: Harvard University Press.

Sapolsky, Robert M., 1998, *Why Zebras Don't Get Ulcers: An Updated Guide to Stress, Stress-related Diseases, and Coping*. New York: W. H. Freeman and Company.

Sapolsky, Robert, M., and the Department of Neurology and Neurological Sciences, Stanford University School of Medicine, 2003, "Glucocorticoids and Hippocampal Atrophy in Neuropsychiatric Disorders."

Savage, Leonard J., 1972, *The Foundations of Statistics*. New York: Dover.

Schacter, Daniel L., 2001, *The Seven Sins of Memory: How the Mind Forgets and Remembers*. Boston: Houghton Mifflin.

Schelling, Thomas, 1971, "Dynamic Models of Segregation." *Journal of Mathematical Sociology* 1: 143–186.

———, 1978, *Micromotives and Macrobehavior*. New York: W. W. Norton and Company.

Scheps, Ruth, ed., 1996, *Les sciences de la prévision*. Paris: Éditions du Seuil.

Schroeder, Manfred, 1991, *Fractals, Chaos, Power Laws: Minutes from an Infinite Paradise*. New York: W. H. Freeman and Company.

Schumpeter, Joseph, 1942, *Capitalism, Socialism and Democracy*. New York: Harper.

Seglen, P. O., 1992, "The Skewness of Science." *Journal of the American Society for Information Science* 43: 628–638.

Sextus Empiricus, 2000, *Outline of Scepticism*, edited by Julia Annas and Jonathan Barnes. New York: Cambridge University Press.

———, 2005, *Against the Logicians*, translated and edited by Richard Bett. New York: Cambridge University Press.

Shackle, G.L.S., 1961, *Decision Order and Time in Human Affairs*. Cambridge: Cambridge University Press

———, 1973, *Epistemics and Economics: A Critique of Economic Doctrines*. Cambridge: Cambridge University Press.

Shanteau, J., 1992, "Competence in Experts: The Role of Task Characteristics." *Organizational Behavior and Human Decision Processes* 53: 252–266.

Sharpe, William F., 1994, "The Sharpe Ratio." *Journal of Portfolio Management* 21(1): 49–58.

———, 1996, "Mutual Fund Performance." *Journal of Business* 39: 119–138.

Shiller, Robert J., 1981, "Do Stock Prices Move Too Much to Be Justified by Subsequent Changes in Dividends?" *American Economic Review* 71(3): 421–436.

———, 1989, *Market Volatility*. Cambridge, Mass.: The MIT Press.

———, 1990, "Market Volatility and Investor Behavior." *American Economic Review* 80(2): 58–62.

———, 1995, "Conversation, Information, and Herd Behavior." *American Economic Review* 85(2): 181–185.

———, 2000, *Irrational Exuberance*. Princeton, N.J.: Princeton University Press.

———, 2003, *The New Financial Order: Risk in the 21st Century*. Princeton, N.J.: Princeton University Press.

Shizgal, Peter, 1999, "On the Neural Computation of Utility: Implications from Studies of Brain Simulation Rewards." In D. Kahneman, E. Diener, and N. Schwarz, eds., 1999.

Sieff, E. M., R. M. Dawes, and G. Loewenstein, 1999, "Anticipated Versus Actual Reaction to HIV Test Results." *American Journal of Psychology* 122: 297–311.

Silverberg, Gerald, and Bart Verspagen, 2004, "The Size Distribution of Innovations Revisited: An Application of Extreme Value Statistics to Citation and Value Measures of Patent Significance," www.merit.unimaas.nl/publications/rmpdf/2004/rm2004–021.pdf.

———, 2005, "Self-organization of R&D Search in Complex Technology Spaces," www.merit.unimaas.nl/publications/rmpdf/2005/rm2005–017.pdf.

Simon, Herbert A., 1955, "On a Class of Skew Distribution Functions." *Biometrika* 42: 425–440.

———, 1987, "Behavioral Economics." In J. Eatwell, M. Milgate, and P. Newman, eds., 1987.

Simonton, Dean Keith, 1999, *Origins of Genius: Darwinian Perspectives on Creativity.* New York: Oxford University Press.

———, 2004, *Creativity.* New York: Cambridge University Press.

Sloman, S. A., 1993, "Feature Based Induction." *Cognitive Psychology* 25: 231–280.

———, 1994, "When Explanations Compete: The Role of Explanatory Coherence on Judgments of Likelihood." *Cognition* 52: 1–21.

———, 1996, "The Empirical Case for Two Systems of Reasoning." *Psychological Bulletin* 119: 3–22.

———, 1998, "Categorical Inference Is Not a Tree: The Myth of Inheritance Hierarchies." *Cognitive Psychology* 35: 1–33.

———, 2002, "Two Systems of Reasoning." In T. Gilovich, D. Griffin, and D. Kahneman, eds., 2002.

Sloman, S. A., B. C. Love, and W. Ahn, 1998, "Feature Centrality and Conceptual Coherence." *Cognitive Science* 22: 189–228.

Sloman, S. A., and B. C. Malt, 2003, "Artifacts Are Not Ascribed Essences, Nor Are They Treated as Belonging to Kinds." *Language and Cognitive Processes* 18: 563–582.

Sloman, S. A., and D. Over, 2003, "Probability Judgment from the Inside and Out." In D. Over, ed., *Evolution and the Psychology of Thinking: The Debate.* New York: Psychology Press.

Sloman, S. A., and L. J. Rips, 1998, "Similarity as an Explanatory Construct." *Cognition* 65: 87–101.

Slovic, Paul, M. Finucane, E. Peters, and D. G. MacGregor, 2003a, "Rational Actors or Rational Fools? Implications of the Affect Heuristic for Behavioral Economics." Working Paper, www.decisionresearch.com.

———, 2003b, "Risk as Analysis, Risk as Feelings: Some Thoughts About Affect, Reason, Risk, and Rationality." Paper presented at the Annual Meeting of the Society for Risk Analysis, New Orleans, La., December 10, 2002.

Slovic, P., M. Finucane, E. Peters, and D. G. MacGregor, 2002, "The Affect Heuristic." In T. Gilovich, D. Griffin, and D. Kahneman, eds., 2002.

Slovic, P., B. Fischhoff, and S. Lichtenstein, 1976, "Cognitive Processes and Societal Risk Taking." In John S. Carroll and John W. Payne, eds., *Cognition and Social Behavior.* Hillsdale, N.J.: Lawrence Erlbaum.

———, 1977, "Behavioral Decision Theory." *Annual Review of Psychology* 28: 1–39.

Slovic, P., B. Fischhoff, S. Lichtenstein, B. Corrigan, and B. Combs, 1977, "Preference for Insuring Against Probable Small Losses: Implications for the Theory and Practice of Insurance." *Journal of Risk and Insurance* 44: 237–258. Reprinted in P. Slovic, ed., *The Perception of Risk.* London: Earthscan.

Slovic, Paul, 1987, "Perception of Risk." *Science* 236: 280–285.

———, 2001, *The Perception of Risk.* London: Earthscan.

Sniezek, J. A., and R. A. Henry, 1989, "Accuracy and Confidence in Group Judgement." *Organizational Behavior and Human Decision Processes* 43(11): 1–28.

Sniezek, J. A., and T. Buckley, 1993, "Decision Errors Made by Individuals and Groups." In N. J. Castellan, ed., *Individual and Group Decision Making.* Hillsdale, N.J.: Lawrence Erlbaum.

Snyder, A. W., 2001, "Paradox of the Savant Mind." *Nature* 413: 251–252.

Snyder A. W., E. Mulcahy, J. L. Taylor, D. J. Mitchell, P. Sachdev, and S. C. Gandevia, 2003, "Savant-like Skills Exposed in Normal People by Suppression of the Left Fronto-temporal Lobe. *Journal of Integrative Neuroscience* 2: 149–158.

Soll, J. B., 1996, "Determinants of Overconfidence and Miscalibration: The Roles of Random Error and Ecological Structure." *Organizational Behavior and Human Decision Processes* 65: 117–137.

Sornette, D., F. Deschâtres, T. Gilbert, and Y. Ageon, 2004, "Endogenous Versus Exogenous Shocks in Complex Networks: An Empirical Test." *Physical Review Letters* 93: 228701.

Sornette, D., and K. Ide, 2001, "The Kalman-Levy Filter," *Physica D* 151: 142–174.

Sornette, Didier, 2003, *Why Stock Markets Crash: Critical Events in Complex Financial Systems.* Princeton, N.J.: Princeton University Press.

———, 2004, *Critical Phenomena in Natural Sciences: Chaos, Fractals, Self-organization and Disorder: Concepts and Tools,* 2nd ed. Berlin and Heidelberg: Springer.

Sornette, Didier, and Daniel Zajdenweber, 1999, "The Economic Return of Research: The Pareto Law and Its Implications." *European Physical Journal B* 8(4): 653–664.

Soros, George, 1988, *The Alchemy of Finance: Reading the Mind of the Market.* New York: Simon & Schuster.

Spariosu, Mihai I., 2004, *The University of Global Intelligence and Human Development: Towards an Ecology of Global Learning.* Cambridge, Mass.: The MIT Press.

Spasser, Mark A., 1997, "The Enacted Fate of Undiscovered Public Knowledge." *Journal of the American Society for Information Science* 48(8): 707–717.

Spencer, B. A., and G. S. Taylor, 1988, "Effects of Facial Attractiveness and Gender on Causal Attributions of Managerial Performance." *Sex Roles* 19(5/6): 273–285.

Sperber, Dan, 1996a, *La contagion des idées.* Paris: Odile Jacob.

———, 1996b, *Explaining Culture: A Naturalistic Approach.* Oxford: Blackwell.

———, 1997, "Intuitive and Reflective Beliefs." *Mind and Language* 12(1): 67–83.

———, 2001, "An Evolutionary Perspective on Testimony and Argumentation." *Philosophical Topics* 29: 401–413.

Sperber, Dan, and Deirdre Wilson, 1995, *Relevance: Communication and Cognition,* 2nd ed. Oxford: Blackwell.

———, 2004a, "Relevance Theory." In L. R. Horn, and G. Ward, eds., *The Handbook of Pragmatics.* Oxford: Blackwell.

———, 2004b, "The Cognitive Foundations of Cultural Stability and Diversity." *Trends in Cognitive Sciences* 8(1): 40–44.

Squire, Larry, and Eric R. Kandel, 2000, *Memory: From Mind to Molecules.* New York: Owl Books.

Stanley, H. E., L. A. N. Amaral, P. Gopikrishnan, and V. Plerou, 2000, "Scale Invariance and Universality of Economic Fluctuations." *Physica A* 283: 31–41.

Stanley, T. J., 2000, *The Millionaire Mind.* Kansas City: Andrews McMeel Publishing.

Stanley, T. J., and W. D. Danko, 1996, *The Millionaire Next Door: The Surprising Secrets of America's Wealthy.* Atlanta, Ga.: Longstreet Press.

Stanovich, K., and R. West, 2000, "Individual Differences in Reasoning: Implications for the Rationality Debate." *Behavioral and Brain Sciences* 23: 645–665.

Stanovich, K. E., 1986, "Matthew Effects in Reading: Some Consequences of Individual Differences in the acquisition of literacy." *Reading Research Quarterly* 21: 360–407.

Stein, D. L., ed., 1989, *Lectures in the Sciences of Complexity.* Reading, Mass.: Addison-Wesley.

Sterelny, Kim, 2001, *Dawkins vs. Gould: Survival of the Fittest.* Cambridge, England: Totem Books.

Stewart, Ian, 1989, *Does God Play Dice? The New Mathematics of Chaos.* London: Penguin Books.

———, 1993, "Chaos." In Leo Howe and Alan Wain, eds., 1993.

Stigler, Stephen M., 1986, *The History of Statistics: The Measurement of Uncertainty Before 1900.* Cambridge, Mass.: The Belknap Press of Harvard University.

———, 2002, *Statistics on the Table: The History of Statistical Concepts and Methods.* Cambridge, Mass.: Harvard University Press.

Stiglitz, Joseph, 1994, *Whither Socialism.* Cambridge, Mass.: The MIT Press.

Strawson, Galen, 1994, *Mental Reality.* Cambridge, Mass.: The MIT Press.

———, 2004, "Against Narrativity." *Ratio* 17: 428–452.

Strogatz, S. H., 1994, *Nonlinear Dynamics and Chaos, with Applications to Physics, Biology, Chemistry, and Engineering.* Reading, Mass.: Addison-Wesley.

Strogatz, Steven H., 2001, "Exploring Complex Networks." *Nature* 410: 268–276.

———, 2003, *Sync: How Order Emerges from Chaos in the Universe, Nature, and Daily Life.* New York: Hyperion.

Suantak, L., F. Bolger, and W. R. Ferrell, 1996, "The Hard–easy Effect in Subjective Probability Calibration." *Organizational Behavior and Human Decision Processes* 67: 201–221.

Suddendorf, Thomas, 2006, "Enhanced: Foresight and Evolution of the Human Mind." *Science* 312(5776): 1006–1007.

Sullivan, R., A. Timmermann, and H. White, 1999, "Data-snooping, Technical Trading Rule Performance and the Bootstrap." *Journal of Finance* 54: 1647–1692.

Sunstein, Cass R., 2002, *Risk and Reason: Safety, Law, and the Environment.* Cambridge: Cambridge University Press.

Surowiecki, James, 2004, *The Wisdom of Crowds.* New York: Doubleday.

Sushil, Bikhchandani, David Hirshleifer, and Ivo Welch, 1992, "A Theory of Fads, Fashion, Custom, and Cultural Change as Informational Cascades." *Journal of Political Economy* 100(5): 992–1026.

Sutton, J., 1997, "Gibrat's Legacy." *Journal of Economic Literature* 35: 40–59.

Swanson, D. R., 1986a, "Fish Oil, Raynaud's Syndrome and Undiscovered Public Knowledge." *Perspectives in Biology and Medicine* 30(1): 7–18.

———, 1986b, "Undiscovered Public Knowledge." *Library Quarterly* 56: 103–118.

———, 1987, "Two Medical Literatures That Are Logically but Not Bibliographically Connected." *Journal of the American Society for Information Science* 38: 228–233.

Swets, J. A., R. M. Dawes, and J. Monahan, 2000a, "Better Decisions Through Science." *Scientific American* (October): 82–87.

———, 2000b, "Psychological Science Can Improve Diagnostic Decisions." *Psychogical Science in the Public Interest* 1: 1–26.

Szenberg, Michael, ed., 1992, *Eminent Economists: Their Life Philosophies.* Cambridge: Cambridge University Press.

Tabor, M., 1989, *Chaos and Integrability in Nonlinear Dynamics: An Introduction.* New York: Wiley.

Taine, Hippolyte Adolphe, 1868, 1905. *Les philosophes classiques du XIXe siècle en France*, 9ème éd. Paris: Hachette.

Taleb, N. N., 1997, *Dynamic Hedging: Managing Vanilla and Exotic Options.* New York: Wiley.

———, 2004a, *Fooled by Randomness: The Hidden Role of Chance in Life and in the Markets.* New York: Random House.

———, 2004b, "These Extreme Exceptions of Commodity Derivatives." In Helyette Geman, *Commodities and Commodity Derivatives.* New York: Wiley.

———, 2004c, "Bleed or Blowup: What Does Empirical Psychology Tell Us About the Preference for Negative Skewness?" *Journal of Behavioral Finance* 5(1): 2–7.

———, 2004d, "The Black Swan: Why Don't We Learn That We Don't Learn?" Paper presented at the United States Department of Defense Highland Forum, Summer 2004.

———, 2004e, "Roots of Unfairness." *Literary Research/Recherche Littéraire* 21(41–42): 241–254.

———, 2004f, "On Skewness in Investment Choices." *Greenwich Roundtable Quarterly* 2.

———, 2005, "Fat Tails, Asymmetric Knowledge, and Decision Making: Essay in Honor of Benoît Mandelbrot's 80th Birthday." Technical paper series, *Wilmott* (March): 56–59.

———, 2006a, "Homo Ludens and Homo Economicus." Foreword to Aaron Brown's *The Poker Face of Wall Street.* New York: Wiley.

———, 2006b, "On Forecasting." In John Brockman, ed., *In What We Believe But Cannot Prove: Today's Leading Thinkers on Science in the Age of Certainty.* New York: Harper Perennial.

———, 2007, "Scale Invariance in Practice: Some Patches and Workable Fixes." Preprint.

Taleb, Nassim Nicholas, and Avital Pilpel, 2004, "I problemi epistemologici del risk management." In Daniele Pace, a cura di, *Economia del rischio: Antologia di scritti su rischio e decisione economica.* Milano: Giuffrè.

Tashman, Leonard J., 2000, "Out of Sample Tests of Forecasting Accuracy: An Analysis and Review." *International Journal of Forecasting* 16(4): 437–450.

Teigen, K. H., 1974, "Overestimation of Subjective Probabilities." *Scandinavian Journal of Psychology* 15: 56–62.

Terracciano, A., et al., 2005, "National Character Does Not Reflect Mean Personality Traits." *Science* 310: 96.

Tetlock, Philip E., 1999, "Theory-Driven Reasoning About Plausible Pasts and Probable Futures in World Politics: Are We Prisoners of Our Preconceptions?" *American Journal of Political Science* 43(2): 335–366.

———, 2005, "Expert Political Judgment: How Good Is It? How Can We Know?" Princeton, N.J.: Princeton University Press.

Thaler, Richard, 1985, "Mental Accounting and Consumer Choice." *Marketing Science* 4(3): 199–214.

Thom, René, 1980, *Paraboles et catastrophes*. Paris: Champs Flammarion.

———, 1993, *Prédire n'est pas expliquer*. Paris: Champs Flammarion.

Thorley, 1999, "Investor Overconfidence and Trading Volume." Working Paper, Santa Clara University.

Tilly, Charles, 2006, *Why? What Happens When People Give Reasons and Why*. Princeton, N.J.: Princeton University Press.

Tinbergen, N., 1963, "On Aims and Methods in Ethology." *Zeitschrift fur Tierpsychologie* 20: 410–433.

———, 1968, "On War and Peace in Animals and Man: An Ethologist's Approach to the Biology of Aggression." *Science* 160: 1411–1418.

Tobin, James, 1958, "Liquidity Preference as Behavior Towards Risk." *Review of Economic Studies* 67: 65–86.

Triantis, Alexander J., and James E. Hodder, 1990, "Valuing Flexibility as a Complex Option." *Journal of Finance* 45(2): 549–564.

Trivers, Robert, 2002, *Natural Selection and Social Theory: Selected Papers of Robert Trivers*. Oxford: Oxford University Press.

Turner, Mark, 1996, *The Literary Mind*. New York: Oxford University Press.

Tversky, A., and D. Kahneman, 1971, "Belief in the Law of Small Numbers." *Psychology Bulletin* 76(2): 105–110.

———, 1973, "Availability: A Heuristic for Judging Frequency and Probability." *Cognitive Psychology* 5: 207–232.

———, 1974, "Judgement Under Uncertainty: Heuristics and Biases." *Science* 185: 1124–1131.

———, 1982, "Evidential Impact of Base-Rates." In D. Kahneman, P. Slovic, and A. Tversky, eds., *Judgment Under Uncertainty: Heuristics and Biases*. Cambridge: Cambridge University Press.

———, 1983, "Extensional Versus Intuitive Reasoning: The Conjunction Fallacy in Probability Judgment." *Psychological Review* 90: 293–315.

———, 1992, "Advances in Prospect Theory: Cumulative Representation of Uncertainty." *Journal of Risk and Uncertainty* 5: 297–323.

Tversky, A., and D. J. Koehler, 1994, "Support Theory: A Nonextensional Representation of Subjective Probability." *Psychological Review* 101: 547–567.

Tyszka, T., and P. Zielonka, 2002, "Expert Judgments: Financial Analysts Versus Weather Forecasters." *Journal of Psychology and Financial Markets* 3(3): 152–160.

Uglow, Jenny, 2003, *The Lunar Men: Five Friends Whose Curiosity Changed the World*. New York: Farrar, Straus & Giroux.

Vale, Nilton Bezerra do, José Delfino, and Lúcio Flávio Bezerra do Vale, 2005, "Serendipity in Medicine and Anesthesiology." *Revista Brasileira de Anestesiologia* 55(2): 224–249.

van Tongeren, Paul, 2002, "Nietzsche's Greek Measure." *Journal of Nietzsche Studies* 24: 5.

Vandenbroucke, J. P., 1996, "Evidence-Based Medicine and 'Medicine d'Observation,' " *Journal of Clinical Epidemiology*, 49(12): 1335–1338.

Varela, Francisco J., 1988, *Invitation aux sciences cognitives*. Paris: Champs Flammarion.

Varian, Hal R., 1989, "Differences of Opinion in Financial Markets." In Courtenay C. Stone, ed., *Financial Risk: Theory, Evidence and Implications: Proceedings of the Eleventh Annual Economic Policy Conference of the Federal Reserve Bank of St. Louis*. Boston: Kitiwer Academic Publishers.

Véhel, Jacques Lévy, and Christian Walter, 2002, *Les marchés fractals: Efficience, ruptures, et tendances sur les marchés financiers*. Paris: PUF.

Veyne, Paul, 1971, *Comment on écrit l'histoire*. Paris: Éditions du Seuil.

———, 2005, *L'Empire gréco-romain*. Paris: Éditions du Seuil.

Vogelstein, Bert, David Lane, and Arnold J. Levine, 2000, "Surfing the P53 Network." *Nature* 408: 307–310.

Voit, Johannes, 2001, *The Statistical Mechanics of Financial Markets*. Heidelberg: Springer.

von Mises, R., 1928, *Wahrscheinlichkeit, Statistik und Wahrheit*. Berlin: Springer. Translated and reprinted as *Probability, Statistics, and Truth*. New York: Dover, 1957.

von Plato, Jan, 1994, *Creating Modern Probability*. Cambridge: Cambridge University Press.

von Winterfeldt, D., and W. Edwards, 1986, *Decision Analysis and Behavioral Research*. Cambridge: Cambridge University Press.

Wagenaar, Willern, and Gideon B. Keren, 1985, "Calibration of Probability Assessments by Professional Blackjack Dealers, Statistical Experts, and Lay People." *Organizational Behavior and Human Decision Processes* 36: 406–416.

———, 1986, "Does the Expert Know? The Reliability of Predictions and Confidence Ratings of Experts." In Erik Hollnagel, Giuseppe Mancini, and David D. Woods, *Intelligent Design Support in Process Environments*. Berlin: Springer.

Waller, John, 2002, *Fabulous Science: Fact and Fiction in the History of Scientific Discovery*. Oxford: Oxford University Press.

Wallerstein, Immanuel, 1999, "Braudel and Interscience: A Preacher to Empty Pews?" Paper presented at the 5th Journées Braudeliennes, Binghamton University, Binghamton, N.Y.

Wallsten, T. S., D. V. Budescu, I. Erev, and A. Diederich, 1997, "Evaluating and Combining Subjective Probability Estimates." *Journal of Behavioral Decision Making* 10: 243–268.

Wason, P. C., 1960, "On the Failure to Eliminate Hypotheses in a Conceptual Task." *Quarterly Journal of Experimental Psychology* 12: 129–140.

Watts, D. J., 2003, *Six Degrees: The Science of a Connected Age*. New York: W. W. Norton and Company.

Watts, D. J., and S. H. Strogatz, 1998, "Collective Dynamics of 'Small-world' Networks." *Nature* 393: 440–442

Watts, Duncan, 2002, "A Simple Model of Global Cascades on Random Networks." *Proceedings of the National Academy of Sciences* 99(9): 5766–5771.

Wegner, Daniel M., 2002, *The Illusion of Conscious Will*. Cambridge, Mass.: The MIT Press.

Weinberg, Steven, 2001, "Facing Up: Science and Its Cultural Adversaries." Working Paper, Harvard University.

Weintraub, Roy E., 2002, *How Economics Became a Mathematical Science*, Durham, N.C.: Duke University Press.

Wells, G. L., and Harvey, J. H., 1977, "Do People Use Consensus Information in Making Causal Attributions?" *Journal of Personality and Social Psychology* 35: 279–293.

Weron, R., 2001, "Levy-Stable Distributions Revisited: Tail Index > 2 Does Not Exclude the Levy-Stable Regime." *International Journal of Modern Physics* 12(2): 209–223.

Wheatcroft, Andrew, 2003, *Infidels: A History of Conflict Between Christendom and Islam*. New York: Random House.

White, John, 1982, *Rejection*. Reading, Mass.: Addison-Wesley.

Whitehead, Alfred North, 1925, *Science and the Modern World*. New York: The Free Press.

Williams, Mark A., Simon A. Moss, John L. Bradshaw, and Nicole J. Rinehart, 2002, "Brief Report: Random Number Generation in Autism." *Journal of Autism and Developmental Disorders* 32(1): 43–47.

Williams, Robert J., and Dennis Connolly, 2006, "Does Learning About the Mathematics of Gambling Change Gambling Behavior?" *Psychology of Addictive Behaviors* 20(1): 62–68.

Willinger, W., D. Alderson, J. C. Doyle, and L. Li, 2004, "A Pragmatic Approach to Dealing with High Variability Measurements." *Proceedings of the ACM SIGCOMM Internet Measurement Conference*, Taormina, Sicily, October 25–27, 2004.

Wilson, Edward O., 2000, *Sociobiology: The New Synthesis*. Cambridge, Mass.: Harvard University Press.

———, 2002, *The Future of Life*. New York: Knopf.

Wilson, T. D., J. Meyers, and D. Gilbert, 2001, "Lessons from the Past: Do People Learn from Experience That Emotional Reactions Are Short Lived?" *Personality and Social Psychology Bulletin* 29: 1421–1432.

Wilson, T. D., D. T. Gilbert, and D. B. Centerbar, 2003, "Making Sense: The Causes of Emotional Evanescence." In I. Brocas and J. Carillo, eds., 2003.

Wilson, T. D., D. B. Centerbar, D. A. Kermer, and D. T. Gilbert, 2005, "The Pleasures of Uncertainty: Prolonging Positive Moods in Ways People Do Not Anticipate." *Journal of Personality and Social Psychology* 88(1): 5–21.

Wilson, Timothy D., 2002, *Strangers to Ourselves: Discovering the Adaptive Unconscious*. Cambridge, Mass.: The Belknap Press of Harvard University.

Winston, Robert, 2002, *Human Instinct: How Our Primeval Impulses Shape Our Lives*. London: Bantam Press.

Wolford, George, Michael B. Miller, and Michael Gazzaniga, 2000, "The Left Hemisphere's Role in Hypothesis Formation." *Journal of Neuroscience* 20: 1–4.

Wood, Michael, 2003, *The Road to Delphi*. New York: Farrar, Straus & Giroux.

Wrangham, R., 1999, "Is Military Incompetence Adaptive?" *Evolution and Human Behavior* 20: 3–12.

Yates, J. F., 1990, *Judgment and Decision Making*. Englewood Cliffs, N.J.: Prentice-Hall.

Yates, J. F., J. Lee, and H. Shinotsuka, 1996, "Beliefs About Overconfidence, Including Its Cross-National Variation." *Organizational Behavior and Human Decision Processes* 65: 138–147.

Yates, J. F., J.-W. Lee, H. Shinotsuka, and W. R. Sieck, 1998, "Oppositional Deliberation: Toward Explaining Overconfidence and Its Cross-cultural Variations." Paper presented at the meeting of the Psychonomics Society, Dallas, Tex.

Yule, G., 1925, "A Mathematical Theory of Evolution, Based on the Conclusions of Dr. J. C. Willis, F. R. S." *Philosophical Transactions of the Royal Society of London, Series B* 213: 21–87.

Yule, G. U., 1944, *Statistical Study of Literary Vocabulary*. Cambridge: Cambridge University Press.

Zacks, R. T., L. Hasher, and H. Sanft, 1982, "Automatic Encoding of Event Frequency: Further Findings." *Journal of Experimental Psychology: Learning, Memory, and Cognition* 8: 106–116.

Zajdenweber, Daniel, 2000, *L'économie des extrèmes*. Paris: Flammarion.

Zajonc, R. B., 1980, "Feeling and Thinking: Preferences Need No Inferences." *American Psychologist* 35: 151–175.

———, 1984, "On the Primacy of Affect." *American Psychologist* 39: 117–123.

Zeki, Semir, 1999, *Inner Vision*. London: Oxford University Press.

Zimmer, A. C., 1983, "Verbal vs. Numerical Processing by Subjective Probabilities." In R. W. Scholz, ed., *Decision Making Under Uncertainty*. Amsterdam: North-Holland.

Zipf, George Kingsley, 1932, *Selective Studies and the Principle of Relative Frequency in Language*. Cambridge, Mass.: Harvard University Press.

———, 1949, *Human Behavior and the Principle of Least Effort*. Cambridge, Mass.: Addison-Wesley.

Zitzewitz, Eric, 2001, "Measuring Herding and Exaggeration by Equity Analysts and Other Opinion Sellers." Working Paper, Stanford University.

Zuckerman, H., 1977, *Scientific Elite*. New York: The Free Press.

———, 1998, "Accumulation of Advantage and Disadvantage: The Theory and Its Intellectual Biography." In C. Mongardini and S. Tabboni, eds., *Robert K. Merton and Contemporary Sociology*. New York: Transaction Publishers.

Zweig, Stefan, 1960, *Montaigne*. Paris: Press Universitaires de France.

INDEX